Matthias Unterhuber

Possible Worlds Semantics for Indicative and Counterfactual Conditionals?

A Formal Philosophical Inquiry into Chellas-Segerberg Semantics

LOGOS

Studien zur Logik, Sprachphilosophie und Metaphysik

Band 21 / Volume 21

Matthias Unterhuber

Possible Worlds Semantics for Indicative and Counterfactual Conditionals?

A Formal Philosophical Inquiry into Chellas-Segerberg Semantics

ontos verlag

Frankfurt | Paris | Lancaster | New Brunswick

Bibliographic information published by Die Deutsche Bibliothek
Die Deutsche Bibliothek lists this publication in the Deutsche Nationalbibliographie; detailed bibliographic data is available in the Internet at http://dnb.ddb.de

North and South America by
Transaction Books
Rutgers University
Piscataway, NJ 08854-8042
trans@transactionpub.com

United Kingdom, Ireland, Iceland, Turkey, Malta, Portugal by
Gazelle Books Services Limited
White Cross Mills
Hightown
LANCASTER, LA1 4XS
sales@gazellebooks.co.uk

Livraison pour la France et la Belgique:
Librairie Philosophique J.Vrin
6, place de la Sorbonne ; F-75005 PARIS
Tel. +33 (0)1 43 54 03 47 ; Fax +33 (0)1 43 54 48 18
www.vrin.fr

P.O. Box 15 41, D-63133 Heusenstamm
www.ontosverlag.com

ISBN: 978-3-86838-184-9

Printed on acid-free paper
ISO-Norm 970-6
FSC-certified (Forest Stewardship Council)
This hardcover binding meets the International Library standard

Printed in Germany
by CPI buchbücher.de GmbH

Contents

Preface

The current book is a revised version of my PhD thesis which I wrote at the Heinrich Heine University of Duesseldorf with Prof. Gerhard Schurz. In this book I will investigate possible worlds semantics for both indicative and counterfactual conditional logics. I will do this from both a formal and a philosophical perspective. In particular, I aim to show that a specific type of possible worlds semantics, namely Chellas-Segerberg (CS) semantics (see Chellas, 1975; Segerberg, 1989), is a philosophically plausible and technically viable semantics for conditionals. To show this I will discuss a range of topics.

In Chapter 1 I will address the problem of why conditional logics are of interest, where conditional logics are defined as alternative formalisms for conditionals as opposed to classical logic. I will in particular discuss normic conditionals, a type of conditionals which is often not taken into account in the philosophical literature on conditionals.

In Chapter 2 I will argue that conditional logics can be fruitfully applied in a range of disciplines. I will, furthermore, discuss the relation of conditional logics and default logics from the non-monotonic logic literature which is often not addressed in the philosophical literature on conditonals.

In Chapter 3 I will first describe standard conditional logic semantics, such as the possible worlds semantics of D. Lewis and Stalnaker and the probabilistic **P** systems semantics for conditionals by Adams, discuss the difference between conditional logics for indicative and counterfactual conditionals, and describe fundamental problems for probabilistic semantics for conditionals. Second, I will defend possible worlds semantics for indicative conditionals against criticism by Bennett (2003) based on the following arguments: (a) D. Lewis' (1976) triviality result and (b) Bennett's Gibbardian stand-off argument, which goes back to Gibbard (1980, p. 231f). Third, I will discuss a range of topics which serve for a positive account for CS semantics in Chapter 7, such as different types of Ramsey test interpretations.

In Chapter 4 I will describe the formal framework for the investigation in Chapters 5–7.

In Chapter 5 I will prove frame correspondence on the basis of CS Semantics for a lattice of conditional logic systems which is described by 29 axioms. I shall also discuss a general translation procedure from axiom schemata to frame conditions which guarantees a frame correspondence result.

In Chapter 6 I will prove soundness and a non-trivial notion of completeness for the aforementioned lattice of conditional logic systems and I will discuss single world frames in CS semantics.

In Chapter 7 I will provide (a) an objective interpretation of CS semantics in line with alethic interpretations of Kripke semantics and (b) a subjective interpretation by means of a modified Ramsey test criterion. Furthermore, I show that a range of indicative and counterfactual conditional logics as described by Kraus, Lehmann, and Magidor (1990), Lehmann and Magidor (1992), D. Lewis (1973b), Adams (1965, 1966, 1977), and Stalnaker (1968) can be accouted for on the basis of CS semantics. I will also discuss monotonic systems with and without bridge principles.

In Chapter 8 I shall conclude this book with a summary and give a list of issues that were not addressed herein but that might be interesting to pursue. Furthermore, I will summarize the advantages of CS semantics as a semantics for indicative and counterfactual conditionals.

Acknowledgements: I am indebted to Prof. Gerhard Schurz for his personal aid and professional advice in the course of the thesis that served as basis for this book. I would also like to thank Prof. Edgar Morscher for his support during my study at the Paris Lodron University of Salzburg and Prof. Manuel Bremer for many helpful comments on the present text. Finally, I thank Hannes Leitgeb, Ludwig Fahrbach, Paul Thorn, Ioannis Votsis, Alexander Hieke, Janine Reinert, and Florian Boge for helpful comments.

I acknowledge support from the following projects: (1) the Project (P1) *Probability Logic and Non-Monotonic Reasoning from an Evolutionary Perspective* of the LogiCCC Project *the Logic of Causal and Probabilistic Reasoning in Uncertain Environments (LcpR)* of the EUROCORES program of the ESF and DFG, (2) the Project *(B2) Neuroframes: A Neuro-Cognitive Model of Situated Conceptualization* of the DFG Research Project *Functional Concepts and Frames*, and (3) the Project *(P1) Causality and Explanation* of the Research Group *Causation | Laws | Dispositions | Explanation*. Furthermore, I acknowledge a doctoral stipend from the faculty of cultural and social science at the University of Salzburg on a related but distinct topic.

I Foundational Issues

Chapter 1

Arguments for Conditional Logics

The present book is about the logic and philosophy of conditionals. While for a range of other philosophical topics it seems clear to philosophers that they are of interest, this is not so much the case for this area. I am often asked by philosophers who neither focus on the logic nor the philosophy of conditionals why we need conditional logics. While it seems obvious to experts in the field, there is surprisingly little comprehensive argumentation for a treatment of conditionals in a *conditional logic* framework as opposed to a treatment in terms of *classical logic* (propositional calculus, short: p.c.; first-order logic, short: f.o.l.).

The aim of the present chapter is to address this problem and summarize arguments for a conditional logic project. The reader is however cautioned not to expect a full answer to the above question. I cannot properly address a range of issues here which are intimately connected to this problem, such as the formalization of natural language arguments (e.g., Brun, 2003) and the status of naturalism in logic and philosophical analysis of natural language (e.g., Stein, 1996), which would require book-length treatments on their own.

To argue for a conditional logic project I shall discuss five inference schemata in this chapter that are valid in classical logic for an material implication analysis of conditionals. I will argue that these inference schemata are counter-intuitive for indicative conditionals (see Section 1.2), counterfactual conditionals (see 1.3), and propositional normic conditionals (see Section 1.4.3). Furthermore, in Section 1.4.1 I will show that an account of generic normic conditionals in terms of classical logic does not work, based on the qualification problem. Finally, in Section 1.5 I will argue that neither Gricean maxims of conversational implicature can fully explain why these five inference schemata seem counter-inuitive. Observe in that context that the issue of normic conditionals is often not addressed in the philosophical literature on conditionals (e.g., Bennett, 2003).

Before I focus on the argument for a conditional logic project, let me however first say a few words regarding the general framework of this book.

1.1 The Framework of My Investigation

In this chapter I will argue against the representation of conditionals in terms of classical logic. I use this discussion of as a starting point for the investigation of alternative formalisms for conditionals in general (Chapter 3) and Chellas-Segerberg (CS) semantics in particular (Chapters 4–7). My aim is not – as one might suppose – a fully fledged theory of natural language analysis, as for example carried out by Quine (1960). For that reason I will not address problems of such a theory, such as whether it should be naturalistic – by taking into account empirical results of conditionals from psychology and linguistics and weigh them against theoretical constraints such as coherence and simplicity – or alternatively stay non-naturalistic (cf. Stein, 1996, Chapter 5; Goodman, 1983, p. 63f; Bremer, 2005b, pp. 40–48). Note that a range of questions adressed in Chapters 2–7 seems largely independent from these problems. I will, however, briefly comment on these issues in Section 7.4.

My aim here is much more modest. I intend to make only the following points: (i) The traditional analysis of conditionals in terms of classical logic has severe problems and should, therefore, be replaced (or at least supplemented) by an alternative approach. (ii) A viable account among the alternatives is CS semantics (cf. Chapter 8).

1.2 Indicative Conditionals

In this section I will argue against the representation of indicative conditionals by means of the material implication from classical logic, where indicative conditionals are conditionals such as 'Thomas goes for a walk if the weather is nice' (cf. Section 3.4). To do so I will argue against five inference schemata (see below) which are classically valid but which seem to be counter-intuitive.

In this section (i) I will first discuss a simple natural language argument and show how one typically finds out whether such a natural language argument is valid. (ii) I will, then, explain the notions of intuitive correctness and completeness and (iii) introduce the aforementioned inference schemata. Furthemore, (iv) I shall describe C. I. Lewis' criticism of the material implication, (v) outline a relevance logic's edge regarding this problem, and (vi) describe the conditional logic project. Finally, (vii) I shall defend the conditional logic project against an objection by a conditional logic opponent.

Let us focus on (i). Consider the following natural language argument: (a) 'Thomas goes for a walk if the weather is nice', (b) 'the weather is nice' (premises). Therefore: (c) 'Thomas goes for a walk' (conclusion).

Suppose we want to find out whether (c) follows from (a) and (b). The standard procedure for this problem is the following: Represent the sentences in classical logic and accept the natural language inference as valid if the formal translation is valid in classical logic, where conditionals such as (a) are analyzed in terms of the material implication ('$\rightarrow$'). So, in the above example (a) is translated into a formula of the form '$p \rightarrow q$', where 'p' and 'q' stand for 'the weather is nice' and 'Thomas goes for a walk', respectively. One regards, then, the natural language argument valid if the translation of the natural language argument is valid in classical logic, which is that given the premises are true the conclusion has to be true as well.

Let us now discuss point (ii). If we take the above approach and use the material implication to analyze natural language conditionals, the logical properties of the material implication are imposed on conditionals in natural language. We can thereby distinguish two properties: (A) intuitive correctness and (B) intuitive completeness (Blau, 1978, p. 4; see also Bremer, 2005a, Chapter 3).[1] A logical system is intuitively correct iff no natural language argument that is judged invalid turns out to be valid by its formal evaluation (as described above). A logical system is intuitively complete iff all natural language arguments judged valid are also valid according to their formal evaluations.[2]

Blau (1978, p. 4) correctly observed that intuitive completeness is too much to ask, since the notion of inuitive validity is most often too vague for that purpose. In contrast, intuitive correctness seems to be a fairly clear adequacy criterion for logical systems (Blau, 1978, p. 4). Note that the counterintuitive inference schemata of classical logic discussed throughout this chapter violate the intuitive incorrectness and not the intuitive completeness criterion.

Let us focus on (iii). Consider the following inferences, where examples E1–E4 are essentially taken from Adams (1965, p. 166):

E1 John will arrive on the 10 o'clock plane. Therefore, if John does not arrive on the 10 o'clock plane, he will arrive on the 11 o'clock plane.

E2 John will arrive on the 10 o'clock plane. Therefore, if John misses his plane in New York, he will arrive on the 10 o'clock plane.

E3 If Brown wins the election, Smith will retire to private life. If Smith dies before the election, Brown will win it. Therefore, if Smith dies before the election, then he will retire to private life.

E4 If Brown wins the election, Smith will retire to private life. Therefore, if Smith dies before the election and Brown wins it, Smith will retire to private live.

[1] In Åqvist's (2002, p. 173) framework the properties *intuitive correctness* and *completeness* correspond to right-to-left and left-to-right adequacy, respectively.

[2] The properties *intuitive correctness* and *completeness* naturally also depend on an adequate translation of one's natural language expression in one's formal language. I will abstract from this problem here.

E5 If Andrea wins the lottery, she will donate 500,000$ to UNICEF. Therefore, if Andrea does not donate 500,000$ to UNICEF, then she will not win the lottery.

Although E1–E5 are counter-intuitive they are are valid according to classical logic. In order to see that more clearly, let us formalize examples E1–E5 in terms of inferences of the following sort, where inferences of sort S1 and S2 are often called 'fallacies of material implication' (Adams, 1975, p. 4 and p. 11; cf. also Weingartner & Schurz, 1986, p. 10f):

S1 If p then $\neg p \rightarrow q$ (Ex Falso Quodlibet)
S2 If q then $p \rightarrow q$ (Verum Ex Quodlibet)
S3 If $p \rightarrow q$ and $q \rightarrow r$ then $p \rightarrow r$ (Transitivity)
S4 If $p \rightarrow r$ then $p \wedge q \rightarrow r$ (Monotonicity)
S5 If $p \rightarrow q$ then $\neg q \rightarrow \neg p$ (Contraposition)

The formal inferences S1–S5 correspond to E1–E5, respectively. I use the connectives '$\neg$' ("Negation") and '$\wedge$' ("Conjunction") in addition to the material implication '$\rightarrow$' for inferences S1–S5. I shall henceforth also employ the connective '$\vee$' ("Disjunction"). Furthermore, I assume that the connective '$\neg$' binds the strongest, while both '$\wedge$' and '$\vee$' bind stronger than '$\rightarrow$'. (For a more detailed description of the formal language employed in this book see Section 4.2.1.)

It is easy to see that inferences S1–S5 are valid in classical logic. Since S1–S5 are formalizations of inferences E1–E5 respectively, the above criterion gives us that we should regard inferences E1–E5 as valid as well. Intuitively, however, E1–E5 hardly seem acceptable.

In order to establish a formal terminology for a discussion of conditionals which does not presuppose a material implication analysis of conditionals, I will represent E1–E5 in a more neutral way. For that purpose I will employ the two-place conditional connective '$\Box\!\!\rightarrow$'. This conditional connective can – but need not – have the logical properties of the material implication. Note that this symbol is used in D. Lewis (1973b) for counterfactual conditionals. In spite of that I will use the symbol for various types of conditionals, such as indicative and counterfactual conditionals. In this way we can discuss the formal properties of different types of conditionals without presupposing that the properties of the material implication hold for these conditionals.

Let us, accordingly, reformulate S1–S5 in terms of $\Box\!\!\rightarrow$, where $\alpha, \beta, \ldots$ are formula schemata:

S1′	If α then $\neg\alpha \Box\!\!\rightarrow \beta$	(Ex Falso Quodlibet, EFQ)
S2′	If β then $\alpha \Box\!\!\rightarrow \beta$	(Verum Ex Quodlibet, VEQ)
S3′	If $\alpha \Box\!\!\rightarrow \beta$ and $\beta \Box\!\!\rightarrow \gamma$ then $\alpha \Box\!\!\rightarrow \gamma$	(Transitivity)
S4′	If $\alpha \Box\!\!\rightarrow \gamma$ then $\alpha \wedge \beta \Box\!\!\rightarrow \gamma$	(Monotonicity)
S5′	If $\alpha \Box\!\!\rightarrow \beta$ then $\neg\beta \Box\!\!\rightarrow \neg\alpha$	(Contraposition)

We shall now discuss point (iv). C. I. Lewis (1912) already investigated inferences S1′ and S2′. His main motivation was to find a logic of conditionals which is more in accord with ordinary inferences than the classical account (C. I. Lewis, 1912, p. 522). For that purpose, C. I. Lewis focused on inferences S1′ and S2′ rather than on inferences S3′–S5′ (cf. C. I. Lewis, 1912, p. 528f; Hughes & Cresswell, 1996/2003, p. 194f). S1′ and S2′ differ from inferences of type S3′–S5′ insofar as S1′ and S2′ but not S3′–S5′ are bridge principles, viz. principles, which specify a fixed logical relationship between conditional and non-conditional formulas, that is formulas without the conditional operator (cf. Sections 3.4.2 and 4.2.1).

C. I. Lewis (1912, p. 528f) rejected inferences of sort S1′ and S2′: (a) A false proposition does not imply everything (Principle S1, Ex Falso Quodlibet) and (b) a true proposition is not implied by arbitrary propositions (Principle S2, Verum Ex Quodlibet). Translated in our terminology this means that neither the factual falsehood of the premise nor the factual truth of the conclusion guarantees that the respective conditional is true.

I shall now discuss point (v). A second line of research which does not allow for inferences of sort S1′–S5′, is relevance logic (e.g., Weingartner & Schurz, 1986; Schurz, 1991).[3] The basic idea of relevance logic ap-

[3] I will restrict myself here to the approaches of Weingartner and Schurz (1986) and Schurz (1991). See, for example, Anderson, Belnap, and Dunn (1975, 1992),

proaches, such as Weingartner and Schurz (1986) and Schurz (1991), is to exclude irrelevant inferences that is inferences whose conclusion (or premises) contain an irreleveant part. For such a relevance logic approach we can distinguish at least the following three *irrelevance criteria* (with respect to (w.r.t.) conclusions): (i) Formulas in the conclusion exists which are not present in one of the premises (Weingartner & Schurz, 1986, p. 6f). (b) Formulas in the conclusion are replaceable by their negations salva validate, viz. the inference is valid even if we could replace a formula by its negation (Weingartner & Schurz, 1986, p. 9). (c) Formulas in the conclusion can be replaced by *arbitrary* formulas salva validate (Schurz, 1991, Definitions 21 and 22, pp. 409–411).

Let us now see which inferences of sort S1′–S5′ are rejected by the relevance approach described in the preceding paragraph. Criteria (a)–(c) concur in that respect insofar as they render inferences S1′, S2′, and S4′ irrelevant but not inferences S3′ and S5′. For example, the validity of inference S1′ does not depend on the logical form of the formula α. Analogously, the validity of S2′ is independent of the logical form of the formula β. It is interesting to note that the relevance approach also rejects inference S4′, that is the monotonicity property of conditionals (cf. Section 2.2). Note that S3′ and S5′ imply S4′ given minimal conditions (see Sections 7.2.1 and 7.2.6). So, in order to give up S4′ we plausibly have to give up S3′ and S5′ as well.

Let us now inquire into (vi). The starting point of the conditional logic approach employed here is the following observation: To save the inuitive correctness of a logic of conditionals, we take S1′–S5′ at face value insofar as we need to find alternative formalisms – that is conditional logics – which do not validate S1′–S5′. I call such an approach 'conditional logic project'. Thereby it is particularly important to avoid the monotonicity property of conditionals, as we saw in the previous paragraph.

Note that a proponent of the conditional logic project need not argue for an alternative formal framework for all types of conditionals. D. Lewis (1973b),

Dunn and Restall (2002), Priest (2008, Chapter 10), and Sperber and Wilson (1995) for further relevance (logic) approaches.

for example, proposes conditional logic systems which were intended only for an analysis of counterfactual conditionals (D. Lewis, 1973b, p. 1). Despite that fact D. Lewis still counts as a proponent of a conditional logic project. In contrast, an opponent of a conditional logic project would argue that classical logic suffices and no conditional logic is needed for a philosophically adequate analysis of conditionals.[4]

I shall now discuss point (vii) and defend the conditional logic project against an objection by its opponent. Such an opponent might be completely unimpressed by the above examples and might argue – in the tradition of Tarski – that these examples merely appear to be paradoxical due to the ambiguity and elusiveness of natural language. Once the conditionals are properly analyzed, their paradoxical nature disappears. She might suggest that we paraphrase the conclusion of E1 as 'John will arrive at the 10 o'clock plane or he will arrive on the 11 o'clock plane'. It seems much less paradoxical to infer the latter conclusion from 'John will arrive on the 10 o'clock plane'.[5]

The response of the conditional logic project's opponent, however, begs the question. If we presuppose that the truth conditions of conditionals (of the form 'if α then β') agree with the truth conditions of disjunctions of a certain sort – that is disjunctions of the form 'not α or β' – it is no wonder that E1–E5 only appear to be paradoxical, since we reduce the truth conditions of conditionals to truth conditions of disjunctions. It is, however, this reduction of conditionals to disjunctions which proponents of the conditional logic project reject, since this reduction implies S1′–S5′ based on minimal conditions (see Section 7.3.4).

[4] Some opponents of a conditional logic project admit that some type of conditionals cannot be analyzed in terms of classical logic, such as counterfactual conditionals. Since these opponents presuppose an account of conditionals in terms of classical logic, their view implies that counterfactuals cannot be sensibly formalized in any logical system. I find such a conclusion hard to accept.

[5] The relevance logician might object and argue that the resulting inference (of the form 'if α then $\alpha \vee \beta$') is counter-intuitive, since it is irrelevant according to the above (ir)relevance criteria.

1.3 Counterfactual Conditionals

In this section (i) I will argue against a material implication analysis of counterfactual conditionals based on inference schemata S1′–S5′. Then, (ii) I will briefly highlight the role of counterfactuals in the sciences, the humanities, and everyday reasoning.

Let me address point (i) by arguing that inferences of sort S1′–S5′ are also counter-intuive for counterfactual conditionals. For that purpose consider the following natural language inferences, where Example E3′ is taken from (D. Lewis, 1973b, p. 33):

E1′ The Titanic sank. Therefore, if the Titanic had not sunk, Kennedy would not have been assassinated.

E2′ The Titanic sank. Therefore, if the Titanic had not been built, then it would have sunk.

E3′ If J. Edgar Hoover had been born a Russian, he would have been a Communist. If J. Edgar Hoover had been a Communist, he would have been a traitor. Therefore, if he had been born a Russian, he would have been a traitor.

E4′ If Hitler had not been born, there would have been no World War II. Therefore, if Hitler had not been born and a similarly vile dictator emerged in Germany in the 1930s, there would have been no World War II.

E5′ If Kennedy had not been assassinated by Oswald, Oswald would still have been arrested. Therefore, if Oswald had not been arrested, Kennedy would have been assassinated by Oswald.

E1′–E5′ certainly seem unacceptable but would be rendered valid by a material implication analysis of counterfactual conditionals on the basis of inference schemata S1′–S5′, respectively. [6]

[6] From a linguistic perspective the discussion is complicated by the tense of the clauses and the tendency that true consequents are indicated by modifiers such as ‘still’ as in E5′ (cf. Declerck & Reed, 2001). I will abstract from these difficulties here.

In particular E1′ and E2′ seem to be problematic. However, according to S1′ any counterfactual conditional with a false antecedent would be true, no matter what the consequent is. Analogously, by S2′ any counterfactual with a true consequent would hold, whatever the antecedent. In constrast, it is a generally accepted feature of counterfactual conditionals that they are not vacuously true even if their antecedent is false or their consequent is true: If the antecedent of a counterfactual conditional is false, it also depends on the consequent whether the counterfactual conditional is true. Moreover, if the consequent of a counterfactual conditional is true, it need not be that the counterfactual conditional is true as well.

Let me now discuss point (ii). Counterfactual conditionals play an important role in the sciences, the humanities, and everyday reasoning. For example, counterfactuals such as the ones described above allow for causal inferences (cf. D. Lewis, 1973a, p. 557). Furthermore, questions as 'What would be the case if ...?' lie at the heart of philosophy and the scientific method in general and essentially employ reasoning with counterfactual conditionals. Moreover, in everyday reasoning we also employ questions of this type rather frequently. For example, a car driver might ask herself after a car accident: 'What would have happened if I had reacted differently?' Observe that further examples of counterfactual conditionals from the sciences, the humanities, and everyday reasoning are easily found.

1.4 Normic Conditionals

In this section I will distinguish between propositional and generic normic conditionals and I will argue that both types of normic conditionals cannot adequately be represented by means of classical logic. The discussion of propositional normic conditionals shows that it is the conditional aspect of normic conditionals rather than their generic aspects which leads to the problems of an analysis of normic conditionals in terms of classical logic.

To make my case I will discuss normic conditionals and contrast them with a classical analysis of (universally) quantified statements in Section

1.4.1. In Section 1.4.2 I shall argue against an analysis of generic normic conditionals in terms of classical logic based on the qualification problem, which is closely connected to Lange's (1993) dilemma from the literature on the ceteris paribus laws (cf. Reutlinger, Schurz, & Hüttemann, 2011, §4). In Section 1.4.3 I shall discuss the inference schemata S1′–S5′ for propositional normic conditionals and in Section 1.4.4 I will argue that normic conditionals play an important role in the sciences, the humanities, and everyday reasoning.

1.4.1 The Generic Case

In this section I shall describe generic normic conditionals and contrast them with universally quantified conditionals as described in classical logic. Before I turn to examples for generic normic conditionals, let me first describe the standard account of universally quantified conditionals in first-order logic. This will prove useful, since generic normic conditionals are naturally formulated in what can be considered a quantified form.

Consider, for instance, the following statement:

E6 All fish are cold-blooded.[7]

Conditionals such as E6 are traditionally analyzed in classical logic the following way: For all *x* holds, if *x* is a fish, then *x* is cold-blooded. The very same analysis is not restricted to sentences of the above type but extends to sentences of the following sort:

E7 Fish are cold-blooded.

Both E6 and E7 are thus treated as universally quantified conditionals (short: uq conditionals).[8]

Consider the following examples of generic normic conditionals:

[7] Most but not all fish are cold-blooded. For example, some species of sharks and tunas are exceptions to this rule insofar as they can raise their temperature significantly above ambient temperature (Helfman, Collette, Facey, & Bowen, 2009, p. 96f).

[8] In the literature there is quite some controversy in which way natural language sen-

E8 99.9% of fish are cold-blooded.
E9 Most fish are cold-blooded.
E10 Fish are probably cold-blooded.
E11 Fish are normally cold-blooded.

E8–E11 also qualify as a sort of quantified normic conditionals and I will also call this type of conditionals quasi unversally quantified (quq) conditionals.[9] E8 indicates an exact numeric value of the respective frequency or probability, while E9–E11 do not. Furthermore, E8–E10 suggest an interpretation in terms of frequencies or probabilities (see Section 3.5.1), whereas E11 does not immediately do so. For E11 at least two interpretations can be distinguished: 'normally' might either be understood in the sense of prototypical or statistical normality (Schurz, 2001b, p. 478) as we shall see shortly. Furthermore, E8–E10 and E11 are examples of probabilistic conditionals and normic conditionals (in the narrow sense), respectively (cf. Schurz, 2008, p. 89f; Schurz, 2001b, p. 476f).

Although E7–E11 are related to E6, they qualify as quq conditionals rather than uq conditionals. They differ from E6 insofar as they allow for exceptions, that is there might be fish that are not cold-blooded and that fact does not falsify the respective quq conditional. Nevertheless, exceptions can accumulate in such a way that they become the rule rather than the exception. So, if the majority of fish are not cold-blooded, then E8 and E9 cannot hold. On the basis of Schurz's argument described below, this line of reasoning also extends to conditionals of type E11.

Since an analysis of conditionals in terms of classical logic (cf. example

tences such as E7 should be analysed (Pelletier, 1997; Nickel, 2009, 2010; Leslie, 2007). I will follow here the classical approach in assuming that generics can be appropriately represented by quantification over conditional structures.

[9] There are approaches other than conditional logics for quq conditionals which also reject an analysis of generic (normic) conditionals in terms of the material implication, such as generalized quantifier accounts (e.g., Barwise & Cooper, 1991; Peters & Westerståhl, 2008). Time and space restrictions do not allow me to discuss these approaches here.

E6) does not allow for any exceptions/counterexamples, conditionals such as E7–E11 are prima facie excluded from a traditional f.o.l. analysis.[10]

Let us now focus on quq conditionals of the E11 sort. In a statistical normality interpretation, E11 states that the probability of x being cold-blooded, given that x is a fish, is high. Hence, E11 is roughly synonymous with E9 or E10 in this interpretation. In a prototypical normality interpretation E11 rather says that prototypical fish are cold-blooded.

Both sorts of conditionals are by no means independent. There is a implicative relationship between both sorts of conditionals which is based on evolution-theoretic principles. As Schurz (2001a) shows, for an evolutionary system to be successful, the normal case has to be the statistically normal case in the long run (Schurz, 2008, p. 90f; see Schurz, 2001a, p. 495) where statistically normal cases are specified by conditionals of type E9. So, based on evolution-theoretic principles it is reasonable to assume that normic conditional imply statistical conditionals of type E9 (see also Schurz, 2008, p. 90f).

1.4.2 The Qualification Problem

In this section I will contend that quq conditionals – that is generic normic conditionals – can be adequately accounted for in classical logic, based on the qualification problem. For that purpose I will use an example from biology and argue that the opponent of the conditional logic project faces the following dilemma when she tries to account for quq conditionals by means of uq conditionals: She is either forced (a) to replace a quq conditional by a uq conditional that is only equivalent given our current state of knowledge (where the latter might still turn out to be false) or (b) to replace the quq

[10] It might be argued that we can use quantifiers to count elements in the domain and by these means are able to represent a frequentistic interpretation of quq conditionals. Please note that this approach does not work for cases E9–E11. This is due to the fact that we are generally not able to identify a non-arbitrary threshold of "counterexamples", above which one rejects E9–E11. Moreover, for infinite domains even an arbitrary threshold approach is not viable.

conditional by an analytically true uq conditional.

Observe that the qualification problem – as discussed here – is closely related to a problem from the ceteris paribus literature (cf. Schurz, 2002; Earman, Roberts, & Smith, 2002; Cartwright, 2002; Strevens, in press; Schrenk, 2008; Hüttemann, 1998) which is whether ceteris paribus laws such as 'ceteris paribus, planets have elliptical orbits' (cf. Schurz, 2002, p. 352), can be adequately described by uq conditionals. Such an analysis faces Lange's dilemma (Lange, 1993, 235; see Reutlinger et al., 2011, §4) which is a variation of the dilemma above.

Let us now discuss the qualification problem. The qualification problem stems from the non-monotonic logic literature (see Section 2.1.5) and was used to argue for non-monotonic logics as opposed to classical logic (cf. McCarthy, 1980, p. 27; Horty, 2001, p. 340f). I will use this problem to argue against an account of quq conditionals in terms of uq conditionals from classical logic.

Consider the following true quq conditional from biology, where 'viviparous' means that the respective species' females do not lay eggs but give live birth:

E12 Mammals are normally viviparous.

A conditional logic project opponent might argue here that E11 is equivalent to a uq conditional, whose antencedent is specified in such a way that it excludes all (potential) exceptions. [11] Since platypoda (platypuses) and tachyglossa (spiny anteaters) are the only orders of existing species which do not give live birth (McKenna & Bell, 1997, p. 35), one might be tempted to formulate E11 by means of the following uq conditional:

E13 All mammals are viviparous, except for spiny anteaters and platypuses.

[11] Such an approach is most plausible for normic conditionals in the narrow sense. To make the opponent's case as strong as possible, we restrict the present discussion to this type of conditional.

However, E13 is only eqivalent to E12 based on our current state of knowledge and might turn out false after all. Furthermore, E13 is strictly speaking false since there are mammals which are not spiny anteaters or platypuses but which cannot give live birth (e.g., female apes who completely lack the disposition to give birth).[12] Based on both facts we get horn (a) of the dilemma above.

One possible way out of this predicament is to use the subclass of *prototheria*: All species of this subclass are defined as mammals which are non-viviparous (Macdonald et al., 2007). This allows us to reformulate E12 in a knowledge independent way:

E14 All mammals are viviparous, except for prototheria.

Although example E14 is knowledge independent, it cannot be equivalent to E12 for the following reasons: E14 is analytically true, since prototheria are defined to encompass all mammal species which are non-viviparous. Unlike E14, example E12 has no inherent feature which guarantees its truth. Hence, E12 is not analytically true and we face horn (b) of the dilemma above.

I shall provide here further reasons as to why the opponent's strategy – to account for all possible exceptions by restricting the antecedent of a uq conditional – leads to horn (a) of our dilemma. I will thereby argue for a point made by Schurz (2002, p. 359) that for any quq conditional there are a possibly infinite number of heterogeneous exceptions and that for that reason this strategy of the conditional logic opponent does not work.

To keep the argument simple, presuppose that a specimen's genetic code determines which species it belongs to.[13] Note that the genetic code specifies a specimen's metabolism only on the basis of environmental conditions. Furthermore, environmental influences (e.g., medication, toxic substances)

[12] Viviparousness is a disposition of individual specimen rather than species. I will abstract here from the problem of how to relate dispositions of specimen to dispositions of species.

[13] This assumption is controversial among biologists and philosophers of biology (cf. Ereshefsky, 2007).

can alter a specimen's expression of genes without affecting the specimen's genes. In particular, it is not precluded that there is an environmental condition A which inhibits the organism's metabolic processes that lead to the organism's viviparousness but does not affect its genetic code. [14] Condition A might, for example, be the presence of a toxic substance.

It is furthermore not precluded that there is a second environmental condition B which inhibits A's inhibition of the organism's viviparousness and which does not affect the organism's genetic code. Assume furthermore that the organism retains its viviparousness if both conditions A and B apply. Condition B might, for example, be the presence of a chemical substance which neutralizes the toxic substance's effect on the organism.

Moreover, we might have still a third environmental condition C which inhibits B's inhibition of A's inhibition of the organism's viviparousness. Again we presuppose that environmental condition B does not affect the organism's genetic code. Furthermore, if conditions A, B, and C hold, then C effectively inhibits the organism's viviparousness. Environmental condition C might, for example, be the presence of a chemical substance.

We can continue with environmental conditions D, E, F, ... A fortiori here is no reason why we might not produce a chain of inhibitions of arbitrary length. Note that the total environmental condition in which A inhibits the organism's viviparousness differs from the situation in which conditions A, B, and C are present: In the latter situation only the inhibition of B by C allows A to inhibit the organism's viviparousness. So, A is an environmental condition of a different sort than C. Furthermore, cases D, E, F, ... can be constructed analogously.

Let us now come back to an account of E12 in terms of uq conditionals. To make sure that a uq conditoinal is equivalent to a conditional of sort E12 and does not to fall prey to horn (a) of the dilemma above, one has to exclude all environmental conditions which inhibit the organism's viviparousness. Since we can construct an arbitrary number of non-equivalent environmen-

[14] These processes might be extremely unlikely in a natural environment. This does not affect the present argument.

tal conditions as described above, we essentially get an infinite number of environmental conditions that have to be excluded: inhibitions simpliciter, inhibitions of inhibitions of inhibitions, and so on. Since formulas only allow for a finite number of subformulas, we cannot possibly list all those exceptions in the antecedent of the respective uq conditional, especially as they might not follow any clear pattern. So, even under rather simple circumstances as those described here there seems to exist a possibly infinite number of heterogeneous exceptions for quq conditionals such as E12.

We might be tempted to account for quq conditionals such as E12 by uq conditionals whose antecedent excludes exceptions of any sort. However, this step is not viable, since this would render the respective uq conditional analytic and we would again face horn (b) of the dilemma above.

1.4.3 The Propositional Case

Let us now discuss propositional normic conditionals. I will first (i) describe propositional normic conditionals and then (ii) focus on an inference schema which is particularly counter-intuitive for normic conditionals. Finally, (iii) I will discuss why inferences S1′–S5′ are also counter-intuitive for propositional normic conditionals. The discussion of propositional normic conditionals is intended to show that it is not so much the generic aspects of normic conditionals which leads to the problems of account of normic conditionals in terms of classical logic but rather the conditional aspect.

Let us now focus on (i). Consider the following propositional normic conditionals:

E15 If specimen 214 is a fish, then it is probably cold-blooded.
E16 If specimen 214 is a fish, then it is normally cold-blooded.
E17 If Peter pulls the red lever now, the engine will probably start.
E18 If it rains now, the car will normally not start.

E15 and E17 are probabilistic conditionals and E16 and E18 normic conditionals (in the narrow sense). Observe that the modifiers ‘probably’ and ‘normally’ are interpreted to pertain to the whole conditional structure not only to the consequent. E15 and E16 can be regarded as specific instances of the quq conditionals E10 and E11 (in the narrow sense), respectively. Moreover, analogous to the quantificational case (see Section 1.4.1), probabilistic conditionals are arguably a subtype of normic conditionals. Hence, I restrict myself to the discussion of normic conditionals (in the narrow sense). One can easily check that my points also hold for probabilistic conditionals.

Let us now discuss (ii). Analogous to the quq cases, the conditionals E15–E18 do not preclude counterexamples. In E8–E11 "counterexamples" are individuals (fish that are not cold-blooded), in E15–E18 the "counterexamples" are states of affairs in which the antecedent is true, but the consequent is false (e.g., specimen 214 being a fish and not being cold-blooded). This means that the set of formulas $\{\alpha \Box\!\!\rightarrow \beta, \alpha, \neg\beta\}$ is consistent in the propositional case where the propositional normic conditional is represented by the conditional formula $\alpha \Box\!\!\rightarrow \beta$.[15] In other words the following inference does not hold for normic conditionals (see also Section 3.4.2):

S6′ if $\alpha \Box\!\!\rightarrow \beta$ then $\alpha \rightarrow \beta$ (MP)

In contrast, if we analyze conditionals in terms of the material implication from classical logic, MP holds trivially. Given that normic conditionals do not satisfy S6′, this suggests that a material implication analysis is inappropriate for normic conditionals.

Let us now focus on (iii). Consider the following inferences:

E1″ It rains today. Therefore, if it does not rain today, then Fireball will normally win the race tomorrow.

E2″ It rains today. Therefore, if I Fireball wins then race tomorrow, it will normally rain today.

E3″ If Michaela studies into the night today, she normally gets much work done. If Michaela gets much work done today, she normally does not study into the night. Therefore, if Michaela studies into the night today, then she normally does not study into the night.

[15] The quq case is complicated by quantificational issues. In this book we focus only on the propositional case of S6′ below.

E4″ If Tina turns the ignition key now, the car normally starts. Therefore, if Tina turns the ignition key now and the motor does not get any fuel, the car normally starts.

E5″ If Ludwig does not ascend Mt. Everest, then he normally (still) does not win the Nobel Prize for Physics. Therefore, if Ludwig wins the Nobel Prize for Physics, then he normally ascends Mt. Everest.

Examples E1″–E5″ correspond to inferences S1′–S5′, respectively.

The examples E3″ and E4″ are in need of explanation: In E3″ and E4″ it seems more natural to use the expression 'to be expected' rather than 'normally'. Expectancy, however, can be viewed in this context as synonymous to normality, since it allows for an interpretation in terms of both prototypical and statistical normality: I can accept the consequent either since it is (proto)typically expected on the basis of the antecedent or since the consequent is probable given the antecedent (cf. Section 1.4.1). I use here the expression 'normally' in order to allow for a uniform treatment of propositional normic conditionals.

1.4.4 The Role of Normic Conditionals

In this section I shall highlight the role of normic conditionals in the sciences, the humanities, and everyday reasoning. I will furthermore describe sets of assertions (involving normic conditionals) – examples (A) and (B) below – which seem particularly interesting and to which I will refer later in Section 3.4.2.

Let us briefly remark on the role of normic conditionals in the sciences, the humanities, and everyday reasoning. One can hardly imagine scienfic reasoning and reasoning in the humanities and everyday life without reliance on normic conditionals such as E7–E12 and E15–E18. First, probabilistic reasoning is pevasive and almost omnipresent in almost any facet of human life (see also Section 2.1.4). Second, reasoning with prototypes seems to be very natural and pervasive as well for all types of situations.

Furthermore, consider the following sets of statements (cf. Delgrande,

1987, p. 109):

(A) If specimen 213 is a mammal, it is normally viviparous.
If specimen 213 is a platypus, it is normally not viviparous.
Specimen 213 is a platypus and a mammal.

(B) Mammals are viviparous.
Mammals which are platypus are not viviparous.
There are mammals which are platypus.

In the sciences, the humanities, and in everyday reasoning, sets such as (A) and (B) are considered consistent. As it is easy to see, classical logic cannot account sets of statements since it renders both (A) and (B) inconsistent due to the monotonicity principle (S4′) and the rule Modus Ponens (S6′). I will discuss the problem for Modus Ponens in Section 3.4.2.

1.5 Conversational Implicatures

In this section I will discuss the inference schemata S1′–S5′ on the basis of Gricean conversational implicatures, since a conditional logic project opponent can appeal to the pragmatic analysis of natural language by Grice (1989, see also Bennett, 2003, pp. 22–26) to argue against the conditional logic project. Grice held the view that an analysis of conditionals in terms of the material implication is adequate. Inferences of type S1′–S5′ only appear to be counter-intuitive on pragmatic grounds due to the rules that govern natural language use and which are called 'maxims of conversational implicature'.

To defend the conditional logic project against this type of objection, (i) I will describe the basic assumptions of the Gricean approach, (ii) compare it to Schurz's (1991) relevance logic approach from Section 1.2, and (iii) argue that Grice's approach cannot fully explain why inferences S1′-S5′ only appear to be problematic while they are in fact not. For that purpose I will discuss the Gricean approach as described by Bennett (2003). [16]

[16] I cannot include here the more recent account of Gricean maxims of conversational

Let us now focus on (i). Maxims of conversational implicature include the following (Bennett, 2003, p. 22f): be appropriately informative, be truthful, be relevant, be orderly, and be brief. On the basis of Gricean maxims of conversational implicature, the language is then used in such a way that the language user conveys information "without outright asserting it" (Bennett, 2003, p. 22). An example of a Gricean implicature is the following (cf. Bennett, 2003, p. 23): If Jane Doe says 'God exists', then she conveys that she believes that God exists. Despite this fact, the proposition *God exists* does not entail that Jane Doe believes that God exists. However, asserting a proposition without believing it would result in a violation of the maxim of truthfulness. So, in that example Jane's assertion would be counter-intuitive from a pragmatic perspective or at least in need of further explanation.

Let us discuss point (ii). We can interpret Schurz's (1991) relevance approach also in terms of Gricean maxims. This is due to the fact that the exclusion of irrelevant inferences follows a non-redundancy preference, insofar as inferences with irrelevant conclusions elements are explicitly excluded (cf. Section 1.2). Non-redundancy preferences are tapped by Grice's (1975, p. 46) maxim "relation". However, *the main difference* between Schurz's (1991) relevance approach and the Gricean approach described above is that in Schurz's (1991) account, irrelevance inferences are not just "not assertable" but are regarded as normatively deficient (cf. criterion (a) in the next paragraph).

I shall now focus on point (iii). The conditional logic opponent can defend an analysis of natural language conditionals in terms of the material conditional by Gricean maxims along the following lines: (a) A material implication analysis is adequate from a normative stance, and (b) we can fully explain why inferences of type S1′–S5′ strike us as counter-intuitive by reference to Gricean maxims insofar as it is pragmatically counter-intuitive to assert the conclusion rather than the premise(s), viz. we can draw the respective inferences only on pains of violations of Gricean maxims.

Bennett's (2003, p. 24f) discussion shows that inferences of type S1′ and

implicature as, for example, described in Levinson (2000).

S2′ might be considered counter-intuitive for the following reason: In both cases the maxim of brevity and the maxim of informativeness suggest that one should assert the premise rather than the conclusion: First, the premise is shorter (maxim of informativeness). Second, the premise implies the conclusion, but not vice versa. Thus, the premise is also more informative (maxim of informativeness). Hence, the Gricean maxims suggest that one should assert the premises of inferences, such as E1 and E2, rather than their conclusions.[17]

Observe that not all inferences of type S1′–S5′ can be explained by appeal to the two criteria brevity and informativeness equally well. While an explanation of S4′ seems viable in terms of the maxims of brevity and informativeness, an application of both maxims to S3′ and S5′ yields a rather mixed result. Let us first focus on S4′. The premise in E4 is shorter than the conclusion (maxim of brevity) and the premise implies the conclusion, but not vice versa (maxim of informativeness). Hence, both maxims suggest that one should endorse the premise rather than the conclusion in inferences of sort S4′.

In the case of S3′ both maxims conflict with each other: For example, in inferences of sort E3 the premises imply the conclusion, but not vice versa. Hence, the maxim of informativeness suggests that one should assert the premises of E3 rather than its conclusion. The maxim of brevity gives us reasons to assert the conclusion rather than the premises, since the conclusion is shorter than the conjunction of both premises.

In case of S5′, there is even less support: In E5 the conclusion and the premise are logically equivalent on the basis of classical logic. So, the maxim of informativeness is indecisive. Moreover, the conclusion contains an unnegated antecedent and consequent, while the premise employs negated ver-

[17] In my opinion it is less plausible to analyze inferences with counterfactual conditionals, such as, E1′–E5′, and propositional normic conditionals, such as E1″–E5″, in terms of conversational implicature for indicative conditionals of sort E1–E5. To make as strong as possible a case for a Gricean analysis, I restrict the discussion to examples E1–E5.

sions of both subclauses (as consequent and antecedent, respectively). Hence, the conclusion is shorter than the premise and the maxim of brevity suggests that one should assert the conclusion of E5 rather than its premise.

The failure to account for S5′ inferences such as E5 is more severe than it might first seem. This is due to the fact that, given reasonable restrictions, S5′ implies S3′ and S4′, but not vice versa (Kraus et al., 1990, p. 180f; see Section 7.2.6). Observe that the relevance approaches described in Section 1.2 do not agree with the Gricean account concerning the rejection of inferences S1′–S5′: Only S1′, S2′, and S4′ turn out to be irrelevant in relevance logic approach.

Chapter 2

The Conditional Logic Project in an Interdisciplinary Context and Default Logics

The present chapter focuses on two main points. First, I will discuss the interdisciplinary ramifications of the conditional logic project which goes beyond classical logic (see Chapter 1) in Section 2.1. I aim to show by these means that the conditional logic project can be fruitfully applied in a range of disciplines, such as philosophy, linguistics, psychology, and computers science.

Second, I will sketch some assumptions of default logics in Section 2.2 and compare this type of formalism with conditional logics. A discussion of the latter point is instructive insofar as both types of approaches are correctly described as non-monotonic logics, but differ in essential ways. Note that default logics approaches – and also the related conditional logic approaches from the artificial intelligence literature (e.g., Kraus et al., 1990; Lehmann & Magidor, 1992) – are often ignored in the philosophical literature on conditionals (e.g., Bennett, 2003).

2.1 The Conditional Logic Project in an Interdisciplinary Context

In Chapter 1 we defined the conditional logic project as the project which investigates alternative logical formalism for conditionals to the classical construction (in p.c. and f.o.l.) in terms of the material implication. The conditional logic project – the study of conditional logics – started in the philosophical literature.[1] Despite that fact conditional logics are now investigated in a range of disciplines such as linguistics (e.g., Stalnaker, 1975; Kratzer, 1977, 1981; Portner, 2009), computer science,[2] and psychology (e.g., Evans & Over, 2004). In all those disciplines several distinct but related projects emerged. These projects can be roughly characterized the following way:

(0) the conditional logic project,
(1) the linguistics of conditionals project,
(2) the philosophy of conditionals project,
(3) the psychology of reasoning project, and
(4) the non-monotonic logic project.

Projects (1)–(4) are all related to *the conditional logic project* but pursue distinct goals. I will first describe the conditional logic project and, then, discuss projects (1)–(4) and their relation to the conditional logic project.

Note that my categorization of the individual projects is by no means the only one possible.[3] The main aim of this section is, however, to provide a pragmatic overview over the interdisciplinary ramifications of the condi-

[1] See, for example, Adams (1965, 1966), Stalnaker (1968), Stalnaker and Thomason (1970), and D. Lewis (1971, 1973b). Earlier examples are C. I. Lewis (1912) and C. I. Lewis and Langford (1932).

[2] See, for example, Kraus et al. (1990), Lehmann and Magidor (1992), Ginsberg (1986), Delgrande (1987, 1988, 1998), and Nejdl (1992).

[3] See for example Bremer (2005b, Chapter 1) for an alternative characterization of projects in a formal semantics context.

tional logic project and for that purpose my delineation of the individual projects seems well suited.

2.1.1 The Conditional Logic Project

The aim of the *conditional logic project* is to specify logic systems which can describe conditionals more adequately than classical logic. (This motivation is discussed in detail in Chapter 1.) By means of conditional logics the scope of logical analyses is, then, broadened and logical tools are developed, which allow for a more fine-grained (language-based) analysis of concepts and their application. Examples for the application of conditional logic formalisms are D. Lewis' accounts of causation (e.g., D. Lewis, 1973a, 2000; cf. Section 3.4.3) and dispositions (e.g., D. Lewis, 1997). Furthermore, in Chapters 3–7 I will describe a range of conditional logics that are investigated within the conditional logic project.

2.1.2 The Linguistics of Conditionals Project

The linguistics of conditionals project aims to provide an adequate empirical description of conditionals in natural language. This can either be done on a syntactic/grammatical (e.g., Haegeman, 2003), a semantic (e.g., Portner, 2009) or a pragmatic level (e.g., Declerck & Reed, 2001). Its aim is always descriptive in the sense that an account of actual linguistic behavior is aimed at.

Let us now focus on the relation of the conditional logics and the linguistics of conditionals project. Observe in that context that philosophers such as Adams, D. Lewis, and Stalnaker who developed conditional logics also published on clearly linguistic topics (e.g., Adams, 1970; D. Lewis, 1970; Stalnaker, 1975).

I shall now discuss a specific example for the use of conditional logic formalisms in the literature on the linguistics of conditionals, namely Portner's (2009) approach. Portner (2009) gives a semantic account of modalities and conditionals in natural language based on Angelika Kratzer's work (e.g., Kratzer, 1977, 1981), which had itself a strong impact on linguistics. This approach is based on D. Lewis' systems of spheres semantics (D. Lewis, 1973b; see Section 3.2) and standard modal logic (e.g., Hughes & Cresswell, 1996/2003). Portner (2009) is thereby not interested in the logical and meta-logical properties of logic systems, but uses the formal framework to model the meaning of natural language conditionals and modal expressions (Portner, 2009, p. 49f). For that purpose he takes the context in which such expressions occur into account. In Portner's approach modal expressions such as 'must' and 'should' are viewed as indexicals, whose actual meaning depends also on the context. For example, 'must' can be interpreted in a deontic sense ('dog owners must keep their animals indoors'), but also epistemically ('it must be raining outside', Portner, 2009, p. 30). Thus, Portner uses the parameter 'context' to determine the meaning of linguistic entities, where the paramter 'context' specifies the speaker, the addressee, the time of utterance, and the place of utterance (Portner, 2009, p. 49).

According to Portner (2009, pp. 50–52) there is a second contextual parameter called 'conversational background'. This parameter is used to determine the meaning of sentences with modifiers, such as 'in the view of what I know' and 'in the view of the rules of the secret committee' and is used to determine the type of modality employed (Portner, 2009, pp. 50–52). Portner, then, uses both parameters, *conversational background* and *context*, for his semantics for natural language conditionals (p. 81f): Both parameters are employed to determine an ordering of possible worlds in a D. Lewis' type semantics for conditionals, which gives one truth conditions for conditionals.

2.1.3 The Philosophy of Conditionals Project

The aim of *the philosophy of conditionals project* is to provide a plausible and rationally justified theory of conditionals. For that purpose ordinary language often serves as a guide. The philosophy of conditionals project is, however, normatively loaded insofar as a rational justification for the use and interpretation of conditionals is aimed at. This approach is at its core restricted to an informal discussion of these issues and the philosophical discussion in that area can be used to provide a more adequate analysis of philosophical problems from areas such as philosophy of language, epistemology, and philosophy of science (e.g., Schurz, 2008, p. 90f).

The philosophy of conditionals project and the conditional logic project have a strong mutual influence on each other. On the one hand, conditional logics are taken as formal descriptions of normative theories of conditionals (e.g., Bennett, 2003, pp. 163–168). On the other, in the conditional logic project often the philosophical underpinnings of the formalisms are discussed. For a discussion of theses from the philosophy of conditionals project I refer to Chapters 3 and 7 (see in particular Sections 3.4, 3.5, 3.7, and 3.8).

2.1.4 The Psychology of Reasoning Project

A further project associated with the conditional logic project is the *psychology of reasoning project* which aims to provide an empirically adequate account of human reasoning. One of the most active and interesting areas in this project is reasoning with conditionals (cf. Evans & Over, 2004). This project is descriptive in the sense that it aims at an empirically adequate account of human beings' actual reasoning behavior. Contrary to the linguistics of conditionals project it is less concerned with the linguistic representation of conditionals.

One motivation for the study of conditional logics is that these logics are purported to be descriptively more accurate than classical logic in accounting for human beings' reasoning with conditionals. As basis for the construction of new conditional logic formalisms often one's intuitions about human beings' reasoning behavior with conditionals is employed (see Chapter 1) and it is argued that these intuitions largely agree with human beings' actual reasoning behavior (also called 'common sense reasoning'). To provide support for the latter claim however, one cannot restrict oneself to rational analyses but also has to investigate empirically whether these claims hold up (cf. Pelletier, Elio, & Hanson, 2008). One aim of the psychology of reasoning project is to do exactly that.

Traditionally, psychologists took classical logic and syllogistic as their normative standard. The performance of participants on conditional reasoning tasks was, then, judged against the predictions made by a material implication analysis of the task. A still dominant approach originating from this tradition is the mental model theory of Johnson-Laird and Byrne (2002). According to that approach, human beings use semantic means to perform logical reasoning with conditionals, very much like reasoning with truth tables (p. 650) and represent only those lines of the truth table in which the respective (conditional) formulas are assigned the value 'true' (p. 653). Moreover, human beings do not start with a fully fleshed-out mental model but begin with a mental model that makes only those cases explicit in which the antecedent and the consequent are true (p. 654). The context can, then, help participants to flesh out a complete "mental model" and produce the normatively correct solution which agrees with a material implication analysis of conditionals (p. 659).

More recently, the normative standard in this literature seems to have changed. Evans, Handley, and Over (2003) and Oberauer and Wilhelm (2003), for example, demonstrated that subjects interpret conditionals in terms of conditional probabilities (in line with a probabilistic conditional logic semantics) rather than in terms of the material implication. Furthermore, Douven and Verbrugge (2010) investigated empirically whether the Ramsey test discussed in the philosophy of conditionals project (see Section 3.3) is also plausible from the perspective of the psychology of conditionals project. Moreover, Schurz (2007) inquired empirically into reasoning with normic conditionals (cf. Section 1.4) and regular indicative conditionals in the context of conflicting information as described by example (A) in Section 1.4.4 (see also Unterhuber & Schurz, in press). His results show that human reasoning is in line with probabilistic default reasoning and conditional logic approaches rather than with a material implication analysis of conditionals. Pfeifer and Kleiter (2010), moreover, give an overview over a series of probabilistic reasoning experiments in which specific axioms of a System **P** (Gilio, 2002; Biazzo, Gilio, Lukasiewicz, & Sanfilippo, 2002; see Section 3.6) were investigated and contrasted with inferences S3′–S5′ (Transitivity, Monotonicity, and Contraposition, respectively) from Chapter 1. Their results show that subjects' reasoning concurs largely with the inferences predicted by System **P**.

Evans and Over (2004, Chapter 7) specifically discuss D. Lewis' (1973b) systems of spheres semantics regarding the reasoning with counterfactual conditionals (see Section 3.2). Note that counterfactual reasoning is also investigated from a developmental perspective (e.g., Beck, Robinson, Carroll, & Apperly, 2006; Beck & Guthrie, 2011; Rafetseder, Cristi-Vargas, & Perner, 2010; Rafetseder & Perner, 2010).

2.1.5 The Non-Monotonic Logic Project

Starting in the late 70s a number of logic systems have been developed in computer science and in particular in the artificial intelligence community which deviate from classical logic in an essential and more radical way than, for example, modal logic (e.g., Hughes & Cresswell, 1996/2003). Central to this deviation is the notion of a non-monotonic consequence or derivability relation,[4] where a consequence relation specifies which formulas α are logically implied by which sets of formulas Γ (short: $\Gamma \models \alpha$). In classical logic and modal logic (e.g., Hughes & Cresswell, 1996/2003) the consequence relation is monotonic. That means that if $\Gamma \models \beta$, then $\Delta \models \beta$ for any superset Δ of Γ. So, no matter which formulas are added to Γ, α is still a consequence of this new set. This principle does not hold generally for the systems investigated in the computer science literature. Since this seems to be the characteristic feature of this approach, these systems are often called 'non-monotonic logics'.

A key motivation for the study of non-monotonic logics is the observation that the notion of a default (a type of non-monotonic rule) is required in order to provide an adequate account of general intelligence (McCarthy & Hayes, 1969; McCarthy, 1980, 1989). Since this motivation is closely related to the notion of defaults and non-monotonic rules discussed in this literature, I will investigate this issue in more detail in Section 2.2.

[4] One can, alternatively, describe the following by means of a proof-theoretic derivability relation.

Interestingly the impact of the non-monotonic logic literature on the conditional logic literature is rather limited. In constrast the conditional logic approach is well perceived in the non-monotonic logic literature (e.g., Ginsberg, 1986; Delgrande, 1987, 1998, 1988; Kraus et al., 1990; Nejdl, 1992; Lehmann & Magidor, 1992; Schurz, 1998).

2.2 Non-Monotonic Logics, Defaults Logics, and Conditional Logics

In this section I will focus on the relation of non-monotonic logics, default logics, and conditional logics. This point is important, since both default logics and conditional logics are related but pursue different goals and use distinct formal methods. First, I will sketch in Section 2.2.1 a motivation for the study of default logics which stems from the non-monotonic logic project (see Section 2.1.5) on the basis of the well-known cannibals and missionaries problem. Then, in Section 2.2.2 I will describe a particular default logic approach – Reiter defaults – and discuss the relation of defaults and normic conditionals from Section 1.4. In Section 2.2.3 I will then give an alternative characterization of Reiter defaults in terms of non-derivability and consistency conditions, apply these alternative characterizations to the cannibals and missionaries problem, and discuss default logics in the context of inductive inferences. In Section 2.2.4 I shall discuss different types of (non-)derivability relations and the role of the rule of the substitution in default logics. In Section 2.2.5 I will inquire into the model-theoretic and the proof-theoretic description of default theories and their axiomatization. Finally, in Section 2.2.6 I will discuss in which way default logics and conditional logics differ.

2.2.1 A Motivation for the Study of Non-Monotonic Logics

In this section I shall describe a general observation by McCarthy and colleagues based on the well-known cannibals and missionaries problem where the former of which served as a starting point for exploration of non-monotonic logics. McCarthy and colleagues (McCarthy & Hayes, 1969; McCarthy, 1980, 1989) were interested in a general formal account of intelligence that does not depend on a pre-established formal representation format of the problem at hand. They rather focussed on how the representation of the world must be such that "the solution of the problems follows from the facts expresssed in the representation" (McCarthy & Hayes, 1969, p. 466). Their approach is, hence, akin to the logical programming paradigm as in the programming language PROLOG. In this paradigm a program is not told what to do, but is instead told what is true and asked to draw conclusions from these specifications (Clocksin & Mellish, 2003, p. 255).

For a discussion of adequate representations of the world, let us focus on a standard problem from computer science, namely the cannibals and missionaries problem:

> "Three missionaries and three cannibals come to a river. A rowboat that seats two is available. If the cannibals ever outnumber the missionaries on either bank of the river, the missionaries will be eaten. How shall they cross the river[?]" (McCarthy, 1980, p. 29)

For an adequate representation of this problem, we have to specify that there exist three cannibals (c_1, c_2, and c_3), three missionaries (m_1, m_2, and m_3), a boat (b), etc. Let Γ be a set of formulas which describes all these facts. In order for an agent to draw any (relevant) conclusion from the problem situation (e.g., what to do next), we are forced to introduce certain closed-world conditions in addition to the facts described by Γ, namely that there are no more cannibals, missionaries, boats than specified, that the boat is the only means to cross the river, etc. (cf. McCarthy, 1980, p. 30). Let Δ be the set of formulas which describe these closed world assumptions (e.g., that the boat is the only means to cross the river).

We can, then, represent the cannibals and missionaries problem by classical logic in terms of sets Δ and Γ and draw conclusions from these sets as described above. However, such an approach has one great disadvantage. If we change the problem situation slightly – for example by including a bridge, on which both the cannibals and missionaries can cross the river – the representation of the problem situation based on Δ and Γ becomes inconsistent. This is due to the fact that given this change all closed world conditions specified earlier still hold (i.e., that the boat is the only means to cross the river), based on the monotonicity property of classical logic, that is that whenever we can derive a formula α from a set of formulas Γ' (short: $\Gamma' \vdash \alpha$), then we can also do so for any superset Γ'' of Γ'. Thus, if we add some facts or closed world assumptions to Δ or Γ, all formulas are still derivable that were derivable from $\Delta \cup \Gamma$ earlier (cf. Section 2.1.5). So, to represent for example the changed cannibals and missionaries problem we also have to change Δ in order not to arrive at an inconsistency. Hence, for a general account of intelligence as described above this type of representation format seems hardly adequate in terms of classical logic and quite artificial. Especially Δ seems to be arbitrary, the more as one has to introduce a different set of closed world assumptions for each new problem situation, although the problem situation might only differ slightly.

Let us take a closer look at the above representation of the cannibals and missionaries problem. It is not the set of factual formulas Γ that is problematic but the closed world assumptions in Δ. To avoid the rigidity of the classical representation format of the cannibals and missionaries problem above, we can introduce closed world *conditionally* rather than unconditionally.

In the non-monotonic logic literature the notion of a default is used to describe these conditional closed world assumptions in Δ. Defaults, informally stated, are rules of the form "in the absence of any information to the contrary, assume ..." (Reiter, 1980, p. 81). Note that this type of rule explicitly allows for exceptions: If one has information to the contrary, one is not allowed to draw the respective conclusion. In contrast, for a consequence relation to be monotonic (as in classical logic), no such exceptions are allowed to hold. To see this more clearly let us first specify formally what these defaults look like and then later apply them to the cannibals and missionaries problem.

2.2.2 Reiter Defaults

Let us in this section describe Reiter defaults (Reiter, 1980, p. 88f) and compare them with normic conditionals from Section 1.4 based on Schurz's (1994) representation result. Defaults in Reiter's approach have the following form: [5]

$$\frac{\alpha : \mathbf{M}\beta_1, \ldots, \mathbf{M}\beta_n}{\gamma} \quad (\text{for } n \in \mathbb{N})$$

$\mathbb{N}^+$ is the set of positive integers. The formulas α and γ represent the prerequisite of the default and its consequent, respectively (Reiter, 1980, p. 88). The above Reiter default is to be read the following way: If α is the case and $\beta_1, \ldots \beta_n$ are possible, then one is allowed to conclude γ.

Reiter (1980) defaults are interpreted in the context of extensions of default theories. A default theory Θ is an ordered pair $\langle \Delta, \Gamma \rangle$, where Δ is a set of defaults (as specified above) and Γ a set of f.o.l. formulas. An extension of such a default theory extends the set Γ in such a way that it is deductively closed under f.o.l. and closed under the set of defaults Δ. The expressions $\mathbf{M}\beta_1, \ldots, \mathbf{M}\beta_n$ in the Reiter default described above say that $\neg\beta_1, \ldots, \neg\beta_n$ are not in the respective extension (cf. Reiter, 1980, p. 89).

To specify extensions formally, let us denote the set of f.o.l. theorems of a set of formulas Γ by Th(Γ). Then, an extension E of a default theory $\Theta = \langle \Delta, \Gamma \rangle$ is a set of f.o.l. formulas which satisfies the following conditions: (a) $\Gamma \subseteq E$, (b) $E = \text{Th}(E)$, and (c) if $\alpha : \mathbf{M}\beta_1, \ldots, \mathbf{M}\beta_n / \gamma$ is a default in Δ, [6] then it holds that if $\alpha \in E$ and $\neg\beta_1, \ldots, \neg\beta_n \notin E$ then $\gamma \in E$.

To describe the defaults discussed in the previous section and to apply them to normic indicative conditionals (cf. Section 1.4), a less general notion suffices, namely (closed) normal defaults. Reiter (1980, p. 94f) specifies

[5] I focus here exclusively on closed defaults. This type of default contains exclusively closed f.o.l. formulas, that is f.o.l. formulas in which no free individual variables occurs (cf. Reiter, 1980, p. 88).

[6] The expression '/' is the inference indicator as in the Reiter default described above.

(closed) normal defaults the following way:

(D1) $$\frac{\alpha : \mathbf{M}\beta}{\beta}$$

This type of default has only one possibility requirement and this requirement refers, in addition, to the consequent of the default.

Schurz (1994) shows that normal defaults of type (D1) can alternatively be read as normic conditionals of the form $\alpha \,\Box\!\!\rightarrow \beta$. For that purpose Schurz (1994) gives an interpretation of modified default theories $\Theta = \langle \Delta, \Gamma \rangle$ – default theories where Δ contains only normal defaults – in terms of the probabilistic $\mathbf{P}_\epsilon$ semantics of Adams (1975; see Section 3.6.3). In Schurz's framework an extension E of a default theory Θ corresponds to a set of default assumptions which is generated by E (see Schurz, 1994, esp. Theorems 4 and 5, p. 255f): Whenever a formula α is to be added to an extension E of a default theory $\langle \Delta, \Gamma \rangle$ by a normal default $\alpha \,\Box\!\!\rightarrow \beta$, all other facts in Γ must be regarded as irrelevant insofar as they do not imply $\neg\beta$ conjointly with Δ. A set of defaults for an extension E for a default theory Θ encodes, then, the irrelevance assumptions, which are made in the course of the construction of the extension E.[7]

2.2.3 Default Logics, Non-Monotonic Rules, and Inductive Inferences

In this section I will focus on the characterization of default logics and describe the relation of default logics with non-monotonic rules and inductive inferences. For that purpose (i) I will first reformulate the possibility requirement as specified in Reiter defaults in terms of non-derivability and consistency conditions. Then, (ii) I will apply these reformulations to the cannibals and missionaries problem from Section 2.2.1 and (iii) discuss on that basis the notion of default logics and non-monotonic rules. Finally, (iv) I will

[7] Schurz (1997b, p. 545f) also characterizes Poole (1988) defaults in terms of a probabilistic conditional logic semantics.

explore the connexion between default rules and inductive inferences.

Let us now focus on (i). Although from a technical perspective the reformulations of Reiter defaults are quite trivial, these reformulations allow us to see why normal Reiter defaults suffice for the specification of defaults discussed in Section 2.2.1. Furthermore, based on these reformulations we can compare the default logics more easily with conditional logic approaches in Section 2.2.6.

We can now reformulate the possibility requirement $\mathbf{M}\alpha$ in Reiter (1980) defaults the following way:

(C1) $\neg\alpha \notin \mathrm{Th}(\Gamma)$

Condition (C1) makes the meaning of the possibility requirement in (normal) defaults explicit. Both Reiter's original formulation and (C1) are understood here in a syntactic way and, hence, should be regarded as a part of a system's proof theory rather than its model theory.

(C1) has the following equivalent formulation in terms of consistency conditions:

(C2) $\Gamma \cup \{\alpha\}$ is consistent

To see this, consider the following: If $\neg\alpha \in \mathrm{Th}(\Gamma)$, then $\Gamma \cup \{\alpha\}$ is inconsistent. Furthermore, if the set $\Gamma \cup \{\alpha\}$ is inconsistent, then $\neg\alpha \in \mathrm{Th}(\Gamma)$.

As can easily be seen, (C2) is equivalent to the following condition:

(C3) $\Gamma \nvdash \neg\alpha$

Here, $\Gamma \nvdash \alpha$ denotes that α cannot be derived from Γ (where $\Gamma \vdash \alpha$ states that α can be derived from Γ).

Let us now discuss point (ii). In Section 2.2.1 I specified defaults informally as rules of the form: "in the absence of any information to the contrary, assume ..." (Reiter, 1980, p. 81). In the cannibals and missionaries problem we can use defaults such as 'in absence of information to the contrary, assume that the boat is the only means to cross the river'. Defaults of this type can be represented the following way: Conclude α if $\neg\alpha$ is not the case.

This might sound awkward at first glance, although it is not. To show the latter I shall discuss (C1), (C2), and (C3) in the context of the cannibals and missionaries problem.

Let us focus on (C1). For an adequate representation of the cannibals and missionaries problem one can start with a set of facts Γ and apply conditional closed world conditions from Δ to it (cf. Section 2.2.1). For the construction of extensions from the set of facts Γ one can, then, apply defaults such as 'the boat is the only means to cross the river', if the set Γ does not imply that the boat is not the only means to cross the river. More formally, this precondition corresponds to (C1) and the corresponding default has the following form: I add α to Γ if $\neg\alpha$ is not a theorem of Γ. Note that this default is a normal Reiter (1980) default. It is, however, a degenerated one, since this default does not have a prerequisite.

For the construction of set of defaults Δ one does not consider all logically possible normal Reiter defaults as candidates, but only (normal) Reiter defaults which seem acceptable as normic conditionals in the given context. The default rule discussed in the previous two paragraphs is such an acceptable default assumption. An absurd, but logically possible default assumption would be the following: 'In absence of information to the contrary, every cannibal marries a missionary'. Observe that the latter default assumption would in the present context neither hold up as an acceptable indicative normic conditional (cf. Section 1.4).

Let us now focus on (C2). Due to the equivalence of (C1) and (C2) the above default can also be reformulated the following way: We add α to Γ if $\Gamma \cup \{\alpha\}$ is consistent. This formulation shows why defaults are more appropriate for the representation of closed world conditions than the addition of unconditional closed world assumptions. Normal Reiter defaults allow one to add a closed world assumption such as 'the boat is the only means to cross the river' only if the resulting set does not become inconsistent. This procedure, hence, preserves the consistency of a set Γ and default logics as described above can, thus, be regarded as a consistency-based approach.

Let us reformulate this type of default in terms of (C3). Then, one is allowed to add α to Γ given that $\Gamma \nvdash \neg\alpha$ holds. Non-derivability conditions are not a part of the axiomatization of classical logic, since such an axiomatization is strictly based on rules which only refer to derivability conditions. A canonical example for a classical rule is the Modus Ponens rule. According to that rule we are allowed to conclude $\beta \in \Gamma$ for a formula β and a set of formulas Γ if there is a formula $\alpha \in \Gamma$ such that $\alpha \rightarrow \beta \in \Gamma$ holds. In the case of Modus Ponens the addition of formulas can also be described in terms of the classical derivability relation $\vdash$ the following way: If $\Gamma \vdash \alpha \rightarrow \beta$ and $\Gamma \vdash \alpha$, then $\Gamma \vdash \beta$.

Since defaults are designed to go beyond classical logic, a characterization of these inferences in terms of the classical derivability relation $\vdash$ is not adequate. Hence, in the non-monotonic literature the alternative symbol $\mid\!\sim$ is often used. One might describe (C3) the following way according to a default theory $\Theta = \langle \Delta, \Gamma \rangle$: $\Gamma \mid\!\not\sim \neg\alpha$. A set which is closed under the non-motonic derivability relation relative to a default theory Φ is, then, an extension of Φ in the above sense.

We shall now focus on (iii). We can describe Reiter defaults the following way:

$$\frac{\Gamma \mid\!\sim \alpha \;, \Gamma \mid\!\not\sim \neg\beta_1, \ldots, \Gamma \mid\!\not\sim \neg\beta_n}{\Gamma \mid\!\sim \gamma} \quad (\text{for } n \in \mathbb{N})$$

The inference relation $\mid\!\sim$ thus specified is to be understood always relative to a default theory $\Theta = \langle \Delta, \Gamma \rangle$ (cf. Section 2.2.2). Moreover, unlike earlier the premises of the rule employ a non-monotonic inference relation $\mid\!\sim$ for its derivability and non-derivability conditions. The main reason for this change is that $\mid\!\sim$ is used for an alternative characterization of extensions. Accordingly, the non-monotonic inference relation can be employed to specify sets of formulas, which can be derived from the set Γ of a default theory $\Theta = \langle \Delta, \Gamma \rangle$ by the defaults in Δ. The resulting sets are, then, extensions in the above sense.

Moreover, we can also describe normal Reiter defaults the following way:

$$\frac{\Gamma \mid\!\sim \alpha \,, \Gamma \not\mid\!\sim \neg\beta}{\Gamma \mid\!\sim \beta}$$

The inference relation $\mid\!\sim$ in Reiter defaults is thus non-monotonic in the sense explained in Section 2.2.1. The non-monotonicity property results from the reliance of default logics on non-derivability conditions in non-monotonic rules.[8]

Default rules are, accordingly, also called 'non-monotonic rules' (Schurz, 2008, p. 55f; Schurz, 2001a, p. 373; cf. also Section 4.2.3). While monotonic rules only refer to derivability conditions, non-monotonic rules additionally employ at least one (non-trivial) non-derivability condition. Since it is characteristic for default logic approaches to include non-monotonic rules, I will regard any logical system that essentially refers to (non-trivial) non-monotonic rules as a default logic.

Let us now focus on (iv), the relation between default rules and inductive inferences. An inductive inference is, for example, the following (cf. Section 1.4.1): All fishes observed up to now are cold-blooded. Hence, (probably) all fishes are cold-blooded (see Schurz, 2008, p. 47). According to classical logic this inference is logically invalid, since for any finite number of fishes which are cold-blooded, there might still be a larger number of fishes that are not cold-blooded. Hence, the observation that a finite number of fishes is cold-blooded does not guarantee that all fishes or even the majority of fishes are cold-blooded. Nevertheless, one would accept the inference in many practical situations (where the number of observed fishes is large enough).

We shall now see why the inductive inference in the preceding paragraph is non-monotonic. Let Γ be the observation that n fishes ($n \in \mathbb{N}^+$) are cold-blooded. Suppose we are justified in drawing the inductive inference $\Gamma \mid\!\sim \alpha$

[8] Not all default rules make the inference relation non-monotonic. The following normal default rule does not require the derivability relation to be non-monotonic: $\Gamma \mid\!\sim \alpha$, $\Gamma \not\mid\!\sim \top / \Gamma \mid\!\sim \gamma$: As $\top$ is a theorem of any set of formulas whatsoever, this default rule is inapplicable.

for a sufficiently high n, where the formula α represents the state of affairs that all [most] fishes are cold-blooded. Then the inference does not hold up under all circumstances. Let β represent the (unlikely) observation that m fishes ($m \in \mathbb{N}$) with $m > n$ are not cold-blooded. Here it is not the case that $\Gamma \cup \{\beta\} \mid\!\sim \alpha$. Thus the inductive inference is also non-monotonic in the sense specified in Section 2.2.1.

Thus, both non-monotonic rules and inductive inferences share that the truth of the formulas in Γ does not guarantee the truth of the conclusion α, whatever the circumstances. In the case of non-monotonic rules, circumstances can arise which qualify as exceptions, as indicated by non-derivability conditions (see below). In the case of inductive inferences there might still be convincing counter-evidence that blocks the inductive inference.

Furthermore, Carnap (1962, pp. 192–202) went so far as to postulate two different types of logics: deductive logics and inductive logics. Although this terminology is controversial (Schurz, 2008, p. 47), one can identify logical systems which draw only on monotonic rules as as deductive logics, and default logics as described above as inductive logics (cf. Schurz, 2008, pp. 54–56).

Given the above considerations it hardly seems surprising that we are in need of a default logic approach for a general account of intelligence as described in Section 2.2.1. For intelligent behavior, agents have to be able to draw both deductive and inductive inferences. As we saw in this and the preceding sections default logics allow – in contrast to classical logic – for an adequate account of these inductive inferences, as required for a non-rigid account of closed world conditions. In the framework of McCarthy and colleagues the agents are, then, able to draw inferences (e.g., that the boat is the only means to cross the river) from the absence of other facts (e.g., the boat is the only means to cross the river that is *explicitly mentioned* in the description of the problem). Moreover, as non-derivability conditions can also be formulated in terms of consistency requirements, default logics represent a consistency-based approach, which relies alternatively on consistency requirements.

Prima facie these closed world conditions might seem very natural and almost self-evident. From a formal perspective they are by far non-trivial and require a major deviation from classical logic. Furthermore, the exact specification of those closed world conditions is not an easy task. Note that closed world conditions as discussed here seem to be closely connected to Gricean conversational implicatures (see Section 1.5). Again an exact specification of Gricean conversational implicatures in terms of defaults is far from trivial.

2.2.4 Types of Non-Derivability Conditions and the Rule of Substitution

Let us here review the notions of derivability and non-derivability and discuss the role of the rule of substitution in default logics. For that purpose (i) I will start with some basic properties of derivability in classical logic and then (ii) distinguish between (a) derivability simpliciter (derivability of single formulas) and (b) derivability of formulas from sets of formulas. (iii) I will then contrast the use of (non-)derivability conditions of type (b) in systems, such as Reiter's (1980) default logic with (non-)derivability conditions of type (a) in systems such as Adams' (1975) Conditional Logic System $\mathbf{P}_{\epsilon}$ (see Section 3.6.3). (iv) Finally, I will explain why the rule of substitution does not hold for default logics.

Let us begin with (i) and with a discussion of the notion of derivability in classical logic. Derivability of a formula α from a formula set Γ in classical logic (short: $\Gamma \vdash \alpha$) is always reducible to derivability of a finite subset Γ_f of Γ. Furthermore, it holds that $\Gamma \vdash \alpha$ exactly if $\Gamma_f \vdash \alpha$, which holds exactly if $\vdash \bigwedge \Gamma_f \rightarrow \alpha$ (the deduction theorem), where the expression $\bigwedge \Delta_f$ represents the conjunction of all elements in the finite set of formulas Δ_f.

We can describe the non-derivability in classical logic analogously. A formula α is non-derivable from Γ (short: $\Gamma \nvdash \alpha$) exactly if there is no finite subset Γ_f of Γ such that $\Gamma_f \vdash \alpha$. This condition holds exactly if there is no finite subset Γ_f of Γ such that $\vdash \bigwedge \Gamma_f \rightarrow \alpha$.

Let us now focus on point (ii). We can distinguish between two types of (non-)derivability conditions: (a) derivability simpliciter and (b) derivability from a set of formulas. If one uses (a) and considers derivability (non-derivability) conditions of formulas alone, one only has to check whether a given formula α is derivable (non-derivable) simpliciter (i.e., $\vdash \alpha$), without reference to a set Γ (i.e., $\Gamma \vdash \alpha$).

Let us now discuss point (iii). Approach (b) is needed for Reiter defaults, since for the application of this type of default it is important to find out whether a formula α follows from a particular set of formulas or is consistent with it rather than just to investigate whether a formula is (non-)derivable simpliciter (as in approach (a)). This is due to the following fact: In Reiter defaults, sequences of applications of default rules are permitted. In the course of these applications, the supersets of Γ which are used for the construction of extensions in Reiter defaults are referenced repeatedly and, thus, are changed, where Γ is the set of facts in a default theory. Hence, in these cases one cannot replace the (non-)derivability condition of a formula α from a formula set Γ in a rule by a (non-)derivability of single formulas.

There are, however, default logic systems for which (non-)derivability conditions simpliciter suffice (as in approach (a)). For example, in Adams' (1975; Schurz, 1997b; see Section 3.6.3) System $\mathbf{P}_\epsilon$ a rule with a derivability condition simpliciter is employed. To describe this non-monotonic rule let us represent conditionals, as in Chapter 1, in terms of formulas of the form $\alpha \mathrel{\Box\!\!\rightarrow} \beta$, where α and β are the antecedent and the consequent formula, respectively. The non-monotonic rule in Adams (1975, Rule R7, p. 61) gives us that any set of formulas which contains both formulas $\alpha \mathrel{\Box\!\!\rightarrow} \beta$ and $\alpha \mathrel{\Box\!\!\rightarrow} \neg\beta$ is inconsistent, provided α is consistent. In natural language this seems plausible, since it is, for example, natural to regard conditionals such as 'if Anna goes shopping she is happy' and 'if Anna goes shopping she is not happy' as conjointly contradictory (cf. Section 3.6.3).

To account for Adams' (1975) conditional consistency principle, one has to require that the antecedent of such a conditional formula is consistent. Otherwise it would contradict a very intuitive and central axiom of conditional logic called 'Refl' ("Reflexivity"), namely $\alpha \mathrel{\Box\!\!\rightarrow} \alpha$ (see Section 3.6.3). Since this is the only precondition for Adams' rule, one can specify RCNC the following way, where $\nvdash_{p.c.}$ refers to non-derivability in p.c.:

RCNC if $\nvdash_{p.c.} \neg\alpha$ then $\neg((\alpha \mathrel{\Box\!\!\rightarrow} \beta) \wedge (\alpha \mathrel{\Box\!\!\rightarrow} \neg\beta))$

'RCNC' stands for 'Restricted Conditional Non-Contradiction'. That a formula $\neg\alpha$ is non-derivable in p.c. is is tantamount to α to being p.c.-consistent (see Section 2.2.3).

RCNC is a non-monotonic rule, since it refers to a non-trivial non-derivability condition. Thus, Adams' (1975) System $\mathbf{P}_\epsilon$ is a genuine default logic. Observe that System $\mathbf{P}_\epsilon$ is a default logic in a weaker sense than, for example, Reiter's (1980) default logic, since the former refers only to the non-derivability of single formulas (in p.c.). Adams' alternative probabilistic conditional logic systems (Adams, 1965, 1966, 1977, 1986) do not endorse any kind of non-monotonic rule (see Section 3.6.3; cf. Schurz, 1998, p. 84) and are therefore no default logics. Note that Adams' own formulation of RCNC differs from my specification, since he uses a restricted language for that purpose (see Section 3.6.3).

Finally, let us focus on (iv), that is why in default logics the rule of substitution does not hold (cf. Schurz, 2004, pp. 35–38) – unlike in classical logic. In order to see that, let us take a look at the rule RCNC from Adams' (1975) System $\mathbf{P}_\epsilon$. Although $\neg((p \rightarrow q) \wedge (p \rightarrow \neg q))$ is a theorem of System $\mathbf{P}_\epsilon$ by RCNC, $\neg((p \wedge \neg p \rightarrow q) \wedge (p \wedge \neg p \rightarrow \neg q))$ is not, since $p \wedge \neg p$ is inconsistent and so the non-derivability condition of RCNC is not applicable.

The failure of the rule of substitution has the following consequence. If the rule of substitution does not hold, it is not just a matter of logical form whether an inference is licenced. Inferences that are licenced by default logics are, thus, *not* logical (or structural), viz. they do not hold independently of any particular facts.

Note here two things. First, there exists a more restricted rule of substitution, the rule of isomorphic substitution which holds in general in default logics (see Section 2.2.6; cf. Schurz, 2001a, p. 370; Schurz, 2004, p. 38). Second, I will assume in the following chapters that logics are in general closed under substitution (see Sections 3.6.1 and 4.2.2). So, in my use of the term 'logic', default logics are in fact not logics.

2.2.5 Model Theory, Proof Theory, and Axiomatization of Default Logics

In this section I will discuss two points w.r.t. default logics: (i) a model-theoretic as opposed to a proof-theoretic characterization of default logics and (ii) the axiomatization of default logics.

Let us focus on (i). Earlier in this chapter I described default logics, such as Reiter (1980) defaults, in proof-theoretic terms. In the non-monotonic literature the inference relation $\mid\!\sim$, however, is often formalized in terms of a model-theoretic consequence relation rather than a proof-theoretic derivability relation (e.g., Makinson, 1994; Fuhrmann, 1998). Although the authors in the non-monotonic literature are not very specific as to why they use a model-theoretic notion rather than a proof-theoretic one, one important reason seems to be that often no finite (syntactic) procedure exists (as required for a proof-theoretic notion) which describes the set of theorems of these formalisms.

I shall now focus on (ii), the axiomatization of default logics. The following two options exist for the axiomatization of default logics: (a) One can isolate all theorems which require reliance on non-derivability conditions and add them to the set of axioms. Alternatively, (b) one can simultaneously axiomatize the sets of derivable and non-derivable formulas.

I will now discuss approach (a). Let us focus on Adams' (1975) default logic System $\mathbf{P}_\epsilon$, which – as we saw in Section 2.2.4 – is a genuine default logic that employs the non-monotonic rule RCNC. Strategy (a) would require that we add any instance of a formula to the set of axioms of System $\mathbf{P}_\epsilon$ that is derivable based on RCNC. In a variation of Adams' (1975) system, one might include any formula $\neg((\alpha \mathrel{\Box\!\!\rightarrow} \beta) \wedge (\alpha \mathrel{\Box\!\!\rightarrow} \neg\beta))$ where α is an atomic proposition. This procedure, however, does not give us all substitution instances of RCNC. Moreover, the exact specification of axioms to be added is not an easy task, as one can easily see.

In contrast, in the somewhat less complex default logic $\mathbf{C}^{\mathbf{def}}$ ('def' for 'default') of Carnap this procedure is viable (Hendry & Pokriefka, 1985; Schurz, 2001a, p. 374).[9] For other default logic systems such an approach is not (yet) available. It is important to note that this method does not guarantee that the default logics thus axiomatized are also axiomatized in a finite way. On the other hand finite axiomatization is required for a finite, mechanical notion of proof (cf. Enderton, 2001, p. 109).

Let us now focus on approach (b). Schurz (2001a, 1996), for example, provides a full axiomatization of theorems and non-theorems for Carnap's System $\mathbf{C}^{\mathbf{def}}$ and some default logics discussed in the non-monotonic literature (e.g., Nute, 1992), respectively.

An advantage of approach (b) is that the simultaneous axiomatization of the sets of derivable and non-derivable formulas, as envisaged by approach (b), seems to be more natural than approach (a). A disadvantage of approach (b) is that it is not always applicable. To see this, note that the set of theorems and non-theorems encompasses the whole set of formulas. In order to be able to axiomatize arbitrary sets of formulas – as required by our notion of syntactic proof – these sets have to be decidable, viz. there must be a syntactic, mechanic, and finite procedure to check whether an arbitrary formula is a member of either set. This implies that both sets (theorems and non-theorems) have to be decidable. As Kleene's theorem tells us (Enderton, 2001, Theorem 17F, p. 64; cf. Schurz, 2004, pp. 42–44), the latter fact is only the case if the whole set of formulas is decidable. Since f.o.l. is undecidable (Church's Theorem, cf. Enderton, 2001, p. 238), it remains far from clear how f.o.l. default logics such as Levesque's (1990) and Reiter's (1980) systems can be completely (and finitely) axiomatized in terms of approach (b). However, for a specific type of logic, this approach is still viable: In case a logic is a (negation-)complete f.o.l. theory, it is decidable and, hence, allows for a direct axiomatization in terms of approach (b).

[9] I use the name $\mathbf{C}^{\mathbf{def}}$ rather than **C** (Schurz, 2001a, p. 365) to avoid ambiguity, since I discuss also Kraus et al.'s (1990, p. 176) System **C** in Section 7.2.1.

2.2.6 Conditional Logics and Default Logics

In this section I will discuss the difference between default logics and conditional logics. For that purpose I will focus on the conditional logic approach of Kraus et al. (1990) and Lehmann and Magidor (1992) who specify conditionals in terms of a meta-language inference relation rather than an object-language conditional operator.

In traditional conditional logic systems a conditional connective, such as $\Box\!\!\rightarrow$, is used to represent conditional structures from natural language (see Section 1.2). This conditional connective is a two-place (modal) operator which connects two formulas, namely the antecedent formula α and the consequent formula β of a conditional formula $\alpha \Box\!\!\rightarrow \beta$. The formal expression $\alpha \Box\!\!\rightarrow \beta$ is thus an object language formula and can be negated, iterated, nested, and so on in standard conditional logic approaches, such as Stalnaker (1968; see Section 3.3.3) and D. Lewis (1973b; see Section 3.2),. Hence, in these accounts expressions, such as $\neg(\alpha \Box\!\!\rightarrow \beta)$, $\alpha \Box\!\!\rightarrow (\beta \Box\!\!\rightarrow \gamma)$, and $\alpha \Box\!\!\rightarrow (\delta \wedge (\beta \Box\!\!\rightarrow \gamma))$ are object-language formulas (see also Sections 3.5.3 and 4.2.1).

In contrast default logics employ an inference (or derivability) relation $\mid\!\sim$ which is due to non-monotonic rules of default logics (see Section 2.2.3). The inference relation $\mid\!\sim$ is not a part of the object language, it is a meta-language expression which describes the formal relation between sets of object language formulas (i.e., Γ) and object language formulas (i.e., α). In addition, nestings and iterations of $\mid\!\sim$ neither result in object language nor in meta-language expressions. So, default logics are logics which have a non-monotonic inference (derivability) relation, while conditional logics are most typical logics which have a non-monotonic conditional operator.

Note that there are conditional logics that are not default logics in the above sense. Examples for semantics of this sort are Lewis models (D. Lewis, 1973b; see Section 3.2), Stalnaker models (Stalnaker & Thomason, 1970; see Section 3.3.3), Adams' (1965, 1966, 1977, 1986) Systems $\mathbf{P}^*$ and $\mathbf{P}^+$ semantics (see Section 3.6), and Chellas-Segerberg (CS) semantics (Chellas,

1975; Segerberg, 1989; see Chapters 4–7).

On the other hand, there are default logics which are not conditional logics. Examples for default logics of this type are the formalisms of Reiter (1980; see Section 2.2.2) and Levesque (1990). The object language of these systems does not have a conditional operator or the like. Moreover, it is quite implausible to interpret the non-monotonic inference relation directly as a conditional operator of some sort. [10]

Finally, there exist systems which are both conditional logics and default logics in the above sense. Adams' (1975) $\mathbf{P}_{\epsilon}$ semantics (see Section 3.6.3) and System **Z** semantics by Pearl (1990) and Goldszmidt and Pearl (1996; see also Schurz & Thorn, 2012) are particularly prominent examples for such mixed systems' semantics. In addition, there are a number of studies which describe the relation of AGM belief revision (Alchourrón, Gärdenfors, & Makinson, 1985; Gärdenfors, 1988) with conditional logics (e.g., Gärdenfors, 1988, 1986; Rott, 2001; Lindström & Rabinowicz, 1995; Leitgeb & Segerberg, 2007; Grove, 1988; Giordano, Gliozzi, & Olivetti, 2005; Baltag & Smets, 2008). The AGM belief revision system is also a default logic in the above sense since it essentially refers to a consistency condition. Despite this fact the default aspect of these systems is often not taken seriously in the literature. Observe, furthermore, that the relation between conditionals and AGM belief revision is not an easy task due to Gaerdenfors' (1986; see also Gärdenfors, 1988) triviality result.

Finally, let us discuss conditional logics which are formulated in a default logic terminology. Examples for this type of logic are the ones of Kraus et al. (1990), Lehmann and Magidor (1992), and Hawthorne and Makinson (2007). Unlike the default logics described earlier, these formalisms can be directly translated into a conditional logic language as described above. To tell the difference, one has to be able to distinguish between mere default logic terminology and essential default logic assumptions. The discussion of the preceding sections enables us to do that.

[10] See, however, the interpretation of Reiter (1980) defaults in terms of normic conditionals by Schurz (1994) (see Section 2.2.2).

Let us discuss this type of conditional logic based on the approach by Kraus et al. (1990) and Lehmann and Magidor (1992). Kraus et al. (1990) and Lehmann and Magidor (1992) use a non-monotonic inference relation $\mid\!\sim$. However, they do not specify $\mid\!\sim$ in terms of a relation between sets of formulas (i.e., Γ) and formulas (i.e., α), but rather as a relation between formulas (i.e., $\alpha \mid\!\sim \beta$). A so-called conditional assertion $\alpha \mid\!\sim \beta$ is, then, interpreted the following way: if α then normally β (Kraus et al., 1990, p. 173).

Hence, Kraus et al. (1990) and Lehmann and Magidor (1992) do not interpret $\mid\!\sim$ in terms of a formal, logical inference relation but rather a material one. A material inference relation differs from its formal, logical counterpart insofar as it is not closed under isomorphic substitution, viz. it is not closed under the substitution of atomic propositional letters by (other) atomic propositional letters (cf. Schurz, 2001a, p. 370; Schurz, 2004, p. 37f).[11] Hence, in the framework of Kraus et al. (1990) and Lehmann and Magidor (1992), the fact that $p_1 \mid\!\sim q_1$ holds does not guarantee that its isomorphic substitution $p_2 \mid\!\sim q_2$ holds as well.

Note that Lehmann and Magidor (1992, p. 18f) also use a rule with a non-derivability condition for the specification of their System **R** (see also Section 7.2.4), namely the following:

$$\text{RM} \qquad \frac{\alpha \mid\!\sim \gamma \,,\, \alpha \not\mid\!\sim \neg\beta}{\alpha \land \beta \mid\!\sim \gamma}$$

'RM' abbreviates 'Rational Monotonicity'.[12] The rule RM appears to be a non-monotonic rule, since it relies on a non-derivability condition (or its semantic equivalent). Accordingly, System **R** of Lehmann and Magidor (1992) would qualify as a default logic.

Observe here that the expression $\alpha \not\mid\!\sim \neg\beta$ is somewhat ambiguous.[13] A

[11] The rule of isomorphic substitution is weaker than the rule of substitution discussed in Section 2.2.4.

[12] I use here a slightly different but equivalent version of this principle (cf. Lehmann & Magidor, 1992, p. 18).

[13] Kraus et al. (1990) do not explicitly indicate whether $\alpha \not\mid\!\sim \beta$ is an expression of their

semantic interpretation of RM in terms of models makes the interpretation of $\not\mid\!\sim$ more perspicuous: If $\alpha\mid\!\sim\gamma$ holds in a System **R** model and $\alpha\mid\!\sim\neg\beta$ is *not* the case in that model, then $\alpha\wedge\beta\mid\!\sim\gamma$ also holds in that model. Hence, $\alpha\not\mid\!\sim\neg\beta$ should not be interpreted in terms of non-derivability in the sense of default logics discussed earlier. The latter interpretation in terms of "non-derivability" condition would require that the set $\{\alpha,\beta\}$ is consistent in a logical system such as p.c. (see Section 2.2.3). This would not make sense in the present context. Given this analysis, it is much more plausible to interpret the "non-derivability" condition of RM – translated in our conditional logic language – as $\neg(\alpha \,\square\!\!\rightarrow \beta)$ (cf. Section 7.2.4).

These considerations, hence, suggest that conditional assertions (i.e., $\alpha\mid\!\sim\beta$) in the framework of Kraus et al. (1990) and Lehmann and Magidor (1992) correspond to specific conditional formulas (i.e., $\alpha \,\square\!\!\rightarrow \beta$) rather than to a non-monotonic inference relations in the sense of Reiter (1980). This thesis is strengthened by a further fact. One can identify a second-order inference relation for conditional assertions of type $\alpha\mid\!\sim\beta$. This second-order inference relation can again be monotonic, as it is, for example, for Kraus et al.'s (1990) and Lehmann and Magidor's (1992) System **P** (Lehmann & Magidor, 1992, p. 9).

meta-language. Their use of the expression $\not\mid\!\sim$ suggests that they regard $\alpha\not\mid\!\sim\beta$ as a meta-language expression (e.g., p. 184 and p. 188; see also Lehmann & Magidor, 1992, p. 5; see also Section 4.2.1).

Chapter 3

Possible Worlds Semantics and Probabilistic Semantics for Indicative Conditionals: a Survey and a Defense of Possible Worlds Semantics

The present chapter serves three main purposes. First, I will provide an overview of standard formal-philosophical approaches to conditionals.[1] I will start in Section 3.2 with the possible worlds ordering semantics for conditionals of D. Lewis (1973b), Kraus et al. (1990) and related approaches. In Section 3.3 I shall discuss different Ramsey test interpretations and describe Stalnaker's (1968; Stalnaker & Thomason, 1970) possible worlds semantics for conditionals which is proposed in the context of the Ramsey test. In Section 3.4 I will also review different criteria for distinguishing indicative and counterfactual conditionals and investigate which types of conditional logics might be appropriate to account for both types of conditionals. In Sections 3.5 and 3.6 I will describe fundamental issues in probabilistic conditional logics and discuss Adams' standard **P** systems (Adams, 1965, 1966, 1977, 1975, 1986; see also Schurz, 1997b, 1998, 2005b), respectively. In that con-

[1] I will omit here a discussion of paraconsistent systems, such as connexive logics (McCall, 1966; see also Wansing, 2010) due to space and time restrictions.

text the restrictions of the conditional logic languages (see Sections 3.5.3 and 4.2.1) and D. Lewis' (1976) triviality result (see Section 3.7) for probabilistic conditional logics are of particular importance, the latter of which I will slightly extend in Section 3.7.2.

Second, I will defend an account of indicative conditionals on the basis of possible worlds semantics, such as Chellas-Segerberg (CS) semantics (Chellas, 1975; Segerberg, 1989) described in Chapters 4–7. My defense is targeted at the criticism of such philosophers as Adams (1975), Bennett (2003), Edgington (2007), and Gibbard (1980) who oppose truth value accounts of indicative conditionals. These authors argue that indicative conditionals are not propositions and cannot have truth values (NTV: "No Truth Value") (Bennett, 2003, p. 94; see also Section 3.5.5). Since possible worlds semantics, such as CS semantics (and Stalnaker (1968) semantics) for indicative conditionals, are based on the notions of truth and falsehood, these semantics are also prone to the criticism of NTV proponents (cf. Adams, 1975, p. 7; cf. Section 3.5.5).

For my defense of possible worlds semantics for indicative conditionals I will discuss Bennett's NTV thesis (Section 3.5.5) and then address arguments against truth value approaches for indicative conditionals put forth by Bennett (2003): (a) D. Lewis' (1976) triviality result (see Section 3.7) and (b) Bennett's (2003, pp. 83–93) Gibbardian stand-off argument (see Section 3.8) which goes back to Gibbard (1980, p. 231f). Furthermore, in Section 3.1 I will give an overview of my defense of possible worlds semantics for indicative conditionals against arguments (a) and (b). Furthermore, Sections 3.7 and 3.8 will heavily draw on the earlier sections.

Third, the present chapter also serves as basis for a positive account of possible worlds semantics in terms of CS semantics in Chapters 4–7. For my interpretation of CS semantics in Section 7.1 I shall in particular draw on the discussion of conditional logics for indicative and counterfactual conditionals on the one hand and the Ramsey test on the other, in Sections 3.4 and 3.3, respectively. More specifically, I will discuss different interpretations of the Ramsey test in Section 3.3, including Bennett's (2003) probabilistic interpretation and Stalnaker's (1968) original account. I will, furthermore, describe the conditional logic system of Stalnaker (1968; Stalnaker & Thomason, 1970) and outline some important differences between CS semantics and Stalnaker's semantics, which Stalnaker (1968) gives in the context of his specification of the Ramsey test. I will in addition distinguish between possible worlds semantics, which can directly be interpreted in terms of the Ramsey test, such as CS semantics, and ordering semantics, such as D. Lewis (1973b) and Kraus et al. (1990). In the latter approach orderings of possible worlds serve as basis for the semantics rather than the Ramsey test.

3.1 Outline of My Defense of Possible Worlds Semantics for Indicative Conditionals and the Core Idea of Chellas-Segerberg Semantics

In this section I will outline my defense of possible worlds semantics for indicative conditionals and I will also sketch the core idea of Chellas-Segerberg (CS) semantics (Chellas, 1975; Segerberg, 1989) investigated throughout Chapters 4–7. The description of this possible worlds semantics should make clear which type of semantic approach I will defend.

For my defense of possible worlds semantics for indicative conditionals, I will address Bennett's (2003) thesis that indicative conditionals are not propositions and in general do not have truth values (NTV, "No Truth Value") in Section 3.5.5; and I will discuss two arguments against truth value accounts of indicative conditionals put forth by Bennett (2003): (a) D. Lewis' (1976) triviality result and (b) Bennett's (2003, pp. 83–93) Gibbardian stand-off argument. (Bennett (2003) draws on an argument by Gibbard (1980, p. 231f) for (b) regarding pairs of conditionals with mutually inconsistent consequents.) Bennett discusses two further arguments against truth value accounts, (c) one by Edgington (1995; see Bennett, 2003, p. 102) and (d) the other by Bradley (2000; see Bennett, 2003, p. 105). I focus on (a) and (b), since I regard both (a) and (b) as being by far stronger compared to (c) and (d). Bennett seems to agree on that issue, since he dedicates much more space to (a) and (b) than to (c) and (d). (Bennett (2003) discusses points (c) and (d) for less than a page each.)

Before I outline arguments (a) and (b), let me first illustrate the core idea of possible worlds semantics for conditionals. I use CS semantics for this purpose (see also Chapters 4–7). CS semantics and in general possible worlds semantics for conditional logics are based on the notion of truth at possible worlds. A formula α is, hence, not true simpliciter, but true at a possible world. For formulas not containing the conditional operator $\Box\!\!\rightarrow$, truth value assignments are perfectly analogous to p.c. semantics. In CS semantics, the truth conditions of conditional formulas $\alpha \Box\!\!\rightarrow \beta$ at a world w depend on the accessibility relation between possible worlds such as w and w', relative to subsets X of the set of possible worlds W. In other words, the accessibility relation R_X between pairs of possible worlds is relativized to propositions $X \subseteq W$. In possible worlds semantics, for each formula α the set of possible worlds at which α is true is described by $||\alpha||$. A conditional formula $\alpha \Box\!\!\rightarrow \beta$ is, then, true at a possible world w in CS semantics if and only if β is true at all worlds w' which are accessible from w relative to $R_{||\alpha||}$.

Let me now focus on argument (a). To specify (a) adequately, I have to describe some assumptions of subjective probabilistic accounts as, for example, advocated by Adams (1965, 1966, 1977, 1975, 1986), Bennett (2003), and Edgington (2007) first. In such a framework all formulas to which subjective probability values are assigned are regarded as propositions which can either be true or false. A probability value assigned to a formula α is, then, interpreted as the probability that α is true (see Section 3.5.5). Moreover, the semantic approach of Adams employs the Stalnaker thesis, which states that the probability of a conditional $\alpha \Box\!\!\rightarrow \beta$ equals the (subjective) conditional probability $P(\beta|\alpha)$ ("the probability of β given α"). D. Lewis' (1976) triviality result, then, shows that a (subjective) probabilistic semantics for conditionals, as described by Adams (1965, 1966, 1977, 1975, 1986), cannot be extended to a language, which allows either for (i) conjunctions of conditional formulas (e.g., $(\alpha \Box\!\!\rightarrow \beta) \wedge (\gamma \Box\!\!\rightarrow \delta)$) or (ii) conjunctions of conditional formulas and non-conditional formulas (e.g., $(\alpha \Box\!\!\rightarrow \beta) \wedge \alpha$; cf. D. Lewis,

1976, p. 304, see Section 3.7.1). Furthermore, in Section 3.7.2 I shall show that admitting iterations of conditional formulas (e.g., $\alpha \mathrel{\Box\!\!\rightarrow} (\beta \mathrel{\Box\!\!\rightarrow} \gamma)$) also leads to counter-intuitive consequences in Adams' approach.

In addition, I will discuss Schurz's approach (1997b, 1998, 2005b) which neither presupposes the Stalnaker thesis nor implies it and, therefore, does not fall prey to D. Lewis' (1976) triviality result: Schurz only *associates* conditionals with conditional probabilities in the sense that $\alpha \mathrel{\Box\!\!\rightarrow} \beta$ holds only if the respective conditional probability is sufficiently high (Schurz, 1997b, p. 536; Schurz, 1998, p. 85).

One further way out of this dilemma is the approach taken by Adams (1965, 1966, 1977, 1975, 1986): Adams restricts the language in such a way that D. Lewis' triviality result cannot apply. So, Adams does not allow for (i) Boolean combinations involving conditionals (e.g., $(\alpha \mathrel{\Box\!\!\rightarrow} \beta) \wedge \gamma$) and (ii) nestings of conditional formulas (e.g., $\alpha \mathrel{\Box\!\!\rightarrow} (\beta \wedge (\gamma \mathrel{\Box\!\!\rightarrow} \delta))$). In other words, given that we accept the Stalnaker thesis (as Adams does) the conditional logic language cannot arguably be a language that allows for either (i) or (ii) – otherwise it is prone to D. Lewis' triviality result or other undesirable consequences (see above). Based on that fact Bennett, Adams, Edgington, and Gibbard argue that Boolean combinations involving conditional formulas and nestings of conditional formulas do not represent propositions. These authors go even one step further and contend that also simple conditional formulas are not propositions (e.g., Bennett, 2003, p. 94; see also Adams, 1975, p. 7) where simple conditionals have the form $\alpha \mathrel{\Box\!\!\rightarrow} \beta$ and α and β are purely propositional formulas. Since in this view only propositions have truth values (see Section 3.5.5), Bennett (2003) concludes that also (simple) conditionals do not have truth values.

I agree that, given one accepts all assumptions made by Bennett Adams, Edgington, and Gibbard, it is plausible to conclude that conditionals do not in general have truth values. It is, however, quite another question whether (i) one should accept Bennett, Adams, Edgington, and Gibbard's assumptions and (ii) whether these assumptions also apply to possible worlds semantics. Point (ii) is much less clear than it might seem first. This is due to the fact that possible worlds semantics, such as CS semantics, does not draw on the notion of probability, which is essential for D. Lewis' (1976) triviality result. So, how can D. Lewis' triviality result apply to possible worlds semantics such as CS semantics? I will address the latter question in detail in Sections 3.6.4 and 3.7.3.

Let us now focus on argument (b). Bennett presupposes for his Gibbardian stand-off argument (i) a specific type of Ramsey test interpretation for indicative conditionals and (ii) a consistency criterion for indicative conditionals. With respect to point (ii) I distinguish between (A) a weak conditional consistency criterion and (B) a strong conditional consistency criterion. According to (A) conditionals of the form $\top \mathrel{\Box\!\!\rightarrow} \bot$ are not allowed, where $\top$ and $\bot$ are defined as $p \vee \neg p$ and $p \wedge \neg p$, respectively. (The symbols $\neg$, $\wedge$ and $\vee$ represent object-language negation, conjunction and disjunction, respectively.) Bennett (2003) employs the strong conditional consistency criterion (B) according to which conditionals of the form $\alpha \mathrel{\Box\!\!\rightarrow} \beta$ and $\alpha \mathrel{\Box\!\!\rightarrow} \neg\beta$ are contradictory. While (A) is valid in most standard conditional logic semantics for indicative conditionals (e.g., Adams, 1965, 1966, 1977, 1986; Stalnaker, 1968), (B) is not. Bennett concludes from this argument that only a subjective probabilistic semantics allows to adequately account for indicative semantics.

I will discuss Bennett's argument (b) in some detail and inquire in particular whether assumption (i) holds generally and under which condition (ii), as described by (B), is warranted for possible worlds semantics and probabilistic semantics. I conclude that Bennett's stand-off argument is not decisive concerning possible worlds semantics for indicative conditionals, due to assumptions (i) and (ii): First, some semantics for (generic) indicative conditionals, such as the objective frequency-based semantics of Schurz (1997b, 1998, 2005b) (see Section 3.5), do not refer to the Ramsey test. Second, the only conditional logic systems discussed in this chapter which endorses a (weaker) variant of the conditional consistency criterion (B) are the probabilistic semantics of Adams (1975) and Schurz (1997b).

3.2 Possible Worlds Ordering Semantics for Conditionals: D. Lewis' and Kraus et al.'s Semantics and Related Approaches

I will proceed in this section as follows: I will first give a general characterization of what I call "ordering semantics" and then describe particular instances of this type of semantics, namely D. Lewis' (1973b) systems of spheres semantics and alternative semantics, which are directly based on an ordering relation, such as D. Lewis' (1973b, pp. 48–50) relational semantics, Kraus et al. (1990) and Lehmann and Magidor (1992). Since Stalnaker (1968) proposed his semantics in the context of his Ramsey test formulation, I will postpose a discussion of Stalnaker's (1968) possible worlds semantics for conditionals to Section 3.3 where I will discuss different Ramsey test interpretations. Furthermore, my discussion of ordering semantics will serve as basis for a comparison between ordering semantics on the one hand and Stalnaker (1968; Stalnaker & Thomason, 1970) semantics, set selection semantics, and CS semantics (Chellas, 1975; Segerberg, 1989; see Section 3.3.4) on the other in Section 3.3.5.

Before I describe the ordering semantics of D. Lewis, let me first say a few words regarding different types of ordering semantics. We can distinguish at least three types of ordering semantics. All approaches share that they employ rankings of possible worlds. These are either (i) rankings of truth valued possible worlds, such as in D. Lewis (1973b), Kraus et al. (1990), and Lehmann and Magidor (1992), (ii) rankings of possible worlds according to the numerical values of probability functions (cf. Goldszmidt & Pearl, 1996, p. 58) or (iii) semantics for which the ranks are taken as fundamental (Spohn, 2012). Due to time and space constraints I will focus here exclusively on approach (i).

Regarding (i) we can distinguish between quite different interpretations. Whereas D. Lewis (1973b) and Burgess (1981) interpret their semantics in terms of counterfactual conditionals (see Section 1.3), Kraus et al. (1990)

and Lehmann and Magidor (1992) employ their semantics to account for normic/indicative conditionals (see Section 1.4). Interestingly, both types of approaches draw on essentially the same semantic intuition (cf. Makinson, 1993; Nejdl, 1992).

Let me now describe the systems of spheres semantics of D. Lewis (1973b). D. Lewis gives an account of his systems of spheres semantics in terms of the full language $\mathcal{L}_{\text{CL}}$ ('CL for 'Conditional Logic'), where the formulas of the full language $\mathcal{L}_{\text{CL}}$ are purely propositional formulas plus the additional clause that if α and β are formulas of language $\mathcal{L}_{\text{CL}}$ so is $\alpha \mathrel{\Box\!\!\rightarrow} \beta$, where $\Box\!\!\rightarrow$ is the conditional operator (see Section 4.2.1). Language $\mathcal{L}_{\text{CL}}$, hence, allows for Boolean combinations involving conditional formulas (e.g., $(\alpha \mathrel{\Box\!\!\rightarrow} \beta) \wedge \gamma)$ and nestings of conditional formulas (e.g., $\alpha \mathrel{\Box\!\!\rightarrow} (\beta \wedge (\gamma \mathrel{\Box\!\!\rightarrow} \delta))$). In addition, for any conditional logic language $\mathcal{L}$ we identify $\mathcal{L}$ with the set of all formulas of language $\mathcal{L}$.

Lewis frames (D. Lewis, 1973b, p. 13f and p. 119) can then be defined the following way:

Definition 3.1. *$\mathcal{F}_L = \langle W, \$ \rangle$ is a Lewis frame iff*
(a) W is a non-empty set of possible worlds and
(b) \$ is a function such that $\$: W \rightarrow \text{Pow}(\text{Pow}(W))$, where $\$_w$ is the set which is assigned to w by \$, and for all worlds $w \in W$ the following holds:

(i) $\varnothing \notin \$_w$ *and*
(ii) $\$_w \subseteq \text{Pow}(W)$ *such that for all $S, S' \in \$_w$ holds: $S \subseteq S'$ or $S' \subseteq S$.*

The expression Pow(X) denotes the power set of a set X. I call $\$_w$ as specified by a Lewis frame $\mathcal{F}_L = \langle W, \$ \rangle$ 'system of spheres (of world w)'. The elements $S, S', \ldots \in \$_w$ are, then, spheres (of world w).

D. Lewis (1973b) requires in his definition of Lewis frames, in addition to the specifications in Definition 3.1, that all systems of spheres $\$_w$ are (a) closed under union and (b) closed under (non-empty) intersection. I omit criteria (a) and (b), since D. Lewis (1973b, Footnote on p.119) admits that they are neither needed for his soundness results nor for his completeness results. Moreover, I require that any system of spheres $\$_w$ in a Lewis frame does not contain the empty set. This condition is not included in D. Lewis (1973b). The exclusion of the empty set as a member of systems of spheres $\$_w$ for Lewis frame $\mathcal{F}_L = \langle W, \$ \rangle$ does not impact the truth conditions w.r.t. Lewis frames, which are described below (cf. D. Lewis, 1973b, p. 15). Moreover, D. Lewis (1973b) admits that allowing for the inclusion of the empty set in systems of spheres $\$_w$ is not intuitive and that he does so only for the sake of technical convenience (p. 15).

Let us now focus on systems of spheres in Lewis frames. Note that the spheres in a system of spheres (for w) $\$_w$ determine an ordering of possible worlds. We can express those orderings also in terms of ranks where ranks in systems of spheres are determined the following way: For a system of spheres $\$_w$ world $w' \in W$ is of strictly lower rank than world $w'' \in W$ iff there exist spheres S and S' in $\$_w$ such that $w' \in S'$, $w'' \in S$, $S' \subset S$, and $w'' \notin S'$. Moreover, for a system of spheres $\$_w$ worlds w' and $w'' \in W$ are of equal rank iff there exists a sphere S in $\$_w$ such that $w', w'' \in S$ and there exists no sphere S', such that the following holds: (a) $S' \subset S$ and (b) either (i) w' in S' and not w'' in S' or (ii) w'' in S' and not w' in S', where $\subset$ is the proper subset relation.

In Lewis frames the ordering of possible worlds in a system of spheres $\$_w$ is based on the similarity of possible worlds in W to the world $w \in W$. A system of spheres $\$_w$ in a Lewis frame $\mathcal{F}_L = \langle W, \$ \rangle$ describes a ranking of possible worlds w.r.t. similarity to the (actual) world w (cf. D. Lewis, 1973b, pp. 13–16), where the lower the rank of a world w' in $\$_w$ the more similar w' is to w. In D. Lewis' semantics the systems of spheres $\$_w$ for $w \in W$ need not encompass all possible worlds in W for a Lewis frame $\mathcal{M}_L = \langle W, \$ \rangle$.

In order to allow for a more substantive interpretation of systems of spheres $\$_w$ in terms of similarity of possible worlds, D. Lewis (1973b) adds the following condition to Lewis frames:

$C_{\text{Centering}}$ $\forall w \in W(\{w\} \in \$_w)$

The condition $C_{\text{Centering}}$ is called 'Centering Condition'. Provided $C_{\text{Centering}}$ holds, conditions b.i and b.ii of Definition 3.1 then imply that $\{w\}$ is the innermost sphere of any such Lewis model. This condition seems reasonable given D. Lewis' (1973b) interpretation of systems of spheres $\$_w$ in terms of similarity orderings: The world w is always maximally similar to itself and is the only such world (D. Lewis, 1973b, p. 14f).

I shall now define the notion of Lewis models, where $\mathcal{AV}$ is the set of atomic propositional variables of language $\mathcal{L}_{CL}$:

Definition 3.2. *Let $\mathcal{F}_L = \langle W, \$\rangle$ be a Lewis frame. Then, $\mathcal{M}_L = \langle W, \$, V\rangle$ is a Lewis model iff V is a valuation function such that $V: \mathcal{AV} \times W \rightarrow \{0, 1\}$.*

The values '1' and '0' are interpreted as 'true' and 'false', respectively. I shall now define extensions of Lewis models:

Definition 3.3. *Let $\mathcal{M}_L = \langle W, \$, V\rangle$ be a Lewis model. Then V^* is an extension of V (to arbitrary formulas of $\mathcal{L}_{CL}$) iff*
(a) $\forall p \in \mathcal{AV}\, \forall w \in W$: $V^*(p, w) = V(p, w)$, and
(b) for all formulas α, β and $w \in W$ holds:

$$\models_w^{\mathcal{M}_L} \neg\alpha \quad \text{iff} \quad \not\models_w^{\mathcal{M}_L} \alpha, \qquad (V_\neg)$$

$$\models_w^{\mathcal{M}_L} \alpha \vee \beta \quad \text{iff} \quad \models_w^{\mathcal{M}_L} \alpha \text{ or } \models_w^{\mathcal{M}_L} \beta, \text{ and} \qquad (V_\vee)$$

$$\models_w^{\mathcal{M}_L} \alpha \mathrel{\Box\!\!\rightarrow} \beta \quad \text{iff} \quad (a)\ \neg\exists S \in \$_w \exists w' \in S\, (\models_{w'}^{\mathcal{M}_L} \alpha) \text{ or}$$
$$(b)\ \exists S \in \$_w \forall w' \in S\, (\models_{w'}^{\mathcal{M}_L} \alpha \rightarrow \beta). \qquad (V_{\Box\!\!\rightarrow})$$

The symbols $\neg$, $\exists$, and $\forall$ represent the negation, the existential quantifier, and the universal quantifier in the meta-language, respectively. Moreover, $\rightarrow$ is the material implication (object language) and the expressions $\models_w^{\mathcal{M}_L} \alpha$ and $\not\models_w^{\mathcal{M}_L} \alpha$ abbreviate $V^*(\alpha, w) = 1$ and $V^*(\alpha, w) \neq 1$, respectively. I will, henceforth, refer to Lewis models and extensions of Lewis models indiscriminately. Let me call a world w in a Lewis model $\mathcal{M}_L = \langle W, \$, V\rangle$ an α-world iff α is true at w. Moreover, I refer to a sphere S in $\$_w$ as an α-permitting sphere iff there exists a world $w' \in S$ such that w' is an α-world. Then, according to Definition 3.3 a conditional formula $\alpha \mathrel{\Box\!\!\rightarrow} \beta$ is true at a world w in a Lewis model $\mathcal{M}_L = \langle W, \$, V\rangle$ iff either (a) there is no α-permitting sphere in $\$_w$ or (b) there exists an α-permitting sphere S in $\$_w$ such that $\alpha \rightarrow \beta$ is true at every world in S (cf. D. Lewis, 1973b, p. 16).

Before I continue with a comparison of D. Lewis' sphere semantics with alternative characterizations of ordering semantics, let me, first, consider the following restrictions for Lewis frames:

C_{MP} $\forall w \in W(\$_w \neq \varnothing \wedge \forall S' \in \$_w(w \in S'))$

C_{CS} $\forall w \in W(\$_w \neq \varnothing \Rightarrow \{w\} \in \$_w)$

$C_{\text{P-Cons}}$ $\forall w \in W(\$_w \neq \varnothing)$

The symbols $\varnothing$, $\wedge$, $\Rightarrow$ are the meta-language expressions for the empty set, the conjunction, and the material implication, respectively. Furthermore, 'P-Cons', 'MP', and 'CS' stand for 'Probabilistic Consistency', 'Modus Ponens', and 'Conjunctive Sufficiency', respectively. The following principles correspond to the foregoing frame conditions:

MP $(\alpha \,\Box\!\!\rightarrow \beta) \rightarrow (\alpha \rightarrow \beta)$

E $\alpha \wedge \beta \rightarrow (\alpha \,\Box\!\!\rightarrow \beta)$

P-Cons $\neg(\top \,\Box\!\!\rightarrow \bot)$

By correspondence I mean that whenever condition C_α holds for a Lewis frame $\mathcal{M}_L = \langle W, \$\rangle$ then α is true at all worlds of all Lewis models $\mathcal{M}_L = \langle W, \$, V\rangle$ (for arbitrary valuation functions V), and vice versa.

I describe here conditions C_{MP} and C_{CS}, since both are conjointly equivalent to the centering condition $C_{\text{Centering}}$. C_{MP} is, moreover, sometimes called 'weak centering condition' (D. Lewis, 1973b, p. 120). Condition C_{MP} ensures that there exists a sphere S in $\$_w$ and that w is in all spheres in $\$_w$. This condition ensures that w is in the innermost sphere in $\$_w$ if there exists any. The Principle MP is a variant of the modus ponens rule from p.c., where the conditional $\alpha \,\Box\!\!\rightarrow \beta$ replaces the material implication $\alpha \rightarrow \beta$: We can infer β from $\alpha \,\Box\!\!\rightarrow \beta$ and α. The condition C_{CS} on the other hand makes sure that if there exists a sphere S in $\$_w$ for some world w in a model $\mathcal{M}_L = \langle W, \$, V\rangle$, then $\{w\}$ is in $\$_w$. The syntactic Principle CS allows one to infer $\alpha \,\Box\!\!\rightarrow \beta$ from the conjunction $\alpha \wedge \beta$. Note that both Principles MP and CS are valid in p.c. I will discuss MP and CS in more detail in Section 3.4.

Principle P-Cons and Frame Condition $C_{\text{P-Cons}}$ are interesting insofar as P-Cons seems to be a plausible minimal conditional consistency criterion in the following sense: One should not arrive at a contradiction $\bot$, when one considers any parts of an ordering of worlds in a systems of spheres $\$_w$, which is determined by a tautology $\top$ (Lewis models give us just that), since $\top$ is true at all possible worlds in all Lewis models. Despite this intuition P-Cons is in general not valid in Lewis models. Condition $C_{\text{P-Cons}}$ indicates where it fails: $\top \Box\!\!\rightarrow \bot$ and in fact all formulas of $\mathcal{L}_{\text{CL}}$ are true at a world w in a Lewis model $\mathcal{M}_L = \langle W, \$, V \rangle$ if $\$_w$ is the empty set. In this case the truth conditions of Definition 3.3 hold trivially for any conditional formula. I will postpone a discussion of Principle P-Cons and other types of conditional consistency criteria to Section 3.3.6.

Let me now describe an alternative characterization of systems of spheres by means of an accessibility relation (relational Lewis models) suggested also by D. Lewis (1973b, pp. 48–50). My alternative characterization corresponds to the notion of Lewis frames and models characterized in Definitions 3.1–3.3 without the centering condition. Variants of the relational Lewis semantics are also presented by Kraus et al. (1990), Lehmann and Magidor (1992), Burgess (1981), and Delgrande (1998). Kraus et al. (1990) and Burgess (1981), in addition, investigate weaker semantics than relational Lewis semantics.

Let me now define relational Lewis frames, models, and extensions thereof (cf. also Lehmann & Magidor, 1992, p. 21):

Definition 3.4. *The triple* $\mathcal{F}_{rL} = \langle W, X, S \rangle$ *is a relational Lewis frame (short: rL-frame) iff*
(a) W *is a non-empty set of possible worlds,*
(b) X *is a function such that* $W \rightarrow \text{Pow}(W)$ *where* X_w *is the set of possible worlds assigned to* w *by* X*, and*
(c) S *is a three-place relation on* W *such that for all* $w \in W$ *holds that (A) for all* $w' \notin X_w$ *there is no world* $w'' \in W$ *such that either* $w' S_w w''$ *or* $w'' S_w w'$*, and (B) for all* $w', w'', w''' \in X_w$ *the following is the case:*

(i) $w' S_w w'$ *(Reflexivity),*
(ii) if $w' S_w w''$ *and* $w'' S_w w'''$ *then* $w' S_w w'''$ *(Transitivity), and*
(iii) if $w' S_w w''$ *then* $w''' S_w w''$ *or* $w' S_w w'''$ *(Modularity).*

Condition c.iii of Definition 3.4 is taken from Lehmann and Magidor (1992,

p. 21) and the term 'modularity' is due to the analogous formal property of modular lattice structures (p. 21). One can easily see that condition c.iii is equivalent to the following condition, as employed by D. Lewis (1973b, p. 48), provided Reflexivity and Transitivity hold for S:

Connectivity $\forall w, w', w'' \in W(w' S_w w'' \vee w'' S_w w')$

Observe that Definition 3.4 makes sure that all possible worlds which are not in a given set of possible worlds X_w can neither see any world nor can be seen by any world. So, they are so to speak "outside the ordering". That way truth valuations of formulas at worlds not in X_w do not impact the truth values of (conditional) formulas at world w.

D. Lewis' (1973b, p. 48) specification of relational Lewis frames differs from Definition 3.4 insofar as D. Lewis does not use a function, such as X. D. Lewis rather employs a three-place relation on the total set of possible worlds W and distinguishes between maximal worlds (relative to a world w) and non-maximal worlds. The maximal worlds w' (for a world w) are in my terminology exactly those worlds which are in W but not in X_w. On the other hand D. Lewis' maximal worlds w' in W (w.r.t. world w) are those worlds $w' \in W$ in a Lewis frame $\mathcal{M}_L = \langle W, \$ \rangle$ which are not in $\$_w$. I use Definition 3.4 rather than D. Lewis' (1973b) characterization of relational Lewis frames, since the former is somewhat more perspicuous.

The accessibility relation S gives us, then, an ordering of possible worlds. In my terminology world $w' \in W$ is of (strictly) lower rank than world $w'' \in W$ (w.r.t. world $w \in W$) iff $w' S_w w''$ and not $w'' S_w w'$, whereas w' and $w'' \in W$ are of equal rank (w.r.t. world $w \in W$) iff $w' S_w w''$ and $w'' S_w w'$. Moreover, a world $w' \in W$ is accessible (w.r.t. world $w \in W$) iff there exists a world $w'' \in W$ such that either $w' S_w w''$ or $w'' S_w w'$.

Relational Lewis frames $\mathcal{M}_{\mathrm{rL}} = \langle W, X, S \rangle$ correspond insofar to Lewis frames $\mathcal{M}_L = \langle W, \$ \rangle$, as one can reconstruct the system of spheres function $\$$ in terms of the accessibility relation S and vice versa. When one has $S' \subseteq S''$ for $S', S'' \in \$_w$, then for all worlds $w' \in S'$ and $w'' \in S''$ holds that $w' S_w w''$. Furthermore, for all w, w' holds that $S'_{w'} \in \$_w$ where $S'_{w'} =_{\mathrm{df}} \{w'' \mid w'' S_w w'\}$.

The type of ordering described in Definition 3.4 is also called 'ranked' as opposed to preferential orderings of possible worlds (Lehmann & Magidor, 1992, p. 21), where in preferential orderings all specifications of Definition 3.4 except point c.iii hold (Lehmann & Magidor, 1992, p. 21).

Let me now define relational Lewis models and their extensions:

Definition 3.5. *Let $\mathcal{F}_{rL} = \langle W, X, S \rangle$ be a relational Lewis frame. Then, $\mathcal{M}_L = \langle W, X, S, V \rangle$ is a relational Lewis model (rL-model) iff V is a valuation function such that $V: \mathcal{AV} \times W \rightarrow \{0,1\}$.*

Definition 3.6. *Let $\mathcal{M}_{rL} = \langle W, X, S, V \rangle$ be a relational Lewis model. Then, V^* is an extension of V to arbitrary formulas of language $\mathcal{L}_{\mathrm{CL}}$ iff*
(a) $\forall p \in \mathcal{AV} \forall w \in W$: $V^*(p,w) = V(p,w)$ and
(b) for all formulas α, β and $w \in W$ holds:

$$\models_w^{\mathcal{M}_{rL}} \neg\alpha \quad \text{iff} \quad \not\models_w^{\mathcal{M}_{rL}} \alpha, \qquad (V_\neg)$$

$$\models_w^{\mathcal{M}_{rL}} \alpha \vee \beta \quad \text{iff} \quad \models_w^{\mathcal{M}_{rL}} \alpha \text{ or } \models_w^{\mathcal{M}_{rL}} \beta, \text{ and} \qquad (V_\vee)$$

$$\models_w^{\mathcal{M}_{rL}} \alpha \,\Box\!\!\rightarrow \beta \quad \text{iff} \quad (i)\ \neg\exists w'(w' \in X_w \wedge \models_{w'}^{\mathcal{M}_{rL}} \alpha) \text{ or } (ii)\ \exists w'(w' \in X_w \wedge \models_{w'}^{\mathcal{M}_{rL}} \alpha \wedge \forall w''(w'' S_w w' \Rightarrow \models_{w''}^{\mathcal{M}_{rL}} \alpha \rightarrow \beta)). \qquad (V_{\Box\!\rightarrow})$$

The symbol $\rightarrow$ is again the material implication (object language). The expressions $\models_w^{\mathcal{M}_L} \alpha$ and $\not\models_w^{\mathcal{M}_L} \alpha$ abbreviate $V^*(\alpha, w) = 1$ and $V^*(\alpha, w) \neq 1$, respectively. I will henceforth refer to relational Lewis models and their extensions indiscriminately. A conditional $\alpha \,\Box\!\!\rightarrow \beta$ is – according to Definition 3.6 – true at a world w iff either (i) there is no α-world w', which is accessible w.r.t. w, or (ii) there exists an α-world w' that is accessible w.r.t. w and for all worlds of equal or lower rank than w' holds that $\alpha \rightarrow \beta$ (cf. D. Lewis, 1973b, p. 49).

I will now focus on the difference between D. Lewis's orderings of possible worlds semantics and Kraus et al.'s and Lehmann and Magidor's version. First, Kraus et al. (1990) and Lehmann and Magidor (1992) do not interpret the relation S in their semantics in terms of similarity of possible worlds (in an objective way), as D. Lewis (1973b) does, but in terms of ranks of normality according to an agent's beliefs (Kraus et al., 1990, p. 169f; see also Section 1.4) where the lower the rank of a world w the more normal w is (w.r.t. the agent's beliefs). The ranks of normality are, then, used to determine expectancies of agents: What is most normal – given the information an agent has – is to be expected by the agent (see also Kraus et al., 1990, p. 173f).

Second, in contrast to D. Lewis (1973b) and Burgess (1981), Kraus et al. (1990) and Lehmann and Magidor (1992) use an absolute ordering relation S: In the latter approach S is a two-place relation, which is not relativized to a world w so that $w'Sw''$ indicates that w' is of lower or equal rank than world w'' without further qualification (Kraus et al., 1990, p. 181f; Lehmann & Magidor, 1992, p. 7f, p. 21; see also Makinson, 1993, p. 347f).[2] Kraus et al. (1990) and Lehmann and Magidor (1992) do not use the full language $\mathcal{L}_{\mathrm{CL}}$ but languages which do not allow for nestings of conditionals (e.g., $p \mathbin{\Box\!\!\rightarrow} (q \mathbin{\Box\!\!\rightarrow} r)$ is not a formula of the language $\mathcal{L}_{\mathrm{rCL}^*}$ used in their approach; see Section 4.2.1). Furthermore, in Kraus et al. (1990) and Lehmann and Magidor (1992) at maximum only negations of non-nested conditionals (e.g., $\neg(p \mathbin{\Box\!\!\rightarrow} q)$) are considered formulas, where Boolean combinations of non-nested conditional formulas are called "first degree formulas" when the language does not allow for nestings of conditionals (Makinson, 1993, p. 348). Interestingly, the use of an absolute ordering relation S compared to a relativized but otherwise identical ordering relation S' renders no additional first-degree formulas logically valid (Makinson, 1993, p. 348f). By valid I mean true at all worlds in all models of the pre-specified type.

Kraus et al. (1990) and Burgess (1981) also investigate weaker types of models than described by Definitions 3.1–3.3 on the one hand and Definitions 3.4–3.6 on the other. They use weaker notions of relational Lewis frames (see Definition 3.4) by giving up parts of the restriction on the accessibility relation in point c of Definition 3.4. In addition, Kraus et al. (1990, p. 182 and p. 187) also investigate relational Lewis models which do not rely on orderings of possible worlds, but rather on orderings of non-empty sets

[2] Although Kraus et al. (1990) use the symbol $<$ to describe the relation S, they allow for worlds $w, w' \in W$ that both wSw' and $w'Sw$ hold (Kraus et al., 1990, p. 181, Definition 3.6 and p.183, Definition 3.14).

of possible worlds.

Finally, let me discuss a characteristic aspect of D. Lewis' (1973b) systems of spheres semantics which is how D. Lewis (1973b) deals with infinite descending sequences of α-permitting spheres. For mnemonic reasons, I shall first repeat the informal summary of the truth conditions described in Definition 3.3: A conditional formula $\alpha \mathrel{\Box\!\!\rightarrow} \beta$ is true in a Lewis model $\mathcal{M}_L = \langle W, \$, V \rangle$ at a world w iff either (a) there is no α-permitting sphere in $\$_w$ or (b) there exists an α-permitting sphere S in $\$_w$ such that $\alpha \rightarrow \beta$ is true at every world in S. In case that (b) holds the inclusion requirement b.ii of Definition 3.1 guarantees, then, that $\alpha \rightarrow \beta$ is true at all worlds in spheres S' such that $S' \subseteq S$. This property allows D. Lewis (1973b) to also deal with infinite descending sequences of α-permitting spheres. A system of spheres $\$_w$ contains an infinite sequence of α-permitting spheres $S_1, S_2, \ldots$ in $\$_w$ in a Lewis model $\mathcal{M}_S = \langle W, \$, V \rangle$ iff $S_1, S_2, \ldots$ are α-permitting spheres and it holds that $S_1 \supset S_2 \supset \ldots$, where $\supset$ is the proper subset relation (cf. Schurz, 1998, p. 83). In such a system of spheres $\$_w$, hence, there exists no smallest α-permitting sphere, since for any α-permitting sphere S there exists an α-permitting sphere S' such that $S' \subset S$. D. Lewis' semantics allows one to deal with these cases in an intuitive way: We saw above that if there exists in that sequence of α-permitting spheres a sphere $S \in \$_w$ for some world w such that $\alpha \rightarrow \beta$ is true at all worlds w' in S, then $\alpha \rightarrow \beta$ is true at all worlds in all spheres $S' \subseteq S$. So, even if there is an infinite sequence of α-permitting spheres as described before, the conditional $\alpha \mathrel{\Box\!\!\rightarrow} \beta$ is true at w provided there is a sphere S for all spheres $S' \in \$_w$ so that for all $S' \subseteq S$ holds that $\alpha \rightarrow \beta$ is true at all worlds in S'.

Infinite descending sequences of α-permitting spheres occur only in certain types of Lewis models but not in others. Adams (1977), for example, proves the equivalence of entailment for the probabilistic semantics of the Conditional Logic System $\mathbf{P}^*$ in Adams (1965, 1966, 1977; see Section 3.6) on the one hand and D. Lewis' (1973b) system of spheres semantics on the other. Adams (1977), however, restricts himself in his investigation to finite Lewis models (where a Lewis models $\mathcal{M}_L = \langle W, \$, V \rangle$ is finite iff W contains only a finite number of worlds). In finite Lewis models trivially no infinite descending sequence of α-permitting spheres can occur. [3]

D. Lewis (1973b) also discusses the so-called "limit assumption". This assumption makes sure that no infinite descending sequence of α-permitting spheres can occur. This is done by requiring for any formula α in language $\mathcal{L}_{\mathrm{CL}}$ and any world w in any Lewis model $\mathcal{M}_L = \langle W, \$, V\rangle$ that there always exists a smallest α-permitting sphere in $\$_w$, if an α-permitting sphere exists at all. Formally, the limit assumption amounts to the following: $\forall \alpha \in \mathcal{L}_{\mathrm{CL}} \forall w \in W(\exists S(S \in \$_w \wedge \exists w'(w' \in S \wedge \models_{w'}^{\mathcal{M}_L} \alpha)) \Rightarrow \exists S(S \in \$_w \wedge \exists w'(w' \in S \wedge \models_{w'}^{\mathcal{M}_L} \alpha) \wedge \neg\exists S' \exists w''(S' \in \$_w \wedge w'' \in S' \wedge \models_{w''}^{\mathcal{M}_L} \alpha \wedge S' \subset S)))$.

If one adds the limit assumption to the definition of Lewis models, one can simplify the truth conditions of $\alpha \mathrel{\Box\!\!\rightarrow} \beta$ at a world w in a Lewis model $\mathcal{M}_L = \langle W, \$, V\rangle$ by requiring for $\alpha \mathrel{\Box\!\!\rightarrow} \beta$ to be true at $w \in W$ that (i) either there is no α-permitting sphere in $\$_w$ or (ii) if there is an α-permitting sphere in $\$$, then $\alpha \rightarrow \beta$ is true at all worlds w' in the smallest α-permitting sphere in $\$_w$ (D. Lewis, 1973b, p. 19f). Note that the limit assumption does not make additional formulas valid in Lewis models (D. Lewis, 1973b, p. 121).

D. Lewis (1973b) argues that the limit assumption is not warranted: Consider, for example, the conditional "If there were a line which is strictly longer than 2cm, then X would be the case". The limit assumption would require that there is no infinite descending sequence of α-permitting spheres. According to D. Lewis such a sequence would be needed to represent this case adequately since for any length Y that is greater than 2cm but unequal to 2cm there is a strictly smaller length Z which is strictly smaller than Y and also greater than 2cm (D. Lewis, 1973b, p. 20). D. Lewis (1973b) argues that the limit assumption does not allow for an adequate solution of this problem and that one, therefore, should not accept it.

On the other hand the semantics of Kraus et al. (1990) and Lehmann and

[3] For his account of D. Lewis' ordering semantics Adams (1977) uses the restricted language $\mathcal{L}_{\mathrm{CL}^-}$, which allows only for p.c. formulas that do not contain the conditional operator, or alternatively conditional formulas $\alpha \mathrel{\Box\!\!\rightarrow} \beta$, where α and β are formulas which do not contain the conditional operator (cf. Sections 3.5.3 and 4.2.1).

Magidor (1992) explicitly employ a variant of the limit assumption. Kraus et al. (1990, p. 182) and Lehmann and Magidor (1992, p. 7) call this variant of the limit assumption 'smoothness condition'. (Makinson, 1994, p. 73 calls this property 'stopperedness'.) Accordingly, they employ a variant of the simplified version of D. Lewis' (1973b) truth conditions described above.

Kraus et al. (1990) and Lehmann and Magidor (1992), however, have good reasons to use the limit assumption. First, they intend to describe indicative normic conditionals rather than counterfactual conditionals and assume for that purpose that ranks of normality are assigned to worlds (or sets of worlds) by agents and that these ranks determine the agents' expectancies (see above). Since it seems hardly plausible to presuppose that real agents represent infinite descending sequences of α-permitting spheres, the limit assumption (which excludes these cases) is rationally justified for their semantic approach.

Second, even if one considers a conditional logic semantics which is not interpreted in terms of agents' subjective states, the limit assumption might still be warranted from a pragmatic perspective: In order to arrive at a computationally tractable account of conditionals one cannot allow for infinite descending sequences of α-worlds, for otherwise the respective algorithm might not terminate.

3.3 The Ramsey Test, Stalnaker Semantics, and a General Ramsey Test Requirement

In this section I will discuss the core features of the Ramsey test and inquire whether the conditional consistency requirement and the probabilistic interpretation in Ramsey's original account are an essential feature of the Ramsey test idea. I will then describe Stalnaker's (1968) possible worlds semantics for conditionals, which is formulated in the context of his Ramsey test account, and explain the difference between Stalnaker semantics and related semantics, such as set selection semantics and CS semantics. Then, I will discuss the difference between ordering semantics described in the previous section and the Ramsey test, outline Bennett's (2003) probabilistic version of the Ramsey test, and present a general Ramsey test requirement that any reasonable interpretation of conditional logic semantics has to be satisfied in order to count as a Ramsey test interpretation.

3.3.1 Ramsey's Original Proposal

The Ramsey test received its name due to a footnote by Ramsey (1965, p. 247). In this footnote Ramsey (1965) expressed an intuition that is employed by others, such as Stalnaker (1968) and Bennett (2003, p. 28), to justify a subjective probabilistic approach to conditionals. There are several versions of the Ramsey test, which differ quite strongly from each other. One can distinguish at least the following three: (i) a probabilistic version (Adams, 1975, p. 3; Bennett, 2003, p. 28; see Section 3.3.7), (ii) a possible worlds version (Stalnaker, 1968, p. 101f; see Section 3.3.2), and (iii) a belief revision version (Gärdenfors, 1986; Gärdenfors, 1988; cf. Lindström & Rabinowicz, 1995, p. 147f; Leitgeb, 2010). All three versions refer to Ramsey's footnote, but differ in as much as they use different means to make Ramsey's original proposal more precise.

Furthermore I distinguish between Ramsey test interpretations and ordering approaches (cf. Makinson, 1993; see also Goldszmidt & Pearl, 1996; Spohn, 2012). In an ordering approach conditionals are analyzed on the basis of rankings of entities. Conditionals receive truth values (probability values) based on those ranking of entities. In the literature both approaches are often not clearly distinguished. For example, Stalnaker (1968) describes a Ramsey test interpretation for possible worlds semantics and despite that fact interprets his semantics rather in terms of orderings of possible worlds (p. 104f).

Ramsey test interpretations refer, at least indirectly, to the above-mentioned footnote by Ramsey (1965, p. 247). To determine whether a conditional holds, Ramsey (1965) suggested the following procedure in that footnote:

> If two people are arguing 'If p will q?' and are both in doubt as to p, they are adding p hypothetically to their stock of knowledge and arguing on that basis about q; so that in a sense 'If p, q' and 'If p, $\bar{q}$' are contradictories. We

can say they are fixing their degrees of belief in q given p (Ramsey, 1965, p. 247).

The expression $\bar{p}$ stands in the above quote for the negation of p (Ramsey, 1965, p. xviii). The passage can be divided into the following three parts: (a) *If two people are arguing 'If p will q?' and are both in doubt as to p, they are adding p hypothetically to their stock of knowledge and arguing on that basis about q*, (b) *so that in a sense 'If p, q' and 'If p, $\bar{q}$' are contradictories*, and (c) *we can say they are fixing their degrees of belief in q given p*.

Part (a) seems to describe the basic idea of the Ramsey test and might be roughly summarized the following way: To evaluate a conditional, first, hypothetically suppose that the antecedent holds and then test whether the consequent holds according to that supposition, too.[4] In contrast, passages (b) and (c) seem to refer to optional components. Part (b) describes a consistency requirement for conditionals. According to that requirement one is not allowed to arrive at both 'If p, q' and 'If $p, \neg q$'. Hence, according to the procedure described before, one is not permitted to conclude when supposing the antecedent p that both q and $\neg q$ hold. This suggests a (strong) conditional consistency criterion that renders conditionals of the form $\alpha \mathrel{\Box\!\!\rightarrow} \beta$ and $\alpha \mathrel{\Box\!\!\rightarrow} \neg\beta$ inconsistent rather than the weak version, which merely does not allow for conditionals of the form $\top \mathrel{\Box\!\!\rightarrow} \bot$, where $\top$ and $\bot$ abbreviate $p \vee \neg p$ and $p \wedge \neg p$, respectively (cf. Section 3.1). I shall postpose a further discussion of point (b) to Section 3.3.6.

Moreover, passage (c) suggests a (subjective) probabilistic interpretation, since the notion of a degree of belief (of an agent in a proposition) indicates a subjective agent-relative probability assessment of a specific statement (see Section 3.5). This reference to subjective agent-relative probability is not surprising, since Ramsey is also regarded as one of the forefathers of a subjective approach to probability (Schurz, 2008, p. 101). The probabilistic conditional logic systems of Adams (1965, 1966, 1977, 1975, 1986) follow Ramsey insofar as they use a subjective probabilistic framework. Adams also

[4] Although the basic intuition underlying the Ramsey test seems to be clear, an adequate formulation is not easy to achieve (Bennett, 2003, p. 28f).

employs the Stalnaker thesis, viz. he defines the probability of a conditional as the respective conditional probability (interpreted as the respective degree of belief of an agent; cf. Section 3.1).

Observe that not all probabilistic conditional logic semantics rely on subjectively intepreted probabilities. Schurz (1997b, 1998, 2005b), for example, rather uses objective frequency-based probabilities as basis for his conditional logic semantics. As a consequence the Ramsey test, which is inherently subjective, is not directly applicable to Schurz's (1997b, 1998, 2005b) probabilistic semantics.

In the next few subsections I will discuss Stalnaker's (1968) and Bennett's (2003) version of the Ramsey test. Both versions of the Ramsey test incorporate element (a), but differ with respect to elements (b) and (c). Stalnaker (1968, p. 101f) focuses greatly on component (b), but ignores component (c). This is not surprising, since he aims to justify a possible-worlds semantics rather than a probabilistic semantics by the Ramsey test. Bennett (2003, pp. 28–30) focuses on element (c) and does not discuss element (b) in that context. The latter fact is surprising, since Bennett (2003) also draws on a conditional consistency requirement, as formulated in (b), for his Gibbardian stand-off argument against a truth value semantics for indicative conditionals and refers explicitly to the Ramsey test for that purpose (see Section 3.8).

3.3.2 Stalnaker's Version of the Ramsey Test

I will now describe Stalnaker's (1968) version of the Ramsey test. Stalnaker (1968) suggests that one's deliberations about a conditional should follow the following lines:

> [A]dd the antecedent (hypothetically) to your stock of knowledge (or beliefs), and then consider whether or not the consequent is true. Your belief about the conditional should be the same as your hypothetical belief, under this condition, about the consequent (Stalnaker, 1968, p. 101).

Stalnaker's characterization of the Ramsey test corresponds to part (a) of the Ramsey test in the preceding section. Stalnaker (1968), in addition, provides the following more detailed account of the Ramsey test procedure:

> First, add the antecedent (hypothetically) to your stock of beliefs; second, make whatever adjustments are required to maintain consistency (without modifying the hypothetical belief in the antecedent); finally, consider whether or not the consequent is then true (Stalnaker, 1968, p. 102).

Stalnaker's latter version of the Ramsey test differs from his former account insofar as the latter, in addition, includes the conditional consistency criterion (b). Since Stalnaker (1968) presents the Ramsey test in connexion with his formal semantics, I will describe Stalnaker's semantics in the next section and compare it to the closely related set selection semantics and CS semantics (see Chapters 4–7) before I come back to Stalnaker's Ramsey test interpretation.

3.3.3 Stalnaker Models

I will now specify Stalnaker models and extensions of Stalnaker models (see Stalnaker, 1968, p. 103f; Stalnaker & Thomason, 1970, pp. 25–29). As with Lewis models, I will use here the full conditional logic language $\mathcal{L}_{\mathrm{CL}}$, where $\mathcal{AV}$ is the set of atomic propositions of $\mathcal{L}_{\mathrm{CL}}$:

Definition 3.7. *$\mathcal{M}_{St} = \langle W, \lambda, f, V \rangle$ is a Stalnaker model (short: St-model) iff*
(a) W is a set of possible worlds which has at least one possible world $w \neq \lambda$ as an element and contains, in addition, the absurd world λ,
(b) f is a function such that $f\colon \mathcal{L}_{\mathrm{CL}} \times W \backslash \{\lambda\} \to W$,
(c) V is a valuation function such that $V\colon \mathcal{AV} \times W \backslash \{\lambda\} \to \{0,1\}$ and $V(p,\lambda) = 1$ for all $p \in \mathcal{AV}$, and
(d) for all formulas $\alpha, \beta \in \mathcal{L}_{\mathrm{CL}}$ and worlds $w \in W\backslash\{\lambda\}$ holds:

(i) $f(\alpha, w) \subseteq ||\alpha||^{\mathcal{M}_{St}}$,
(ii) *if* $w \in ||\alpha||^{\mathcal{M}_{St}}$, *then* $f(\alpha, w) = w$, *and*
(iii) *if* $f(\alpha, w) \subseteq ||\beta||^{\mathcal{M}_{St}}$ *and* $f(\beta, w) \subseteq ||\alpha||^{\mathcal{M}_{St}}$, *then* $f(\alpha, w) = f(\beta, w)$.

The values '1' and '0' are interpreted as 'true' and 'false', respectively. Furthermore, $\mathcal{L}_{\mathrm{CL}}$ is the the set of all formulas of language $\mathcal{L}_{\mathrm{CL}}$ (cf. Section 4.2.1). The parameter λ refers to the absurd world, which is a distinguished element among the worlds in W. Moreover, for Stalnaker models $\mathcal{M}_{\text{St}} = \langle W, \lambda, f, V \rangle$ the proposition $||\alpha||^{\mathcal{M}_{\text{St}}}$ is defined as $\{w \in W \mid V(\alpha, w) = 1\}$. Note that $||\alpha||^{\mathcal{M}_{\text{St}}}$ refers to all worlds in W, including the absurd world λ. For the specification of Stalnaker models Stalnaker (1968, p. 103f) and Stalnaker and Thomason (1970, pp. 25–29) employ an accessibility relation R in addition to the parameters W, λ, f, and V described in Definition 3.7. This accessibility relation R is, however, formally ineffective (cf. D. Lewis, 1973b, p. 78) and is, hence, not used in the above definition. Moreover, I restrict the function f to sets of ordered pairs $\mathcal{L}_{\text{CL}} \times W \backslash \{\lambda\}$ rather than sets of ordered pairs $\mathcal{L}_{\text{CL}} \times W$, since I will specify below extensions of St-models – in line with Stalnaker (1968) and Stalnaker and Thomason (1970) – in such a way that the truth conditions for formulas α w.r.t. the absurd world do not draw on the world selection function f. This is the case, since all formulas α of language $\mathcal{L}_{\text{CL}}$ are true at the absurd world λ.

Let me now define extensions of Stalnaker models:

Definition 3.8. *Let $\mathcal{M}_{St} = \langle W, \lambda, f, V \rangle$ be a Stalnaker model. Then, V^* is an extension of V to arbitrary formulas of $\mathcal{L}_{\text{CL}}$ iff*

(a) $\forall p \in \mathcal{AV}\, \forall w \in W$: $V^*(p, w) = V(p, w)$,

(b) $\forall \alpha \in \mathcal{L}_{\text{CL}}$: $V^*(\alpha, \lambda) = 1$, and

(c) for all $\alpha, \beta \in \mathcal{L}_{\text{CL}}$ and $w \in W \backslash \{\lambda\}$ holds:

(i) $\models_w^{\mathcal{M}_{St}} \neg\alpha$ iff $\not\models_w^{\mathcal{M}_{St}} \alpha$, $(V_{\neg})$

(ii) $\models_w^{\mathcal{M}_{St}} \alpha \vee \beta$ iff $\models_w^{\mathcal{M}_{St}} \alpha$ or $\models_w^{\mathcal{M}_{St}} \beta$, and $(V_{\vee})$

(iii) $\models_w^{\mathcal{M}_{St}} \alpha \mathbin{\Box\!\!\rightarrow} \beta$ iff $\models_{f(\alpha,w)}^{\mathcal{M}_{St}} \beta$. $(V_{\Box\!\rightarrow})$

The expressions $\models_w^{\mathcal{M}_{\text{St}}} \alpha$ and $\not\models_w^{\mathcal{M}_{\text{St}}} \alpha$ abbreviate $V^*(\alpha, w) = 1$ and $V^*(\alpha, w) = 0$, respectively, for world w in model $\mathcal{M}_{\text{St}}$. I shall henceforth use both V and V^* indiscriminately in the context of Stalnaker models and their extensions.

The absurd world λ is needed in order to deal with conditionals with inconsistent antecedents. For example, the formula $\bot \mathbin{\Box\!\!\rightarrow} \bot$ is true at all worlds in all Stalnaker models by point d.i of Definition 3.7: Since $||\bot||^{\mathcal{M}_{\text{St}}} = \{\lambda\}$ we have $f(\bot, w) = \lambda$, and the value $f(\bot, w)$ in any Stalnaker model $\mathcal{M}_{\text{St}} = \langle W, \lambda, f, V \rangle$ must be the absurd world λ. Furthermore, according to point (b) of Definition 3.8 every formula is true at λ including $\bot$. Thus, $\bot \mathrel{\Box\!\!\rightarrow} \bot$ turns out to be true at any world w in any Stalnaker model by condition $V_{\Box\!\rightarrow}$ of Definition 3.8. Furthermore, also conditionals of the form $\alpha \mathrel{\Box\!\!\rightarrow} \bot$ can be true at a possible world w in a Stalnaker model $\mathcal{M}_{\mathrm{St}} = \langle W, \lambda, f, V\rangle$ even in the case when α is an S-consistent formula where S is Stalnaker's conditional logic system. Such a formula $\alpha \mathrel{\Box\!\!\rightarrow} \bot$ is true at a world w in a Stalnaker model $\mathcal{M}_{\mathrm{St}}$ iff f assigns the absurd world λ to the ordered pair α and w.

Let me now interpret Stalnaker's (1968) formal semantics of conditionals in terms of the Ramsey test. Definition 3.7 gives one – in line with Stalnaker (1968) and Stalnaker and Thomason (1970) – that the selection function f in a Stalnaker model $\mathcal{M}_{\mathrm{St}} = \langle W, \lambda, f, V\rangle$ determines for each formula α and world w in $W - \{\lambda\}$ a world w' in W, which can be interpreted as resulting from hypothetically putting the antecedent to one's stock of beliefs. The conditional $\alpha \mathrel{\Box\!\!\rightarrow} \beta$ is, then, true at w if and only if β is true at world w', which is selected by f for the ordered pair α and w.

As we saw above, Stalnaker (1968) gives two versions of his Ramsey test interpretation of conditionals: One includes the conditional consistency criterion (B) and the other does not. We, furthermore, distinguished in Section 3.1 between two conditional consistency criteria: the weak and the strong conditional consistency criterion. The weak and the strong conditional consistency criteria are (A) $\neg(\top \mathrel{\Box\!\!\rightarrow} \bot)$ and (B) $\neg((\alpha \mathrel{\Box\!\!\rightarrow} \beta) \wedge (\alpha \mathrel{\Box\!\!\rightarrow} \neg\beta))$, respectively, for given formulas α and β and where $\top$ and $\bot$ are defined as $p \vee \neg p$ and $p \wedge \neg p$, respectively. Interestingly, Stalnaker's semantics does not make the strong conditional consistency criterion (B) valid. This is due to the absurd world λ, which can be assigned to a standard world w in a Stalnaker model $\mathcal{M}_{\mathrm{St}} = \langle W, \lambda, f, V\rangle$ and an arbitrary antecedent α. Since f is interpreted as providing the world, which results from hypothetically putting the antecedent to one's stock of beliefs, and all formulas, including contradictory ones, are true at the absurd world, Stalnaker's semantics does not render the strong conditional consistency criterion (B) valid. (The weak conditional consistency criterion (A) is, however, valid in Stalnaker semantics by point d.ii of Definition 3.7.) As a result, conditional formulas of the form

$\alpha \mathrel{\Box\!\!\rightarrow} \bot$ are true at some worlds in some Stalnaker models and some, such as $\bot \mathrel{\Box\!\!\rightarrow} \bot$, are even true at all worlds in all Stalnaker models.

3.3.4 Stalnaker Semantics, Set Selection Semantics, and Chellas-Segerberg Semantics

There are variants of Stalnaker's semantics which avoid the inclusion of the absurd world λ (e.g., Nute & Cross, 2001, p. 9f). In my eyes the most elegant one is the following: Instead of assigning a single possible world to ordered pairs of formulas of language $\mathcal{L}_{\mathrm{CL}}$ and possible worlds by a function f, one can assign *sets* of possible worlds to those pairs which are either singletons or otherwise empty. The truth condition for conditional formulas from Definition 3.8 is, then, changed so that $\alpha \mathrel{\Box\!\!\rightarrow} \beta$ is true at a possible world w in a Stalnaker model $\mathcal{M}_{\mathrm{St}} = \langle W, f, V\rangle$ iff β is true at all worlds, which f selects for the ordered pair α and w (cf. D. Lewis, 1973b, p. 58).

One can generalize this alternative semantics not just to empty and singleton sets of possible worlds, but also to arbitrary sets of possible worlds. The modified function f' then assigns *sets* of possible worlds to ordered pairs of possible worlds and formulas of $\mathcal{L}_{\mathrm{CL}}$. D. Lewis (1973b, p. 58) calls functions of type f' 'set selection functions'. Set selection models can, then, be formally defined the following way:

Definition 3.9. *$\mathcal{M}_{set} = \langle W, f', V\rangle$ is a set selection model iff*
(a) W is a non-empty set of possible worlds,
(b) f' is a function such that $f' : \mathcal{L}_{\mathrm{CL}} \times W \to \mathrm{Pow}(W)$, and
(c) V is a valuation function such that $V : \mathcal{AV} \times W \to \{0, 1\}$.

The expression Pow(W) refers to the power set of W. Extensions of set selection models are, then, defined the following way:

Definition 3.10. *Let $\mathcal{M}_{set} = \langle W, f', V\rangle$ be a set selection model. Then, V^* is an extension of V to arbitrary formulas of $\mathcal{L}_{\mathrm{CL}}$ iff*
(a) $\forall p \in \mathcal{AV}\, \forall w \in W$: $V^*(p, w) = V(p, w)$ and
(b) for all formulas $\alpha, \beta \in \mathcal{L}_{\mathrm{CL}}$ and $w \in W$ holds:

$$\models_w^{\mathcal{M}_{set}} \neg\alpha \quad \textit{iff} \quad \not\models_w^{\mathcal{M}_{set}} \alpha, \qquad (V_\neg)$$
$$\models_w^{\mathcal{M}_{set}} \alpha \vee \beta \quad \textit{iff} \quad \models_w^{\mathcal{M}_{set}} \alpha \text{ or } \models_w^{\mathcal{M}_{set}} \beta, \text{ and} \qquad (V_\vee)$$
$$\models_w^{\mathcal{M}_{set}} \alpha \mathrel{\Box\!\!\rightarrow} \beta \quad \textit{iff} \quad \forall w'(w' \in f'(\alpha,w) \Rightarrow \models_{w'}^{\mathcal{M}_{set}} \beta). \qquad (V_{\Box\rightarrow})$$

Henceforth, I will use V and V^* indiscriminately.

Set selection models as specified by Definition 3.10 use essentially the same truth conditions for conditional formulas as in the modified semantics for Stalnaker models described above. Observe, furthermore, that the notion of set selection models is not only more general than Stalnaker models in terms of allowing for assigning arbitrary sets of possible worlds to ordered pairs of formulas and possible worlds, but also by omitting the specifications in terms of point (d) of Definition 3.7.

Set selection semantics is highly similar to CS semantics (Chellas, 1975; Segerberg, 1989) described and investigated in Chapters 4–7 (see also Section 3.1). There are two main differences between set selection semantics on the one hand and CS semantics on the other. The less important difference lies in the fact that in CS semantics an *accessibility relation* rather than a set selection *function* is used, as specified in Definition 3.9. It is not too difficult to see that one can replace the set selection function f' in set selection models by a three-place accessibility relation R' between pairs of possible worlds and formulas, and vice versa, without changing the truth values of formulas at any possible world: Suppose that f' in a given set selection model $\mathcal{M}_{\text{set}} = \langle W, f', V\rangle$ assigns the set of possible worlds X to the ordered pair α and w. One can, then, encode that information by an accessibility relation R' in such a way that all and only the worlds in the set X are accessible from w relative to α. Furthermore, let R' be an accessibility relation as specified above. Then, R' gives us for arbitrary formulas α and worlds w in W the sets of worlds $X \subseteq W$ which are accessible from w by the relation R'. Hence, one can encode the information in the three-place accessibility relation R' by a selection function f', which assigns the sets of possible worlds X to the ordered pair α and w.

The second, more important difference between set selection semantics and CS semantics is that in CS semantics, propositions rather than formulas are used, where a proposition $||\alpha||^{\mathcal{M}}$ is identified as the set of possible worlds, at which α is true in a particular model $\mathcal{M}$. So, in CS semantics a three-place accessibility relation between pairs of possible worlds and sets of possible worlds X rather than between formulas and pairs of possible worlds is employed. In CS semantics a conditional formula $\alpha \mathrel{\Box\!\!\rightarrow} \beta$ is true at a possible world w iff β is true at all possible worlds w' which are accessible from w by R relative to $||\alpha||^{\mathcal{M}}$ (see also Section 3.1). Although the use of formulas rather than propositions for the specification of the accessibility relation seems to make only a small difference (cf. D. Lewis, 1973b, p. 60), it has a strong impact on the formal properties of the semantics, insofar as a characterization in terms of structural as opposed to model-based conditions is only possible for CS semantics and not for set selection semantics (cf. Chapters 4–6).

3.3.5 Contrasting Ramsey Test Interpretations of Conditionals and Ordering Semantics

In this section I will contrast Ramsey test interpretations, such as described in Section 3.3.1, with the core idea of ordering semantics. I will argue that CS semantics (Chellas, 1975; Segerberg, 1989; see also Section 3.1 and Chapters 4–7) can be more naturally interpreted in terms of the Ramsey test – and to a lesser degree Stalnaker semantics (1968; Stalnaker & Thomason, 1970) and set selection semantics – while ordering semantics, as for example described in D. Lewis (1973b), Kraus et al. (1990), and Lehmann and Magidor (1992) – is based on a different semantic intuition. Despite that fact both types of semantics are often referred to quite indiscriminately (e.g., D. Lewis, 1973b, p. 57f and p. 77; Gärdenfors, 1979, p. 381; Makinson, 1993, p. 340, p. 343).

The core idea that underlies the ordering semantics of D. Lewis (1973b), Kraus et al. (1990), and Lehmann and Magidor (1992) is that possible worlds (used in the semantics) are ranked according to a criterion, such as similarity to the actual world (D. Lewis, 1973b) or normality preferences of agents (Kraus et al., 1990). While in D. Lewis' (1973b) approach a new system of spheres $\$_w$ is used for each possible world w, Burgess (1981), Kraus et al. (1990), and Lehmann and Magidor (1992) employ an absolute ordering relation between possible worlds according to which all possible worlds (or sets of possible worlds) are ranked on the basis of a global criterion. To determine the truth value of a conditional, one only draws on the relevant segment of the whole ordering of possible worlds.

The Ramsey test, on the other hand, does not require rankings of possible worlds. Rather each conditional is determined by putting the antecedent to one's stock of beliefs and by specifying the truth value of a conditional (at a possible world) on that basis (see Section 3.3.1). Hence, the Ramsey test allows for a step-by-step process rather than a global one-time ranking, as an ordering semantics presupposes. To find out whether a conditional is true at a possible world w in an ordering semantics, in all cases one has to provide a general ordering of possible worlds (w.r.t. the actual world w), independently of the antecedent of the respective conditional. In that sense a Ramsey test approach is much more parsimonious, since for the Ramsey test for a conditional (w.r.t. possible worlds semantics) one only has to take into account the possible worlds which are relevant in the context of the antecedent of that conditional. In an ordering semantics the ranking of possible worlds (w.r.t. the actual world w) gives us the truth value of any conditional (at world w). Except for trivial cases such an ordering of possible worlds involves by far more information than required to assess a conditional by means of the Ramsey test. Furthermore – as we shall see in Section 7.1 – the basic CS semantics (e.g. Section 3.1) allows for a plausible and minimal Ramsey interpretation, but not so much for an interpretation in terms of ordering semantics. The basic Stalnaker semantics (that is Definition 3.7 without condition d) and set selection semantics (see Definition 3.9) seem to be good candidates as well – at least prima facie – but are ultimately inferior to CS semantics, as I will argue below.

Prima facie, one might suspect that Stalnaker (1968) suggests an interpretation for Stalnaker semantics purely in terms of the Ramsey test, the more as he introduces this idea into the modern discussion. Surprisingly,

Stalnaker (1968) does not do that. In the interpretation of his semantics, Stalnaker (1968) mixes elements of a Ramsey test interpretation with elements of ordering semantics: In the informal description of his semantics, Stalnaker (1968, p. 104) essentially draws on the notion of rankings of possible worlds according to their similarity (to the actual world) and argues that his truth conditions require that the world $w' \in W$ must differ minimally from the actual world w, where w' is selected by f for both formula α and $w \in W \backslash \{\lambda\}$ in a Stalnaker model $\mathcal{M}_{St} = \langle W, \lambda, f, V \rangle$ (p. 104). By these means Stalnaker motivates conditions d.ii and d.iii of Definition 3.7, since these ensure – conjointly with the rest of Definition 3.7 – that possible worlds are ordered according to some criterion, such as similarity (Stalnaker, 1968, p. 104f).

I admit that Stalnaker's semantics (see Definitions 3.7 and 3.8) cannot solely be justified on the basis of the Ramsey test, even if one omits point (d) in Definition 3.7. The main reason is that Stalnaker (1968) presupposes that "a possible world is the ontological analogue of a stock of hypothetical beliefs" (p. 102). Accordingly, the world selection function f in a Stalnaker model $\mathcal{M}_{St} = \langle W, \lambda, f, V \rangle$ chooses for each formula α and world $w \in W \backslash \{\lambda\}$ a single world $w' \in W$. Since at any standard possible world w' (a world that is not the absurd world λ), either β or $\neg\beta$ is true for all formulas β, it follows that in Stalnaker's account stocks of hypothetical beliefs must be negation-complete, in the sense that for every formula β either β or else $\neg\beta$ is (hypothetically) believed. The only exception is the case in which f chooses the absurd world λ. Since every formula is true at λ, the absurd world λ represents an inconsistent stock of hypothetical beliefs.

It is hardly plausible that hypothetically putting the antecedent α to one's stock of beliefs results in all cases either in an inconsistent or a negation-complete stock of hypothetical beliefs. This would, for example, imply – in order to find out whether an arbitrary conditional is true – that one has (A) to hypothetically believe that either (i) Goldbach's conjecture is true or (ii) that Goldbach's conjecture is not true, or otherwise (B) arrive at an inconsistent stock of hypothetical beliefs.

On the other hand set selection semantics (see Definitions 3.9 and 3.10) and CS semantics do not make the assumption that hypothetically putting a formula to one's stock of hypothetical beliefs always results either in an inconsistent stock of beliefs or a negation-complete stock of beliefs. Thus, both semantics are more plausible candidates for a pure interpretation of conditionals in terms of the Ramsey test. Furthermore, CS semantics is more promising than set selection semantics, since the former but not the latter uses propositions (sets of possible worlds) rather than formulas. For that reason CS semantics allows a structural as opposed to a model-based characterization of the respective semantics (cf. Section 3.3.4; see also Section 4.1). I will not enquire into this issue any further in this context and postpone an interpretation of CS semantics in terms of a modified Ramsey test to Section 7.1.

3.3.6 Stalnaker Semantics, Conditional Consistency Criteria, and the Principle of Conditional Excluded Middle

In this section I will discuss different conditional consistency criteria (see criterion (b) of the Ramsey test in Section 3.3.1). I shall first describe different versions of conditional consistency criteria and then discuss the role of consistency criteria for conditionals in Stalnaker semantics (Stalnaker models, see Definitions 3.7 and 3.8). For that purpose I use the full conditional logic language $\mathcal{L}_{\mathrm{CL}}$ (see Sections 3.2 and 4.2.1).

Let me now focus on the notion of conditional consistency. A set of formulas Γ is inconsistent (simpliciter) if all formulas of the language are theorems of Γ. By conditional consistency I mean something different, namely that (i) certain types of conditionals are not allowed to have consequent formulas which are inconsistent, or (ii) that two or more consequents of conditional formulas with the same antecedent contradict each other. In both cases (i) and (ii), if a formula set Γ contains such formulas, then Γ becomes inconsistent. A weak conditional consistency criterion of sort (i) is the Principle P-

Cons ("Probabilistic Consistency") that we encountered in Sections 3.1 and 3.2 (criterion (A)). For mnemonic reasons I shall now repeat this principle here:

P-Cons $\neg(\top \,\Box\!\!\rightarrow \bot)$

So, why should one consider P-Cons a conditional consistency criterion? The answer is that P-Cons makes sure – given reasonable assumptions – that whenever all formulas of the form $\alpha \,\Box\!\!\rightarrow \beta$ can be inferred, all non-conditional formulas of the language can also be inferred from it, including $\bot$ (see below). This is not the case in all conditional logics. P-Cons is, for example, not valid in the class of all Lewis models (see Definitions 3.1–3.3): So, for instance, any Lewis model $\mathcal{M}_L = \langle W, \$, V \rangle$ such that $\$_w = \varnothing$ makes all conditionals $\alpha \,\Box\!\!\rightarrow \beta$ true at the world $w \in W$, while $\bot$ is always false at any $w \in W$.

The above-mentioned reasonable assumptions are that the following rules and axiom schemata are valid:

LLE if $\vdash \alpha \leftrightarrow \beta$ and $\alpha \,\Box\!\!\rightarrow \gamma$, then $\beta \,\Box\!\!\rightarrow \gamma$
RW if $\vdash \alpha \rightarrow \beta$ and $\gamma \,\Box\!\!\rightarrow \alpha$, then $\gamma \,\Box\!\!\rightarrow \beta$
CM $(\alpha \,\Box\!\!\rightarrow \gamma) \wedge (\alpha \,\Box\!\!\rightarrow \beta) \rightarrow (\alpha \wedge \beta \,\Box\!\!\rightarrow \gamma)$

'LLE', 'RW', and 'CM' abbreviate 'Left Logical Equivalence', 'Right Weakening', and 'Cautious Monotonicity', respectively. Furthermore, Principles LLE, RW, and CM or variants are valid in conditional logic semantics, such as Stalnaker models (Definitions 3.7 and 3.8), Lewis models (Definitions 3.1–3.3) and the semantics of Adams' **P** systems (1965, 1966, 1977, 1975, 1986; cf. Schurz, 1997b, 1998; see Section 3.6).

Let us now prove the following lemma, where a conditional logic system S is assumed to be closed under propositional consequence except possibly for the rule of substitution (cf. Section 4.2.5), where Γ is a set of formulas of $\mathcal{L}_{\mathrm{CL}}$:

Lemma 3.11. *Let S be a conditional logic system that contains LLE, RW, and CM. Then, $\Gamma \vdash_S \top \,\Box\!\!\rightarrow \bot$ iff for all formulas α and β it holds that $\Gamma \vdash_S \alpha \,\Box\!\!\rightarrow \beta$.*

Proof. "⇐": The proof is trivial.
"⇒": By Lemma 3.12 we obtain the left-to-right direction of Lemma 3.11.

□

Lemma 3.12. *LLE+RW+CM+*$\top \mathrel{\Box\!\!\rightarrow} \bot \Rightarrow \alpha \mathrel{\Box\!\!\rightarrow} \beta$

Proof.

1.	$\top \mathrel{\Box\!\!\rightarrow} \bot$	given
2.	$\top \mathrel{\Box\!\!\rightarrow} \alpha$	1, RW
3.	$\top \mathrel{\Box\!\!\rightarrow} \beta$	1, RW
4.	$\top \wedge \alpha \mathrel{\Box\!\!\rightarrow} \beta$	3, 2, CM
5.	$\alpha \mathrel{\Box\!\!\rightarrow} \beta$	4, LLE

□

Observe that P-Cons is implied by Principles MP ("Modus Ponens") which is the following principle:

MP $(\alpha \mathrel{\Box\!\!\rightarrow} \beta) \rightarrow (\alpha \rightarrow \beta)$

The following lemma gives us that MP implies P-Cons:

Lemma 3.13. *MP ⇒ P-Cons*

Proof.

1.	$(\top \mathrel{\Box\!\!\rightarrow} \bot) \rightarrow (\top \rightarrow \bot)$	MP
2.	$\neg(\top \mathrel{\Box\!\!\rightarrow} \bot)$	1, p.c.

□

The stronger conditional consistency criterion (B) can be formalized in the following ways, where $\nvdash_S \alpha$ indicates non-derivability in a conditional logic system S, which implies closure under p.c. consequence (except possibly for the rule of substitution): [5]

[5] Note that CNC and CNC′ are nothing but variants of Boethius' thesis and Aristotle's thesis, respectively (cf. Routley & Montgomery, 1968, p. 82). The Boethius' thesis is sometimes formulated as 'if $\alpha \mathrel{\Box\!\!\rightarrow} \beta$ then $\neg(\alpha \mathrel{\Box\!\!\rightarrow} \neg\beta)$'. Furthermore, Aristotle's thesis is $\neg(\neg\alpha \mathrel{\Box\!\!\rightarrow} \alpha)$, which is logically equivalent to CNC′ given a logic which contains LLE, RW, AND (i.e., $(\alpha \mathrel{\Box\!\!\rightarrow} \beta) \wedge (\alpha \mathrel{\Box\!\!\rightarrow} \gamma) \rightarrow (\alpha \mathrel{\Box\!\!\rightarrow} \beta \wedge \gamma)$), Refl (i.e., $\alpha \mathrel{\Box\!\!\rightarrow} \alpha$), and closure under propositional consequence (including the rule of substitution) (cf. Section 4.2.5).

CNC $\neg((\alpha \,\Box\!\!\rightarrow \beta) \wedge (\alpha \,\Box\!\!\rightarrow \neg\beta))$
CNC′ $\neg(\alpha \,\Box\!\!\rightarrow \bot)$
RCNC if $\nvdash_S \neg\alpha$ then $\neg((\alpha \,\Box\!\!\rightarrow \beta) \wedge (\alpha \,\Box\!\!\rightarrow \neg\beta))$
RCNC′ if $\nvdash_S \neg\alpha$ then $\neg(\alpha \,\Box\!\!\rightarrow \bot)$

'CNC' and 'RCNC' abbreviate 'Conditional Non-Contradiction' and 'Restricted Conditional Non-Contradiction', respectively. CNC says that any two conditional formulas $\alpha \,\Box\!\!\rightarrow \beta$ and $\alpha \,\Box\!\!\rightarrow \neg\beta$ contradict each other without further qualification. RCNC states that two conditional formulas $\alpha \,\Box\!\!\rightarrow \beta$ and $\alpha \,\Box\!\!\rightarrow \neg\beta$ contradict each other if α is S-consistent according to a conditional logic system S (see before). Furthermore, CNC′ states that no conditional is allowed to have an S-inconsistent consequent, while RCNC′ gives us that no conditional formula with an S-consistent antecedent formula has an S-inconsistent consequent formula. The difference between CNC′ and P-Cons lies in the fact that if CNC′ is a theorem of a logic S, then for a formula set Γ a single conditional with an S-inconsistent consequent formula in Γ suffices to make Γ S-inconsistent. For RCNC′ compared to CNC only conditionals with inconsistent antecedent formulas are exempted.

A variant of Principle CNC is, furthermore, the following axiom schema:

D $(\alpha \,\Box\!\!\rightarrow \beta) \rightarrow (\alpha \,\Diamond\!\!\rightarrow \beta)$.

Note that $(\alpha \,\Diamond\!\!\rightarrow \beta)$ is defined as $\neg(\alpha \,\Box\!\!\rightarrow \neg\beta)$ ($\mathrm{Def}_{\Diamond\!\rightarrow}$, Section 4.2.1). The name D for this principle derives from the fact that it is a generalization of Principle D from Kripke semantics (see Section 3.6.3). We postpone a proof of the logical equivalence of Principle CNC and Principle D to Section 3.6.3 (see Lemma 3.24). Furthermore, one can easily prove that CNC and CNC′ on the one hand and RCNC and RCNC′ on the other are equivalent given reasonable restrictions, where the reasonable restrictions are that RW and the following principle are valid:

AND $(\alpha \,\Box\!\!\rightarrow \beta) \wedge (\alpha \,\Box\!\!\rightarrow \gamma) \rightarrow (\alpha \,\Box\!\!\rightarrow \beta \wedge \gamma)$

Principle AND is valid in a range of conditional logic semantics, such as Lewis models, Stalnaker models, CS semantics, and the probabilistic semantics of Adams' **P** systems (1965, 1966, 1977, 1975, 1986; cf. Schurz, 1997b, 1998; see Section 3.6).

Let me now prove that CNC is equivalent to CNC′ given RW and AND:

Lemma 3.14. *RW+AND ⇒ (CNC ⇔ CNC′)*

Proof. By Lemmata 3.15 and 3.16 we obtain Lemma 3.14. □

Lemma 3.15. *RW+CNC ⇒ CNC′*

Proof.

1.	$\neg((\alpha \mathrel{\Box\!\!\rightarrow} \beta) \wedge (\alpha \mathrel{\Box\!\!\rightarrow} \neg\beta))$	CNC
2.	$\neg\neg(\alpha \mathrel{\Box\!\!\rightarrow} \bot)$	assumption, ind. (indirect) proof
3.	$\alpha \mathrel{\Box\!\!\rightarrow} \bot$	2, p.c.
4.	$\alpha \mathrel{\Box\!\!\rightarrow} \beta$	3, RW
5.	$\alpha \mathrel{\Box\!\!\rightarrow} \neg\beta$	3, RW
6.	$(\alpha \mathrel{\Box\!\!\rightarrow} \beta) \wedge (\alpha \mathrel{\Box\!\!\rightarrow} \neg\beta)$	4, 5, p.c.
7.	$\neg(\alpha \mathrel{\Box\!\!\rightarrow} \bot)$	2–6, 6, 1, ind. proof

□

Lemma 3.16. *RW+AND+CNC′ ⇒ CNC*

Proof.

1.	$\neg(\alpha \mathrel{\Box\!\!\rightarrow} \bot)$	CNC′
2.	$\neg\neg((\alpha \mathrel{\Box\!\!\rightarrow} \beta) \wedge (\alpha \mathrel{\Box\!\!\rightarrow} \neg\beta))$	assumption, ind. proof
3.	$(\alpha \mathrel{\Box\!\!\rightarrow} \beta) \wedge (\alpha \mathrel{\Box\!\!\rightarrow} \neg\beta)$	2, p.c.
4.	$\alpha \mathrel{\Box\!\!\rightarrow} \beta \wedge \neg\beta$	3, AND
5.	$\alpha \mathrel{\Box\!\!\rightarrow} \bot$	4, RW
6.	$\neg((\alpha \mathrel{\Box\!\!\rightarrow} \beta) \wedge (\alpha \mathrel{\Box\!\!\rightarrow} \neg\beta))$	2–5, 5, 1, ind. proof

□

The proof for the equivalence of RCNC and RCNC′ is perfectly analogous to the proof of Lemma 3.14. Due to Lemma 3.14 I will henceforth often not specifically refer to the variants RCNC′ and CNC′. Furthermore, I will discuss both CNC and RCNC in more detail in Sections 3.6.3 and 3.8.

Let me now focus on conditional consistency criteria in Stalnaker's (1968; Stalnaker & Thomason, 1970) semantics. As I noted in Section 3.3.3, the conditional consistency criterion (b) of the Ramsey test corresponds to CNC, CNC′, RCNC, and RCNC′ rather than to P-Cons and P-Cons′. More specifically, Stalnaker argues that "one cannot legitimately reach an impossible conclusion from a consistent assumption" (Stalnaker, 1968, p. 104). This suggests a conditional consistency criterion in line with RCNC or RCNC′.

Interestingly, as we observed in Section 3.3.3 none of the Principles CNC, CNC′, RCNC, and RCNC′ is valid in Stalnaker's (1968; Stalnaker & Thomason, 1970) semantics (see Section 3.3.3) Despite that fact Stalnaker (1968) seems to include such a conditional consistency criterion in his semantics. Stalnaker uses an accessibility relation R in his definition of Stalnaker models to "make sure" that the impossible world λ – at which all formulas, including $\bot$, are true – is chosen only if there is no world that is accessible at which the antecedent formula is true (cf. Section 3.3.3). Stalnaker's informal interpretation of the accessibility relation R is, however, not warranted, since R is simply ineffective in his semantics. Hence, in Section 3.3.2 I followed D. Lewis (1973b) in omitting this parameter in my specification of Stalnaker models.

Furthermore, the majority of standard conditional logic systems makes neither RCNC nor CNC valid (e.g., Stalnaker, 1968; Stalnaker & Thomason, 1970; D. Lewis, 1973b; Adams, 1965, 1966, 1977, 1986; Schurz, 1998) including the majority of conditional logic systems investigated based on CS semantics in Chapters 4–7. Notable exceptions are the semantics of Adams (1975) and Schurz (1997b) which render variants of Principles RCNC and RCNC′ valid (see Section 3.6.3). As the precondition of these variants of RCNC or RCNC′ also refer to non-derivability conditions, the reliance on this type of conditional consistency criterion makes the resulting system a default logic in the sense described in Section 2.2.

It is, finally, instructive to also compare Principle CNC to the following characteristic principle of Stalnaker's (1968) conditional logic:

CEM $(\alpha \,\Box\!\!\rightarrow \beta) \vee (\alpha \,\Box\!\!\rightarrow \neg\beta)$

'CEM' abbreviates here 'Conditional Excluded Middle'. Two variants of Principle CEM are the following:

CEM′ $(\alpha \diamondrightarrow \beta) \rightarrow (\alpha \boxrightarrow \beta)$

CEM″ $\neg(\neg(\alpha \boxrightarrow \beta) \wedge \neg(\alpha \boxrightarrow \neg\beta))$

Principles CEM′ and CEM″ are logically equivalent to CEM (on the basis of $\mathrm{Def}_{\diamondrightarrow}$).

The difference between CNC and CEM and their variants lies in the following fact: CNC states that $\alpha \boxrightarrow \beta$ and $\alpha \boxrightarrow \neg\beta$ cannot both hold. In constrast, Principles CEM, CEM′, and CEM″ give that at least one of both conditionals $\alpha \boxrightarrow \beta$ and $\alpha \boxrightarrow \neg\beta$ must hold. Note that an interpretation of conditional formulas in terms of the material implication makes CEM, CEM′, and CEM″ valid but not CNC.

3.3.7 Bennett's Version of the Ramsey Test

Let me finally discuss Bennett's (2003) version of the Ramsey test, which explicitly refers to component (c), the degree of belief component of the Ramsey test (see Section 3.3.1). Bennett (2003) specifies his version of the Ramsey test as follows, where the expression $A \rightarrow C$ represents an indicative conditional of the form 'if *A* then *C*' (Bennett, 2003, p. 15):

> To evaluate $A \rightarrow C$, I should (1) take the set of probabilities that constitutes my present belief system, and add to it a probability = 1 for *A*; (2) allow this addition to influence the rest of the system in the most natural, conservative manner; and then (3) see whether what results from this includes a high probability for *C* (Bennett, 2003, p. 29).

Observe that (1) and (3) above describe the core aspect (a) of the Ramsey test (see Section 3.3.5): Hypothetically suppose that the antecedent holds and then test whether the consequent holds according to that supposition. Note that Bennett goes in (1) beyond the core aspect of the Ramsey test, insofar as he uses a subjective probabilistic framework (aspect (c)). Point (1) describes in fact nothing but the (classical) conditionalization of a probability function *P* on a formula *A* (e.g., Talbott, 2008, §2; Howson & Urbach, 1993, p. 99; Bovens & Hartmann, 2003; see also Thorn, in press and Douven & Romeijn, 2011). Particularly interesting here in this context is condition (2) which represents a variant of the optional consistency criterion (b) of the Ramsey test (see Section 3.3.5). First, the adjustment of one's own belief system "in the most natural, conservative manner" hints at an ordering approach. Second, Bennett seems to presuppose that such a unique natural and conservative way of changing one's beliefs is available in all cases and he does not go into further details regarding that issue. I rather believe that one has to say more about that issue than merely to presuppose that a unique way of changing one's beliefs always exists (cf. Section 3.8). Approaches, such as Bennett's own, that employ the conditional consistency requirement in terms of CNC or RCNC have to make sure that the consequent of every conditional (with a consistent antecedent) is also consistent. The problem here is – contrary to what Bennett seems to assume – that in the most typical case there are multiple ways in which one can change one's stock of (hypothetical) beliefs when the antecedent contradicts one's prior beliefs, and in these cases the result of one's change of beliefs is not uniquely determined.

3.3.8 A General Ramsey Test Requirement

The aim of the present subsection is to specify a general Ramsey test requirement that any Ramsey test interpretation of conditionals has to satisfy. For that purpose I will draw on the preceding sections and make more precise in which way the core criterion (a) of the Ramsey test identified in Section 3.3.1 has to be drawn out.

Let me now summarize the results of our discussion of the Ramsey test in the preceding sections. We saw in Sections 3.3.1 and 3.3.2 that the core aspect of the Ramsey test of conditionals amounts to the following: Hypothetically putting the antecedent to one's stock of beliefs and to check whether the consequent holds (aspect (a)). Moreover, according to my analysis two further components of the Ramsey test are optional: (b) a conditional consistency requirement and (c) a probabilistic interpretation of the Ramsey test. Point (b) says that any stock of beliefs which results from adding the antecedent to one's stock of beliefs must be consistent. Furthermore, point (c) requires the Ramsey test to be formulated in terms of subjective agent-relative degrees of belief.

I argued in Section 3.3.5 that there are truth value possible worlds semantics, such as CS semantics (see Chapters 4–7; see also Section 3.1) which allow for a pure Ramsey test interpretation and do not incorporate aspects (b) and (c).

Let me now discuss which further conditions have to satisified in order that component (a) is appropriately specified. First, the Ramsey test is subjective in nature in the sense that it refers to a stock of (hypothetical) beliefs of some agent. In that way it differs from D. Lewis' (1973b) interpretation of his systems of spheres semantics: In Lewis models the basic notion is overall similarity of possible worlds which is objective in nature (see Section 3.2). The Ramsey test, however, shares the subjective component, for example, with the ordering semantics of Kraus et al. (1990) and Lehmann and Magidor (1992), who interpret their semantics in terms of preference normality orderings by agents (see Section 3.2).

Second, the Ramsey test refers to stocks of beliefs. Stocks of beliefs do not generally correspond to single possible worlds in possible worlds semantics, contrary to Stalnaker's (1968, p. 102) view (see Section 3.3.5): Putting the problem of inconsistent stocks of beliefs aside, the representation of stocks of beliefs by single possible worlds would require that stocks of beliefs are negation-complete in the sense that I accept either α or $\neg\alpha$ for any formula α (see Section 3.3.5). Negation completeness, however, does not follow if I represent stocks of beliefs by sets of possible worlds rather than single possible worlds. Hence, the ontological analogue to a stock of (hypothetical) beliefs is a set of possible worlds rather than a single possible world.

The use of stocks of (hypothetical) beliefs implies that I do not only arrive at sets of possible worlds in possible worlds semantics by putting the antecedent hypothetically to my stock of beliefs, but that I also start from a set of possible worlds rather than a single possible world. Note that in possible worlds semantics the evaluation of formulas is done for each possible world and not for sets of possible worlds (cf. Sections 3.2 and 3.3.3). However, to allow for a Ramsey test interpretation of possible worlds semantics one need not change possible worlds semantics, but can reinterpret possible worlds semantics as described here in a specific way: In CS semantics (Chellas, 1975; Segerberg, 1989), for example, one can achieve this by requiring that a conditional formula $\alpha \boxright \beta$ is true w.r.t. a set of possible worlds X just in case $\alpha \boxright \beta$ is true at all possible worlds $w \in X$. I will discuss this issue in more detail in the context of CS semantics in Section 7.1.

3.4 Indicative and Counterfactual Conditionals and Conditional Logics

In this section I will discuss the notions of indicative and counterfactual conditionals and logics for these types of conditionals. In Section 3.4.1 I will describe criteria for distinguishing between indicative and counterfactual conditionals, as already presupposed in Chapter 1, and I will discuss the implications of such a distinction for conditional logics for indicative and counterfactual conditionals in Section 3.4.2. In this context the discussion of bridge principles is particularly important, where bridge principles specify a fixed relationship between conditional and non-conditional formulas. I will argue that one is not justified to assume bridge principles for indicative conditionals and that rather a non-monotonic rule should be employed as described by Delgrande (1988), which is nothing but a variant of Carnap's (1962) and Reichenbach's (1949) Principle of Total Evidence. Finally, in Section 3.4.3 I will discuss in how far indicative and counterfactual conditionals can justifiedly be interpreted in a subjective and an objective way, respectively.

Let me at this point provide reasons as to why the distinction between indicative and counterfactual conditionals is of importance: A central assumption of several approaches to conditionals is that indicative and counterfactual conditionals serve quite different purposes and are thereby assumed to have different assertibility conditions. For example, Adams (1975), Bennett (2003), Edgington (2007), and Gibbard (1980) hold the view that (a) indicative and counterfactual conditionals require distinct theoretical treatments and that (b) indicative conditionals should be analyzed in terms of a subjective agent-relative probabilistic semantics such as Adams (1975; see Section 3.6; cf. Section 3.1). For the analysis of counterfactuals these authors disagree in as much as Adams (1975, Chapter 4) and Edgington (2007, p. 202) argue for an analysis of counterfactuals in terms of a probabilistic semantics, while Bennett (2003, p. 291) prefers a temporal analysis based on D. Lewis (1979).[6] Note here that my defense of possible worlds semantics in the following sections against arguments forwarded by Bennett (2003) is a defense of possible worlds semantics for indicative conditionals rather than for counterfactual conditionals (cf. Section 3.1).

3.4.1 Criteria for Distinguishing Indicative and Counterfactual Conditionals

Before I discuss criteria for distinguishing indicative and counterfactual conditionals, let me first provide a classical example for an indicative and a counterfactual conditional (cf. Adams, 1970, p. 90), where E19 and E20 are the counterfactual and the indicative conditional, respectively:

E19 If Oswald hadn't shot Kennedy in Dallas, then someone else would have.

E20 If Oswald didn't shoot Kennedy in Dallas, then someone else did.

We seem to reject E19, while we seem to accept E20. In the literature pairs of conditionals, such E19 and E20, are often regarded as prototypical examples of the difference between indicative and counterfactual conditionals, respectively, since both conditionals have the "same content" but differ w.r.t. their acceptibility conditions. In Section 3.4.3 we will discuss which difference in acceptibility (or assertibility) criteria seems to underlie indicative and counterfactual conditionals such as E20 and E19, respectively, and I will also inquire into their relation to the Ramsey test.

[6] Not all authors employ a probabilistic semantics for indicative conditionals in terms of subjective probabilities. For instance, Schurz (1997b, 2001b, 2005b) advocates a treatment of (generic) indicative conditionals in terms of objective frequency-based probabilities (see also Pearl, 1988; D. Lewis, 1980; Bacchus, 1990; cf. Section 3.5.1).

There are two basic approaches in the literature to describe the difference between indicative and counterfactual conditionals, such as E20 and E19: (A) by mood (indicative vs. subjunctive mood) or (B) by presupposing that the antecedent of a counterfactual conditional must be "counter to the facts".

According to criterion (A), the main difference between conditionals of type E20 and E19 is that E20 is in the indicative mood, while E19 is in the subjunctive mood. Criterion (B) identifies suppositions about the truth value of the antecedent as the essential difference between E19 and E20: In E19 it is presupposed that the antecedent is "counter to facts" viz. false, while in E20 the antecedent is neither presupposed to be true nor presupposed to be false. It is essential to specify what one means by 'presuppose' here. For example, D. Lewis (1973b) argues in his book *Counterfactuals* that this presupposition may be due to "conversational implicature, without any effect on truth conditions" (p. 3). Accordingly D. Lewis provides truth conditions for counterfactuals, for which the antecedent of a counterfactual is not required to be false or presupposed to be so (p. 26).

In this context the more fine-grained distinction by Declerck and Reed (2001) is useful. In order to provide a general terminology for the categorization of conditionals, Declerck and Reed (2001) distinguish between two types of counterfactual conditionals: (B1) "genuine" counterfactuals and (B2) tentative counterfactuals. Counterfactuals of type (B1) are counterfactuals, for which the antecedent is false or presupposed to be so, while counterfactuals of type (B2) include counterfactuals whose antecedent is merely improbable (p. 13f).

Prima facie (A) might seem to be a well-established criterion in linguistics for the distinction between indicative and counterfactual conditionals. This is, however, not the case (cf. Declerck & Reed, 2001, p. 13). It is, for example, not clear that the subjunctive mood describes 'would' conditionals of type E19 in the English language reliably (Bennett, 2003, p. 11). Moreover, in French and Spanish no subjunctive mood, but rather a conditional tense is used to formulate counterfactuals (Bennett, 2003, p. 11).

So, should one choose (A), (B1), or (B2) as a criterion for distinguishing between indicative and counterfactual conditionals? Bennett (2003, p. 12) favors criterion (A), but implicates that the difference between criteria (A) and (B) is negligible and that he merely uses criterion (A) for labeling two sorts of conditionals (indicative vs. counterfactual conditionals). I disagree, since for philosophers such as Bennett who subscribe to points (a) and (b) from the beginning of this section there are substantial reasons to accept criterion (A) rather than criteria (B1) or (B2). To describe reasons that make criterion (A) preferable from Bennett's view, I first have to outline some basic assumptions of subjective probabilistic approaches to indicative conditionals.

According to point (a) described at the beginning of this section, a subjective probabilistic semantics is in order for indicative conditionals. For example, in the subjective probabilistic framework endorsed by Bennett (2003), all formulas to which probabilities are assigned represent propositions that can either be true or else false. A probability assigned to a formula α is, then, interpreted as the probability that α is true (cf. Bennett, 2003, p. 47f). Truth and falsity in this framework are constructed in an objective way, in contrast to truth and falsity according to an agent's beliefs.

Based on criterion (a), Bennett (2003, p. 12) argues for criterion (A) by objecting against criterion (B1):[7] If one were to use criterion (B1), (i) one's classification of counterfactual conditionals would not so much depend on the fact whether the antecedent is objectively false but rather on the fact whether the speaker believes so. Furthermore, in contrast to counterfactual conditionals (ii) indicative conditionals were not in any specific sense "factual", "pro-factual" or the like. Moreover, (iii) the label 'counterfactual' rather depends on pragmatic assumptions regarding conditionals and, thus, should not be taken too seriously and neither serve as basis for the categorization of indicative and counterfactual conditional.

In my view points (ii) and (iii) are rather weak for the following reasons: First, there exists an objective frequency-based semantics (e.g. Schurz, 1997b, 2001b, 2005b; see also Section 3.5.1) which is clearly targeted for indicative conditionals rather than counterfactual conditionals. So, there is

[7] Bennett does not discuss criterion (B2).

a semantics which renders the former type of conditionals "factual" rather than not (point (ii)). Second, it is far from clear that the assumptions (B1) and (B2) are only made on the pragmatic level as opposed to the semantic level (point (iii)). Even if one presupposes that (B1) and (B2) are only made on a pragmatic level, (B1) and (B2) might still be essential for the application of counterfactual conditionals to real world reasoning and, therefore, be appropriate criteria for distinguishing between indicative and counterfactual conditionals.

Let me now discuss point (i). Given a subjective probabilistic framework as endorsed in (a), point (i) seems particularly plausible. This is due to the fact that this approach only draws on the notions of objective truth and subjective probabilities (see above). In general, however, it is far from clear that one must accept such a subjective probabilistic framework. One might alternatively use a subjective notion of truth and interpret both conditional and non-conditional formulas in terms of truth and falsehood according to an agent's beliefs. In that interpretation a counterfactual can, then, be constructed as being "contrary to the facts the agent believes".

A person who accepts points (a) and (b) from the beginning of this section might, in addition, reject criterion (B2) for the following reasons: Since (B2) states that the antecedent of a counterfactual as opposed to an indicative conditional is regarded as improbable, it seems plausible to describe this probabilistic component also in terms in the subjective probabilistic framework already used for indicative conditionals (point (b)). By point (a) one is not allowed to use the same characterization of indicative and counterfactual conditionals. So, if one accepts criterion (B2) and allows for a subjective probabilistic semantics for counterfactual conditionals, it seems again plausible (I) that there are no counterfactuals with subjectively probable antecedents and (II) that there are no indicative conditionals with subjectively improbable antecedents.

While (I) might be regarded plausible on general grounds, it is much less clear that (II) is adequate from a position in line with points (a) and (b) such as Adams (1975), Bennett (2003), Edgington (2007), and Gibbard (1980).

These authors argue that indicative and counterfactual conditionals, such as E20 and E19, respectively, require a completely different analysis (assumption (a)). To accept (B2) and (II) would deny exactly that. This is due to the fact that conditionals with a low subjective antecedent probability, such as E20, must according to (II) also be regarded as counterfactual conditionals. Hence, both E19 and E20 would be regarded counterfactual conditionals and, thus, should be subjected to the same philosophical analysis. In sum, there are substantial reasons for persons who subscribe to position (a) to prefer an analysis of conditionals in terms of criterion (A) rather than (B1) and (B2).

In the following subsection I will spell out the difference between indicative and counterfactual conditionals further in terms of criteria (B1) and (B2). I opt for (B1) and (B2) rather than (A), since I can hardly see how criterion (A) – which refers solely to syntactical/grammatical features of natural language – can lead to a fruitful discussion which types of logics and semantics are justified for both types of conditionals. In contrast criteria (B1) and (B2) make – from a philosophical and a logical perspective – more substantial claims concerning the difference between both types of conditionals.

3.4.2 Bridge Principles and Logics for Indicative and Counterfactual Conditionals

I will now discuss bridge principles and their plausiblity in the context of logics for indicative and counterfactual conditionals. I will not endorse any particular set of bridge principles for indicative and counterfactual conditionals here but advocate an alternative approach for indicative conditionals by Delgrande (1988) which is in line with Carnap's (1962) and Reichenbach's (1949) Principle of Total Evidence (see also Section 3.5.1).

The following principles are examples of prominent and prima facie plausible bridge principles:

MP $(\alpha \mathrel{\Box\!\!\rightarrow} \beta) \rightarrow (\alpha \rightarrow \beta)$
CS $(\alpha \wedge \beta) \rightarrow (\alpha \mathrel{\Box\!\!\rightarrow} \beta)$

Det $(\top \mathrel{\Box\!\!\rightarrow} \alpha) \rightarrow \alpha$
Cond $\alpha \rightarrow (\top \mathrel{\Box\!\!\rightarrow} \alpha)$

The symbol $\rightarrow$ stands for the material implication. The expressions 'MP', 'CS', 'Det', and 'Cond' abbreviate 'Modus Ponens', 'Conditional Sufficiency', 'Detachment', and 'Conditionalization', respectively. In addition, the following axiom schemata are bridge principles:

VEQ $\beta \rightarrow (\alpha \mathrel{\Box\!\!\rightarrow} \beta)$
EFQ $\neg\alpha \rightarrow (\alpha \mathrel{\Box\!\!\rightarrow} \beta)$

The expressions'VEQ' and 'EFQ' abbreviate 'Verum Ex Quodlibet' and 'Ex Falso Quodlibet', respectively and are nothing but Principles S1 and S2 from Section 1.2, respectively. We saw in Sections 1.2 and 1.3 that in contrast to MP, CS, Det, and Cond, the Principles VEQ and EFQ are far from plausible for indicative and counterfactual conditionals, respectively. [8]

By definition the bridge principles (dicussed in the context of conditional logic systems) assume a fixed relationship between conditional and non-conditional formulas. For example, in the case of MP, $\alpha \mathrel{\Box\!\!\rightarrow} \beta$ must be false, whenever α is true and β is false. I will not further characterize the notion of bridge principle here and will postpone a formal definition of this notion to Section 4.2.1.

I will restrict the discussion of (plausible) bridge principles to Principles MP and CS, since given reasonable assumptions – as for example provided in Kraus et al.'s (1990) System **P** (see Section 7.2.3) and all extensions of D. Lewis' (1973b) basic Conditional Logic System **V** (see Section 7.2.5) – Principles MP and CS are logically equivalent to Det and Cond by Lemmata 7.57 and 7.60, respectively. Note that we encountered Principles MP and CS already in Section 3.2 in the context of Lewis models and saw that they correspond to the weak centering condition C_{MP} and the second centering condition C_{CS}, respectively.

[8] The Principles P-Cons, CNC, and their variants discussed in Section 3.3.6 are also bridge principles of a weak sort, since they imply that when all conditional formulas can be inferred all *non*-conditional formulas (including $\perp$) can be inferred as well.

Let me now focus on Principle MP in the context of normic conditionals, such as the following, which are a particular subtype of indicative conditionals (see Section 1.4):

E15 If specimen 214 is a fish, then it is probably cold-blooded.
E16 If specimen 214 is a fish, then it is normally cold-blooded.

Consider now the following set of assertions which is nothing but a variant of example (A) from Section 1.4.4:

E21 If specimen 213 is a mammal, it is normally viviparous.
E22 If specimen 213 is an anteater, it is normally not viviparous.
E23 Specimen 213 is an anteater and a mammal.

E21–E23 are most plausibly represented by the following set of formulas: $\{\alpha \mathrel{\Box\!\!\rightarrow} \gamma, \beta \mathrel{\Box\!\!\rightarrow} \neg\gamma, \alpha \wedge \beta\}$ and we can now show that this set of formulas is inconsistent given MP holds: From $\alpha \mathrel{\Box\!\!\rightarrow} \gamma$ and $\beta \mathrel{\Box\!\!\rightarrow} \neg\gamma$ follow by MP that $\alpha \rightarrow \gamma$ and $\alpha \rightarrow \neg\gamma$, respectively. The latter two formulas imply γ and $\neg\gamma$ based on $\alpha \wedge \beta$. It is important to remark here that the above set of propositions does not represent a marginal case and that perfectly analogous cases are readily availabe in the sciences, humanities, and in everyday reasoning (see Section 1.4.4; cf. Schurz, 2011 and Unterhuber & Schurz, in press).

Furthermore, Principle CS does not seem to be valid for indicative conditionals either such as E15 and E16. Principle CS fails for both examples E15 and E16 for the following reason: If specimen 214 is a fish and cold-blooded, it might still be rather unlikely and surprising (in terms of "normal" expectancy, cf. Section 3.2) that specimen 214 is cold-blooded given it is a fish: Specimen 214 might be an exception which does not concur with what is probable for fish and to be expected of them on rational grounds.

Do, then, Principles MP and CS hold for counterfatual conditionals?[9] Let me first inquire into Principle MP. If one uses the narrower definition (B1)

[9] I will discuss Principles MP and CS in in terms of assumptions of truth and falsity. One can, however, make a perfectly analogous case in terms of the notions of acceptibility or assertibility.

then MP is valid. This is due to the fact that the antecedent of a counterfactual by (B1) is always assumed to be false. Hence, $\alpha \rightarrow \beta$ is always true and MP must hold. This is not the case if I instead use criterion (B2) or (A). In these cases the antecedent of the counterfactual is merely improbable or is formulated in the counterfactual mood and therefore does not imply the truth or falsity of Principle MP.

What about Principle CS in the context of counterfactual conditionals? Given criterion (B1), the antecedent α of counterfactual conditional $\alpha \Box\!\!\rightarrow \beta$ is always false and, hence, $\alpha \wedge \beta$ is always false for any counterfactual conditional. It follows that CS is trivially valid for all counterfactual conditionals interpreted in terms of (B1). Criterion (B2) does not give us that result, since it only requires that the antecedent of a counterfactual is improbable. Furthermore, (A) neither implies that CS is valid or invalid for counterfactual conditionals.

The present discussion, hence, shows that MP and CS do not hold for a logic of indicative conditionals. Despite that fact Principles CS and MP are valid in logics which are specifically targeted for indicative conditionals, such as Adams' Systems $\mathbf{P}^*$ (Adams, 1965, 1966, 1977), $\mathbf{P}_\epsilon$ (Adams, 1975, see also Section 3.6), and Stalnaker's (1968) System **S** (see Sections 3.3.3 and 7.3.3) while they are not valid in other systems, such as System **P** (i.e., Kraus et al., 1990; Lehmann & Magidor, 1992) and Adams' System $\mathbf{P}^+$ (Adams, 1986).

If Bridge Principles MP and CS are not valid for indicative conditionals and if I do not advocate any other bridge principles, which type of logical relationship do I then assume between states of affairs corresponding to indicative conditional formulas and non-conditional formulas? Let me discuss the example above (items E21–E23). Since in that example anteaters are by definition mammals, the unambiguous conclusion is that specimen 213 is not viviparous. To describe this type of inference we can use a non-monotonic rule, which is a variant of the Principle of Total Evidence by Carnap (1962, p. 211) and Reichenbach (1949, §72) (see also Section 3.5.1) as specified by Delgrande (1988, Section 5.1). The basic idea behind this principle is to take all non-conditional facts into account and to allow detachment from conditionals where there is no conditional with a stronger antecedent which would contradict the conclusion. In the above example, E23 represents the conjunction of all non-conditional facts and we can infer from this that specimen 213 is not viviparous, since no conditional in the above set exists for E23 which has a stronger antecedent than E22 (and whose consequent contradicts the consequent of E22). This is due to the fact that to be an anteater implies to be a mammal, but not vice versa. On the other hand, the inference from E21 that specimen 213 can fly is blocked, since this would contradict the detachment of the consequent from E22 where the latter of which has a stronger antecedent than that of E21.

Let me finally summarize the results of the discussion of Principles MP and CS in the context of counterfactual conditionals. These principles are guaranteed to hold for counterfactual conditionals if one presupposes criterion (B1) and they do not hold for either criterion (A) and (B2). Observe here that both principles are valid in D. Lewis' (1973b) logic **VC** (his preferred logic for counterfactuals) and Stalnaker's (1968) logic **S**, where the latter of which is also intended as a logic for counterfactual conditionals (Footnote 3, p. 99).

3.4.3 Subjective and Objective Interpretations of Indicative and Counterfactual Conditionals and the Ramsey Test

We saw in the previous sections that often different philosophical and logical analyses are given for indicative and counterfactual conditionals. I will first provide examples for the differential treatment of indicative and counterfactual conditionals and then explain in more detail why indicative and counterfactual conditionals are often viewed as being subjective agent-relative and objective in nature, respectively.

Let me at this point give three further examples of different theoretical treatments of indicative and counterfactual conditionals. First, although, for example, D. Lewis (1973b, p. 3) admits that the term 'counterfactuals' is too narrowly construed for his investigation of conditional logics, his discussion clearly shows that he focuses on counterfactual conditionals rather than on indicative conditionals. Second, Bennett (2003) accepts a Ramsey test interpretation of indicative conditionals but rejects such an interpretation for counterfactual conditionals. Bennett argues in that context that, for example, the following counterfactual does not pass the Ramsey test (cf. Section 3.3):

> "[I]f Yeltsin had been in control of Russia and of himself, Chechnya would have achieved independence peacefully" (Bennett, 2003, p. 30).

Unfortunately, Bennett (2003) only states that the Ramsey test fails for this particular counterfactual conditional, but does not give general reasons why this would be the case.[10] Third, Adams (1970, p. 92) argues that his probabilistic semantics (1965, 1966, 1977; see also Adams, 1975; Schurz, 1996, 1997b, 1998, 2005b; see Section 3.6) is only applicable to indicative conditionals but not to counterfactual conditionals.

There are more views in the literature regarding the exact criteria for distinguishing counterfactuals on the one hand and indicative conditionals on the other which I did not review in Section 3.4.1 (see Bennett, 2003, Chapters 22 and 23). These approaches do not stop at the criteria (A), (B1), and (B2) discussed at the beginning of this section but in addition try to describe the commonalities and differences of indicative and counterfactual conditionals. I will not survey these approaches here but focus instead on some general observations that seem to underlie these approaches.

We saw in Section 3.4.1 that statement E20 is naturally accepted while E19 is not. What is, then, the difference between E19 and E20 in terms of acceptibility (or assertibility) conditions? In example E20 one seems to take into account one's own knowledge about the assassination of Kennedy and its historic context. Hence, it is natural to read E20 in a subjective agent-relative way. This is not the case for E19. Here one rather focusses

[10] Bennett (2003) provides only the one example and neither discusses this issue any further nor does he refer to other sources.

on the event of the assassination and abstracts from one's knowledge about Kennedy's assassination and its historic context. This difference of interpretation of E19 and E20 seems to underlie Adams' (1970, p. 92) argumentation, since Adams suggests an analysis of conditionals such as example E20 in terms of his subjective probabilistic semantics (Adams, 1965, 1966, 1977) but not of conditionals of type E19.

What does one make, then, of this subjective aspect of indicative conditionals such as E20? We saw in Section 3.3.8 that the Ramsey test has an essentially subjective component by requiring that conditionals are evaluated relative to an agent's (hypothetical) beliefs. It is, hence, natural to provide an interpretation of indicative conditionals in terms of the Ramsey test. Moreover, this subjective element is missing in conditionals of type E19.

A Ramsey test interpretation is, however, not the only possible interpretation of indicative conditionals: For example, Kraus et al. (1990) also suggest an interpretation of (normic) indicative conditionals in terms of agents' subjective orderings of "normality" preferences (without mention of the Ramsey test; see Section 3.2), which is based on a different semantic intuition than the Ramsey test (see Section 3.3.5).

How about interpretations of counterfactual conditionals, such as E19? We saw that E19 does not rely on one's particular assumptions about Kennedy's assassination and its historic context. It is, hence, not surprising that D. Lewis (1973b) suggests an analysis of counterfactuals, such as E19, in terms of objective criteria such as similarity of possible worlds (see Section 3.2).

For D. Lewis this is a part of a larger philosophical project: First, D. Lewis (1973b) bases his analysis of counterfactuals on the notion of similarity between possible worlds. Second, D. Lewis uses his counterfactual semantics in order to provide an analysis of other philosophical concepts, such as causation (D. Lewis, 1973a, 2000) and alethic necessity (D. Lewis, 1973b, p. 22f, see also pp. 137–142), where alethic necessity refers to states of affairs which are not just true but necessarily so, as opposed to necessity in a deontic/normative way ('it ought to be the case that ...') or an epistemic way ('agent *A* knows [believes] that ...'). Observe that one can explicate alethic necessity in different ways, such as necessity according to logical laws, physical laws, etc. (D. Lewis, 1973b, p. 7f; see also Schurz, 1997a, p. 8 and p. 18). These notions of alethic necessity share that they refer to necessity of truth rather than necessity according to beliefs or normative principles.

Causation and alethic necessity are often conceived as being objective, although this is naturally a matter of debate. So, an analysis in terms of counterfactuals, which is in turn based on an objective criterion such as similarity of possible worlds, is in that sense a perfect fit. Moreover, to find out whether an alethic necessity statement such as 'it is necessarily the case that $2+2=4$', is true, one often engages in counterfactual reasoning, such as 'If things were different, would $2+2$ still be 4?' So, by these means D. Lewis (1973b) provides a natural and coherent analysis of alethic necessity and counterfactual conditionals.

3.5 Fundamental Issues of Probabilistic Approaches to Conditional Logic

Before I describe Adams' (1965, 1966, 1977, 1975, 1986) probabilistic **P** systems, I will first focus on the following fundamental issues of probabilistic conditional logics: (i) the difference between subjective and objective frequency-based probabilistic semantics (Section 3.5.1), (ii) taking either non-conditional or conditional (open) formulas as basic and the definition of probabilities of conditionals in terms of the Stalnaker thesis (Section 3.5.2), (iii) the languages in which the conditional logic systems are formulated (Section 3.5.3), (iv) reasons for restricting conditional logic languages (Sections 3.5.3 and 3.5.4), and (v) finally the thesis NTV ("No Truth Value"), which is that conditionals are not propositions and do not in general have truth values (Section 3.5.5).

Point (i), then, aids to emphasize a point made in my discussion of Bennett's Gibbardian stand-off argument in Section 3.8, namely that there are probabilistic semantics for conditionals such as frequency-based semantics which are not based on the Ramsey test. Points (ii)–(iv), then, serve as basis for the discussion of D. Lewis' (1976) triviality result in Section 3.7. Finally, point (v) adds to my defense of possible worlds semantics for indicative conditionals, since Bennett (2003) bases his argument against truth value approaches to indicative conditionals on his NTV thesis (cf. Section 3.1).

3.5.1 Subjective and Objective Probabilistic Semantics and the Principle of Total Evidence

One can distinguish at least two fundamental approaches in probability theory: a subjective agent-relative account and an objective frequency-based account of probabilities (Schurz, 2008, p. 99). Both approaches satisfy the Kolmogorov axioms (Kolmogorov 1950; see Schurz, 2008, p. 101f) but differ crucially in the interpretation of probabilities (cf. Schurz, 2008, p. 99). It is, hence, not surprising that one can distinguish between subjective and objective frequency-based probabilistic semantics for conditional logics as well.

In the subjective probabilistic semantics of Adams' **P** systems (1965, 1966, 1977, 1975, 1986; see Section 3.6) indicative conditionals are analyzed in terms of conditional formulas $\alpha \mathrel{\Box\!\!\rightarrow} \beta$, where α and β stand for the antecedent and the consequent of the natural language conditional, respectively. According to such an analysis, formulas α and β are either (a) formulas of an f.o.l. language which do not contain free individual variables or else (b) formulas of a p.c. language. Approach (a) makes sense only if one also allows for quantification over individual variables in non-conditional formulas. Since this is not presupposed for Adams' **P** systems I use approach (b) here.

In constrast, in an objective frequency-based approach, such as Schurz (1997b, 2001b, 2005b; see also Bacchus, 1990, Chapters 3 and 4), natural language conditionals are represented by conditional formulas of the form $\alpha[x] \mathrel{\Box\!\!\rightarrow} \beta[x]$. The antecedent formula $\alpha[x]$ and consequent formula $\beta[x]$ are formulas which contain only monadic predicates and where x is the only free variable occurring in $\alpha[x]$ and $\beta[x]$ (Schurz, 1997b, p. 538).

So, in subjective and objective frequency-based probabilistic conditional logic accounts, probabilities are assigned to closed and open formulas, respectively. In the subjective approach the assigned probability values express an agent's degree of belief in the proposition described by the formula (cf. Schurz, 2008, p. 99). Such a (conditional) formula does *not* specify a *type* of event (or a type of state of affairs), but rather a particular instance of an event at a particular time at a particular place, in other words a fully determined state of affairs. In the objective frequency-based approach, probabilities which are assigned to open formulas are interpreted as frequencies of the properties expressed by open formulas such as $\alpha[x]$ or their relative frequencies in the long run (cf. Schurz, 2008, p. 99). For this approach to be applicable, the events or instances described by a formula such as $\alpha[x]$ have to be in some sense "repeatable". Hence, the frequency-based approach cannot directly account for single-case probabilities, which refer to specific instances occurring only once at a particular time at a particular place (or which refer to fully determined state of affairs).

Conditional logic systems with a subjective or a frequency-based probabilistic semantics for conditionals do not allow for an equally natural interpretation of all conditional structures. In Section 1.4 two types of normic conditionals were described – quasi-universally quantified (quq) conditionals and their propositional counterparts. Instances E9 ('most fish are cold-blooded') and E15 ('if specimen 214 is a fish, then it is probably cold-blooded') are examples for the quq and the propositional case, respectively. Since propositional cases such as E15 refer to fully determined states of affairs, the propositional case seems best represented in terms of a subjective probabilistic semantics. Analogously, since E9 seems to refer to frequencies, an objective frequency-based interpretation seems best for it.

For my description of subjective and objective frequency-based approaches, I employed a p.c. language and a monadic f.o.l. language, respectively. For both the subjective and the frequency-based approach a full probabilistic f.o.l. version for a conditional logic is, however, not easy to achieve (Schurz, 1997b, p. 538).

Furthermore, the relation between quq conditionals, such as E9, and single-case conditionals, such as E15, is far from trivial. On the one hand there exists no deterministic relationship between both types of conditionals, as for example the quq conditional E9 does not strictly imply its instance E15 (see Section 1.4). On the other hand both are not unrelated, as Adams' (1998, p. 285f) account suggests but there exists a logical relationship between both types of probabilistic conditionals.

Let me focus on statements E9 and E15 to describe this logical relationship. Schurz (2008, p. 101 and p. 115f; see also Schurz, 2005b, p. 42; 1997b, p. 538) suggests that the relationship between subjective probabilities and frequency-based probabilities can be accounted by means of the (non-monotonic) Principle of Total Evidence which goes back to Carnap (1962, p. 211) and Reichenbach (1949, §72). Schurz, however, does not directly specify the relation between subjective probabilistic conditionals, such as E15, and their frequency-based counterparts, such as E9, but rather accounts for the subjective probability of *non-conditional* facts on the basis of *conditional* frequency-based facts.

Schurz's (2008, p. 116; see also Unterhuber & Schurz, in press) solution amounts to the following (see also Section 3.4.2): Given one has a finite set of facts and individuals, one can determine the subjective probability $P_{\text{subj}}(\gamma[a])$ from a set of objective probability conditionals the following way: $P_{\text{subj}}(\gamma[a] \mid \alpha[a] \wedge \beta[b_1, \ldots, b_n] \wedge P(\gamma[x] \mid \alpha[x]) = r) = r$, where $\alpha[a]$ summarizes one's given total factual knowledge about individual a and $\beta[b_1, \ldots, b_n]$ is the complete factual knowledge about the other individuals $b_1, \ldots, b_n$. Alternatively, one can use only the statistically relevant factual knowledge pertaining to the individuals, viz. by omitting all facts which do not bear on the probabilistic inferences drawn.

In other words, in both cases the Principle of Total Evidence determines the probability $P_{\text{subj}}(\gamma[a])$ by the narrowest reference class available on the basis of one's total knowledge. According to this approach, hence, one's degrees of belief in a proposition $\alpha[a]$ is determined by either one's total knowledge or alternatively one's complete statistically relevant knowledge and objective facts described by conditional frequency statements.

3.5.2 The Stalnaker Thesis, the Ramsey Test, and Conditional or Non-Conditional Probabilities as Primitive

In this section I will describe two key assumptions for probabilistic conditional logic semantics, such as the **P** systems of Adams (1965, 1966, 1975, 1977, 1986): (a) the Stalnaker thesis for defining the probability of conditionals and (b) the question whether conditional or non-conditional probabilities should be taken as fundamental. I shall furthermore discuss the relation of (a) to the Ramsey test (see Section 3.3).

Let me first focus on (a). The subjective probabilistic semantics for the **P** systems of Adams (1965, 1966, 1977, 1975, 1986; see Section 3.6) and the objective frequency-based account of Schurz (1997b, 2001b, 2005b) differ in the way they specify probabilities for conditional formulas. In the Adams approach the (unrestricted) Stalnaker thesis is assumed (Bennett, 2003, p. 58; see also Section 3.1) which states that $P(\alpha \mathrel{\Box\!\!\rightarrow} \beta)$ equals $P(\beta \mid \alpha)$ (see Chapter 3.7 for a discussion of variants of the Stalnaker thesis).[11] Although the (unrestricted) Stalnaker thesis seems quite plausible, it gives rise to D. Lewis' (1976) well-known triviality result (see Chapter 3.7). As we will see in Chapter 3.7 it is the restriction of the language that saves probabilistic conditional logic systems with the (unrestricted) Stalnaker thesis from triviality.

The probabilistic semantics of Schurz (1997b, 1998, 2005b) on the other hand is not prone to D. Lewis' (1976) triviality result: Schurz (1997b, 1998, 2005b) does not equate $P(\alpha[x] \mathrel{\Box\!\!\rightarrow} \beta[x])$ with $P(\beta[x] \mid \alpha[x])$, but only *associates* indicative conditionals with conditional probabilities: In Schurz's account a conditional $\alpha[x] \mathrel{\Box\!\!\rightarrow} \beta[x]$ holds only if the respective conditional probability $P(\beta[x] \mid \alpha[x])$ is high (Schurz, 1997b, p. 536; Schurz, 1998, p. 85; Schurz, 2005b, p. 38). Note here that in Schurz's account the occurence of x

[11] Strictly speaking Adams (1965, 1966, 1977, 1975, 1986) does not use the notion of conditional probability in his approach, but defines $P(\alpha \mathrel{\Box\!\!\rightarrow} \beta)$ directly by means of $P(\alpha \wedge \beta)$ and $P(\alpha)$. However, he does so in accordance with the usual definition of the conditional probability (see Sections 3.6.1 and 3.6.3).

in $\alpha[x] \mathrel{\Box\!\!\rightarrow} \beta[x]$ and $P(\beta[x]|\alpha[x])$ is bound by the conditional operator $\mathrel{\Box\!\!\rightarrow}$ and the probability function P, respectively.

Let us now focus on Bennett's (2003, p. 55) thesis that the Stalnaker thesis is the core idea of the Ramsey test (see also Adams, 1975, p. 3; see also Section 3.3.7). Despite that fact the Stalnaker thesis is neither a necessary nor a sufficient element of the Ramsey test. The Stalnaker thesis is not a necessary element of the Ramsey test, since for example Stalnaker (1968) describes a version of the Ramsey test which does not rely on probabilities, but which is targeted for possible worlds semantics (see Section 3.3). Furthermore, CS semantics, a possible worlds semantics for conditionals, can also be interpreted in the Ramsey test, as I will show in Section 7.1.1. Second, the Stalnaker thesis is not a sufficient element of the Ramsey test either. If we add the Stalnaker thesis to an objective frequency-based approaches like Schurz (1997b, 2001b, 2005b), this does not allow for an interpretation in terms of the Ramsey test since in this approach a frequency-based interpretation of conditionals is assumed (see Section 3.5.1).

Let me now address point (b). As we saw earlier, Adams effectively defines conditional probabilities by means of non-conditional probabilities. This is, however, not the only possible approach in the literature. Hawthorne (1996, p. 191), for example, takes conditional probabilities as primitive by means of so-called Popper functions. If one uses Popper functions, one can – but need not – define non-conditional probabilities on the basis of conditional probabilities by defining that $P(\alpha) =_{df} P(\alpha|\top)$ or, alternatively, that $P(\alpha[x]) =_{df} P(\alpha[x]|\top[x])$. Here $\top$ and $\top[x]$ represent a logically true formula and a logically true formula based solely on monadic predicates and with a single free variable x, respectively.

There are two important points to make here. First, the definitions described in the previous paragraph are by no means innocuous, but render – given a sufficiently rich language – additional logical principles valid, that is rule versions of the principles Det and Cond described in Section 3.4.2. Second, to avoid that result one can alternatively (A) give up a fixed relationship between the probability of conditional and non-conditional formulas as described in the preceding paragraph and in addition endorse a variant of the Principle of Total Evidence described in Section 3.5.1. Alternatively, one can (B) restrict the conditional logic language in such a way that that no non-conditional formulas are expressible and that hence such a problem does not occur in the first place (e.g., as in languages $\mathcal{L}_{\mathrm{rCL}}$, $\mathcal{L}_{\mathrm{rCL^*}}$, and $\mathcal{L}_{\mathrm{rrCL}}$; cf. Sections 3.5.3 and 4.2.2; see also the next section).

3.5.3 Languages of a Probabilistic Conditional Logic

In this section I will discuss languages that are used to describe logic systems for (indicative) conditionals, such as Adam's **P** systems (1965, 1966, 1977, 1975, 1986; see Section 3.6). I will describe these languages here informally and postpone the formal definitions to Section 4.2.1.

Before I start with a description of the different languages, let me first describe the key motivation for restricted languages for probabilistic conditional logics. As we already observed in Section 3.5.2, Adams' semantics for the **P** systems employs the Stalnaker thesis and is on these grounds prone to D. Lewis' (1976) triviality result, unless one restricts the language for his systems. Adams (1965) and Bennett (2003) give further reasons for the use of his restricted language, which we will discuss in Section 3.5.4. It should be clear that from today's perspective D. Lewis' (1976) triviality result is still the key motivation for the restriction of the language of the probabilistic conditional logics (see Section 3.7).

Let me begin with the full language $\mathcal{L}_{\mathrm{CL}}$. This language allows the two-place conditional operator '$\Box\!\!\rightarrow$' to recombine in such a way as the material implication in p.c. does (cf. Section 3.2). Language $\mathcal{L}_{\mathrm{CL}}$ is employed by the majority of conditional logics with a possible worlds semantics described here, such as Lewis models (see Section 3.2), Stalnaker models (see Section 3.3.3), and CS semantics (see Chapters 4–7; see also Section 3.1). Noteable exceptions are the semantics of Kraus et al. (1990) and Lehmann and Magidor (1992).

The languages employed by Adams' **P** systems are restricted in the following way: (a) They allow only partially for Boolean combinations (by means of p.c. connectives) involving conditional formulas and (b) do not in general allow for nestings involving conditional formulas, where the formulas $(p_1 \mathrel{\Box\!\!\rightarrow} p_2) \wedge (p_3 \mathrel{\Box\!\!\rightarrow} p_4)$ and $\neg(p_1 \mathrel{\Box\!\!\rightarrow} p_2)$ on the one hand and $p_1 \mathrel{\Box\!\!\rightarrow} (p_2 \wedge (p_3 \mathrel{\Box\!\!\rightarrow} p_4))$ on the other are examples for Boolean combinations and nestings involving conditional formulas, respectively.

Let me describe the language $\mathcal{L}_{\text{CL}^-}$ that is used in Adams' (1965, 1966, 1977) Conditional Logic System $\mathbf{P}^*$. Formulas of $\mathcal{L}_{\text{CL}^-}$ are either purely propositional formulas (without the conditional operator) or contain exactly one conditional operator '$\Box\!\!\rightarrow$' which is the formula's main connective (simple conditionals; cf. Adams, 1965, p. 184; Adams, 1966, p. 270). The symbol '$-$' in '$\mathcal{L}_{\text{CL}^-}$' indicates that $\mathcal{L}_{\text{CL}^-}$ is a restricted version of language $\mathcal{L}_{\text{CL}}$.

Adams (1986, p. 256f) effectively employs language $\mathcal{L}_{\text{rCL}}$ to formulate his Conditional Logic System $\mathbf{P}^+$, where '$\mathcal{L}_{\text{rCL}}$' stands for 'restricted version of language $\mathcal{L}_{\text{CL}}$'. Language $\mathcal{L}_{\text{rCL}}$ allows only for simple conditional formulas of the form $\alpha \mathrel{\Box\!\!\rightarrow} \beta$ (where α and β are purely propositional) and disjunctions of conditional formulas (e.g., $(p_1 \mathrel{\Box\!\!\rightarrow} p_2) \vee (p_3 \mathrel{\Box\!\!\rightarrow} p_4)$). Adams' (1986) exact language differs from language $\mathcal{L}_{\text{rCL}}$ insofar as he limits disjunctions of conditional formulas to conclusions of inferences. Furthermore, Adams (1986) also allows for empty disjunctions (disjunctions with zero disjuncts) and I shall postpose a discussion of this issue to Section 3.6.1.

Observe that Schurz (1998, p. 84f) shows that to allow for inferences with disjunctions in the antecedent is inessential insofar as any valid inference in Adams' (1986) system with disjunctive premises can be transformed into *sets* of inferences, which have only disjunctions in the conclusion (see Section 4.2.1). This result does not imply that one can also express all inferences by *single* inferences in Adams' (1986) restricted language rather than by sets of inferences.

Let us now focus on language $\mathcal{L}_{\text{rCL}^*}$. Language $\mathcal{L}_{\text{rCL}^*}$ results if one reconstructs the conditional logic language of Kraus et al. (1990), Lehmann and Magidor (1992), Hawthorne (1996), and Hawthorne and Makinson (2007) in terms of an object language account. Since these approaches focus on second-order conditional assertions $\alpha \mathrel{|\!\!\sim} \beta$ (and their negations) only (see Sections 2.2.6 and 3.2), one can specify an object language in which accordingly only conditional formulas and negated conditional formulas are allowed. Observe that both languages $\mathcal{L}_{\mathrm{rCL}}$ and $\mathcal{L}_{\mathrm{rCL}^*}$ are equally expressive, as I shall show in Section 4.2.1 (cf. Schurz, 1998, p. 84f).

Finally, I will also draw on the language $\mathcal{L}_{\mathrm{rrCL}}$. The label '$\mathcal{L}_{\mathrm{rrCL}}$' indicates that the language referred to is a very restricted one. The formulas of language $\mathcal{L}_{\mathrm{rrCL}}$ are all simple conditional formulas.

3.5.4 Further Reasons for the Restriction of Languages of Probabilistic Conditional Logics

In the previous section we observed that the key motivation for the use of a restricted conditional logic language is to avoid D. Lewis' (1976) triviality result (see Section 3.7). In this section I will discuss further reasons as to why logics for indicative conditionals employ only a restricted language, such as $\mathcal{L}_{\mathrm{CL}^-}$.

Adams (1965, p. 181f) argues that it is often unclear how to interpret disjunctions of conditionals in natural language (see also Adams, 1975, p. 32). In a similar vain Bennett (2003, p. 95) claims that indicative conditionals cannot be embedded freely in larger structures (as by Boolean combinations and nestings involving conditionals). In the case of Bennett (2003), however, the reasons for his thesis seem to stem again from D. Lewis' (1976) triviality result.

In contrast Adams (1965) gives further reasons for the thesis that indicative conditionals cannot freely be combined. In particular, Adams (1965) argues that disjunctions of conditionals correspond to meta-assertions rather than assertions. However, to show that in general Boolean combinations involving conditional formulas do not lend themselves easily to a probabilistic analysis, Adams relies on a justification of assertions (denials) of conditionals in terms of rational betting behavior and shows that this approach does not directly work for Boolean combinations of conditionals (cf. Adams, 1965, p. 181 and p. 183). Adams' argument for assertions of Boolean combinations of conditionals being meta-assertions thereby presupposes an analysis of indicative conditionals in terms of a subjective probabilistic semantics such as Adams (1965, 1966, 1977).

Note here that a justification of assertions (denials) of conditionals is a specific problem of subjective accounts of probabilities. These approaches have to justify that the application of their formalisms is rational. On the other hand, for objective frequency-based interpretations of probabilities no such justification is needed (Schurz, 2008, p. 114).

As we saw earlier there are quite different alternative analyses of indicative conditionals to the subjective probabilistic accounts (see Sections 3.3 and 3.4) and, thus, there is little reason to assume that, for example, an analysis of indicative conditionals in terms of the betting behavior schema must have a similiar impact on the justification of possible worlds semantics for indicative conditionals. We will in fact see in Section 7.1.1 that the justification of CS semantics, a possible worlds semantics, can be given for indicative conditionals without reference to the betting behavior schema.

I have to admit that I am in general skeptical of the thesis that indicative conditionals do not freely embed in larger structures. It seems plainly true that we can embed indicative conditionals in natural language quite freely and can interpret this structures rationally. Evidence for my skeptical position is, for example, given by Adams (1965, p. 181, Example F10) who also formulates Boolean combinations of conditional sentences despite his philosophical position. (Adams also seems to assume that these expressions can be sensibly interpreted.) In addition, one can easily find examples for nested conditionals in natural language, such as the following: [12]

E24 If the cup broke if it was dropped, it was fragile. (Gibbard, 1980, p. 235)

E25 If a Republican wins the election, then if it's not Reagan who wins it will be Anderson. (McGee, 1985, p. 462)

In sum, I think that there is little reason to suppose that Boolean combi-

[12] Note that Bennett (2003) would not object that E24 and E25 can be rationally interpreted, but would argue that E24 and E25 are in that sense exceptional (pp. 98–102).

nations and nestings involving conditionals cannot in general be rationally accounted for and should, hence, be excluded from an analysis in terms of conditional logics. If anything, this argumentation rather shows that probabilistic approaches to indicative conditionals (with the Stalnaker thesis) are defective insofar as they cannot account for Boolean combinations and nestings involving conditionals in natural language (on pain of triviality based on D. Lewis' (1976) triviality result).

3.5.5 Propositions, NTV ("No Truth Value"), and Conditional Logic Languages

Let me conclude this section with a discussion on the question whether conditionals are propositions and whether they can have (objective) truth values. This point is substantial, since proponents of NTV ("No Truth Value", see Section 3.1), such as Bennett, Edgington, Adams, and Gibbard, argue that conditionals are not propositions and do in general not have truth values (Bennett, 2003, p. 94) and, therefore, should not be treated as such in conditional logics. This argumentation can be seen as a further argument against possible worlds semantics for indicative conditionals which most typically employ the full language $\mathcal{L}_{\text{CL}}$ (see previous section) and, thus, effectively treat conditionals as propositions.

First, observe that Bennett does not explicitly define 'proposition'. His discussion, however, shows that he associates at least three properties with the notion of a proposition: (1) having truth value, (2) being embeddable into more complex structures, for example by means of conjunctions, etc. (Bennett, 2003, p. 95), and (3) being the object of probability assignments in a subjective probabilistic framework. As I shall argue here, points (1)–(3) are not equivalent and should therefore not be treated as such.

I will first address the relation between points (1) and (2). First note that (2) does not imply (1). One can, for example, allow for Boolean combinations of conditionals (point (2)), as it is done in languages $\mathcal{L}_{\text{rCL}}$ (disjunctions) and $\mathcal{L}_{\text{rCL}^*}$ (negations). In such languages simple conditionals are embeddable in more complex structures, viz. either in disjunctions or negations of conditional formulas. A more extreme example is language $\mathcal{L}_{\mathrm{CL}}$, in which all Boolean combinations involving conditionals are allowed. Admitting Boolean combinations of conditionals, however, does not per se imply that conditional formulas have truth values: If I intend to give a (semantic) interpretation of conditionals in a compositional way (cf. Connolly, Fodor, Gleitman, & Gleitman, 2007, p. 2f; Schurz, 2012; Werning, 2012), then it suffices that the meaning of disjunctions of conditionals and negations of conditionals is uniquely determined by the meaning of simple conditionals. This does not imply that in such a case simple conditionals *must* have truth values. Alternatively one could use a probabilistic framework. Note that in a standard probabilistic framework (cf. Section 3.6) $P(\alpha \wedge \beta)$ and $P(\alpha \wedge \neg\beta)$ determine the probability $P(\beta\,|\,\alpha)$ and, hence, simple conditionals are already compositional in that sense (cf. Section 3.5.2).

Furthermore, (1) does not imply (2). In particular, the fact that conditional formulas have truth values does not imply that they can be parts of Boolean combinations or nestings. One can, for example, easily restrict the language of standard p.c. to language $\mathcal{L}_{\mathrm{CL}^-}$ and accept the extensional semantics of p.c. for that language, so that we assign truth values to conditional formulas but do not allow for embeddings of conditionals in more complex structures.

Observe the following important point here: In order to assign truth values to conditionals, one has to refer to expressions of a specific language. One can represents natural language conditionals by conditional assertions in the meta-language of the form $\alpha \mid\!\sim \beta$, as Kraus et al. (1990), Lehmann and Magidor (1992), Hawthorne (1996), and Hawthorne and Makinson (2007) do (cf. Sections 3.2, 3.5.3, and 3.6.2). If one does so, neither an assignment of truth values to conditional assertions is possible in the object language nor can one embed conditional assertions in more complex object language formulas. Hence, in this case both points (1) and (2) are not possible w.r.t. the object language.

It is important to note here that Kraus et al. (1990) and Lehmann and Magidor (1992) employ a truth value based possible worlds semantics for the analysis of indicative conditionals. The latter fact contradicts Bennett's

(2003, p. 95) position, since he argues that it is an advantage of probabilistic semantics for indicative conditionals over truth value based semantics that conditionals cannot be embedded in more complex structures.

Let me now address point (3). Bennett (2003) and Adams (1975) use approach (3) to argue against truth value approaches. Their argument is based on the fact that – given a subjective probabilistic framework as described in Adams (1965, 1966, 1977, 1975, 1986) – the full language $\mathcal{L}_{\mathrm{CL}}$ cannot be used as basis for a conditional logic (see Sections 3.1 and 3.7). Since the formulas to which probabilities are assigned are regarded as propositions, and a non-restricted language – containing arbitrary Boolean combinations involving conditional formulas – is prone to D. Lewis' (1976) triviality result, these authors rather argue that conditional formulas are not propositions and, hence, cannot in general have truth values (cf. Section 3.5.4; see also Section 3.5.3).

Let me, for the sake of the present discussion, presuppose that (2) were tantamount to (1), viz. that any proposition X described by a specific language has a truth value, and vice versa. Even in that case, (3) does not give one that conditionals do not in general have truth values. Such an inference is problematic for at least two reasons. First, I am not aware of any formal result which shows that truth value semantics, such as CS semantics (see Chapters 4–7), presuppose a subjective probabilistic framework as described by Adams (1965, 1966, 1977, 1975, 1986). However, such a result would be needed in order to show that indicative conditionals as described by truth valued possible worlds semantics in general cannot have truth values (cf. Section 3.7).

In contrast D. Lewis' (1976) triviality result gives us only the following: Provided (i) one endorses a subjective probabilistic account as in Adams (1965, 1966, 1977, 1975, 1986; cf. Section 3.6) and (ii) one uses the full language $\mathcal{L}_{\mathrm{CL}}$, then the probabilistic systems become trivial. Hence, D. Lewis' (1976) result does not extend to all approaches for which (ii) does not hold. D. Lewis' (1976) triviality result is, hence, not a genuine problem of truth value approaches but rather of probabilistic approaches. My argument is supported by D. Lewis (1976) who concludes that "[i]t was not the connection between truth and probability that led to my triviality results, but only the application of standard probability theory to the probability of conditionals" (p. 304).

Moreover, a closer look at D. Lewis' (1976) triviality result shows that Adams' (1975, p. 35) and Bennett's (2003, p. 94) NTV approach does not line up as perfectly with D. Lewis' triviality result as Bennett (2003, p. 77) seems to assume. D. Lewis' (1976, p. 304) discussion shows that one does not need to exclude all types of Boolean combinations involving conditionals, but only has to make sure that no conjunctions involving conditional formulas are expressible in the respective language (see Section 3.7). Hence, one can allow either for (i) negations of conditionals (as in language $\mathcal{L}_{rCL^*}$) or (ii) disjunctions of conditionals (as in language $\mathcal{L}_{rCL}$), but not both. It should be clear that NTV alone can hardly explain why disjunctions and negations of conditionals seem unproblematic. Accordingly, Adams (1986, p. 260) justifies his use of disjunctions of conditionals (although only in the meta-language) not by NTV, but by his notion of probabilistic enthymemes, viz. by assuming that tacit premises in inferences exist which assert that something is not very improbable. I shall not go into further detail regarding this issue.

Let me now focus in more detail on NTV. I will argue here that it is quite counter-intuitive to assume NTV even on the basis of a subjective probabilistic semantics. The upshot of my argument is that NTV proponents, such as Bennett, have to justify why that conditional and non-conditional statements have a fundamentally different status, despite the fact that this does not seem to be the case. I presuppose for that purpose – in line with Adams (1975) and Bennett (2003) – (i) a subjective probabilistic framework, as described, for example, by Adams (1965, 1966, 1977, 1975), and (ii) accept NTV. Consider now the following statements:

(A) Specimen 214 is a cold-blooded fish.
(B) If specimen 214 is a fish, then it is probably cold-blooded. (E15)

One can analyze (A) and (B) in a subjective probabilistic framework and assign to both (A) and (B) subjective probabilities according to (i). (ii), then, implies that that statements such (A) have an (objective) truth value, while conditionals such as (B) do not. Consequently, assumptions (i) and (ii) show that conditional statements must have a different ontological status than non-conditional statements, in the sense that (B) makes sense only in an epistemic interpretation and can – contrary to (A) not objectively be the case. Interpreted in terms of the correspondence theory of truth, this means that there are no facts which correspond to conditional statements, although there are always facts which correspond to non-conditional statements. This implication seems, however, hardly acceptable. So, what went wrong? In my view NTV is to blame here. A probabilistic framework as in, for example, Adams (1965, 1966, 1977) alone (without NTV) does not give us that conditionals are not propositions and hence do not have truth values.

3.6 Adams' Probabilistic P Systems

I will now focus on probabilistic Conditional Logic System **P** and its variants. In Section 3.6.1 I shall discuss Systems **P** (Schurz, 2005b; see also Kraus et al., 1990; Lehmann & Magidor, 1992), **P*** (1965, 1966, 1977), and **P**$^+$ (Adams, 1986; Schurz, 1998) and present Schurz's Central **P** Theorem (Theorem 3.23). In Section 3.6.2 I will outline the probabilistic threshold semantics by Hawthorne and Makinson (2007) and Hawthorne (1996) and compare it to Adams' probabilistic validity criterion described in Section 3.6.1. I will then address a further variant of System **P**, that is System $\mathbf{P}_\epsilon$ (Adams, 1975; Schurz, 1997b), in a separate section (Section 3.6.3), since this system is – contrary to Systems **P**, **P***, and **P**$^+$ – also a genuine default logic in the sense of Section 2.2. Finally, in Section 3.6.4 I shall investigate the relationship between possible worlds semantics such as Lewis models (see Definitions 3.1–3.3) and CS semantics (see Chapters 4–7) on the one hand and Adams' **P** systems semantics on the other. The last point will, then, serve as basis for the discussion of the implication of D. Lewis' (1976) possible worlds semantics for indicative conditionals in Section 3.7.

Observe that there are alternative probabilistic validity criteria to Adams' validity criteria described in Sections 3.6.1 and 3.6.3 (see Leitgeb, 2004, Chapter 9) for which quite a few converge, insofar as they support the same type of inferences (Leitgeb, 2004, Theorem 70, p. 178f). Accordingly I will restrict myself here to standard probabilistic semantics along the lines of Adams' probabilistic semantics.

3.6.1 The P Systems: Systems P, P*, and P^{+}

In this section I will describe Adams' **P** Systems, in particular Systems **P** (e.g., Schurz, 2005b), **P*** (e.g., Adams, 1965, 1966, 1977), and **P**$^{+}$ (e.g., Adams, 1986; Schurz, 1998). These probabilistic conditional logics draw on the same probabilistic model theory and differ only in terms of the language the same semantics is formulated for. It is the languages' expressibility that has a profound impact on which proof theory characterizes the respective semantics.

I use the probabilistic semantics of Adams' (1965, 1966, 1977) System **P*** as starting point for my discussion of the model theory of the **P** Systems, that is Systems **P**, **P***, **P**$^{+}$, and **P**$_{\epsilon}$. (See Section 3.6.3 for System **P**$_{\epsilon}$.) Moreover, the proof-theoretic side of System **P**, as described by Kraus et al. (1990) and Lehmann and Magidor (1992), serves a basis for our discussion of the proof theories of the **P** systems. At the end of this subsection I will then discuss the Central **P** Theorem described by Schurz (2005b) for different restrictions of the conditional logic's language.

Let me now focus on the semantics for System **P*** (Adams, 1965, 1966, 1977). To formulate System **P***, Adams (1965, 1966, 1977) uses the restricted language $\mathcal{L}_{\text{CL}^-}$ (see Sections 3.5.3 and 4.2.1). As basis of the semantics serve probability assignments which can be specified the following way, where $\models_{p.c.}$ says that α is a p.c. valid formula:

Definition 3.17. *Function P is a probability function over* $\mathcal{L}_{\text{CL}^-}$ *iff*
(a) $P\colon \mathcal{L}_{\text{CL}^-} \rightarrow \mathbb{R}$
(b) for all formulas $\alpha, \beta \in \mathcal{L}_{\text{prop}} \subseteq \mathcal{L}_{\text{CL}^-}$ *holds:*

(i) $0 \leq P(\alpha) \leq 1$ *and* $P(\top) = 1$,
(ii) *if* $\models_{p.c.} \alpha \rightarrow \beta$, *then* $P(\alpha) \leq P(\beta)$,
(iii) *if* $\models_{p.c.} \neg(\alpha \wedge \beta)$, *then* $P(\alpha \vee \beta) = P(\alpha) + P(\beta)$, *and*
(iv) *if* $P(\alpha) \neq 0$, *then* $P(\alpha \,\Box\!\!\rightarrow \beta) = P(\alpha \wedge \beta)/P(\alpha)$, *and if* $P(\alpha) = 0$, *then* $P(\alpha \,\Box\!\!\rightarrow \beta) = 1$.

Here $\mathbb{R}$ refers to the set of real numbers and $\mathcal{L}_{\text{prop}}$ describes the set of formulas of language $\mathcal{L}_{\text{CL}^-}$ which do not contain an instance of a conditional operator. Definition 3.17 modifies the Kolmogorov axioms of conditional probabilities insofar as it assigns probability one to a conditional formula if the probability of its antecedent formula is zero (Adams, 1965, p. 185; Adams, 1966, p. 272f). In clause b.iv $P(\alpha \,\Box\!\!\rightarrow \beta)$ is directly defined as $P(\alpha \wedge \beta)/P(\alpha)$ in case $P(\alpha) > 0$. The conditional probability $P(\beta|\alpha)$ is customarily defined as $P(\alpha \wedge \beta)/P(\alpha)$ for $P(\alpha) > 0$ (Feller, 1968, p. 115). To capture clause b.iv of Definition 3.17 adequately, one can, then, in turn define the probability $P(\alpha \,\Box\!\!\rightarrow \beta)$ as $P(\beta|\alpha)$ for $P(\alpha) > 0$. Hence, Adams (1965, 1966, 1977) effectively defines the probability of a conditional as its respective conditional probability and, in addition, takes non-conditional probabilities as primitives (see Section 3.5.2). One need not define $P(\alpha \,\Box\!\!\rightarrow \beta)$ of clause b.iv of Definition 3.17 in the way Adams does. One can instead define $P(\alpha \,\Box\!\!\rightarrow \beta)$ as $P(\beta|\alpha)$ and then define $P(\beta|\alpha)$ as $P(\alpha \wedge \beta)/P(\alpha)$ for $P(\alpha) > 0$ and as one for $P(\alpha) = 0$. The latter procedure can, hence, be used to reformulate Adams' (1965, 1966, 1977) approach in such a way that $P(\alpha \,\Box\!\!\rightarrow \beta)$ is defined as $P(\beta|\alpha)$ without further qualification.

Note that Definition 3.17 draws on p.c. validities in clauses b.ii and b.iii (cf. Adams, 1965, p. 184). Since Definition 3.17 relies on the notion of p.c. validity, the definition is, hence, not purely probabilistic. It is one motivation of the so-called Popper functions to remedy this drawback (see Hawthorne 1996, p. 190f; see Section 3.5.2).

Adams (1965, 1966, 1977) then goes on to introduce the following probabilistic validity criterion (short: *p* validity criterion) for his System **P***:

Definition 3.18. *(P Valdity)*
Suppose that $\alpha \in \mathcal{L}_{\mathrm{CL}^-}$, Γ *is a finite set of formulas of language* $\mathcal{L}_{\mathrm{CL}^-}$, ϵ, $\delta \in \mathbb{R}$, *and* P *is a probability function over language* $\mathcal{L}_{\mathrm{CL}^-}$. *Then,* α *is a probabilistic consequence of* Γ *(short:* $\Gamma \models_p \alpha$*) iff* $\forall \epsilon > 0, \exists \delta > 0, \forall P$: *If* $\forall \beta \in \Gamma$ *holds that* $P(\beta) \geq 1-\delta$, *then* $P(\alpha) \geq 1-\epsilon$.

The p validity criterion says that for any assignment of an arbitrarily high probability $1-\epsilon$ (less than one) to the conclusion there exists a threshold $1-\delta$ that if the probability of every single premise is higher than $1-\delta$, then the probability of the conclusion is higher than $1-\epsilon$. So, given an inference is p-valid, one can guarantee arbitrarily high probabilities (less than one) for the conclusion provided sufficiently high probabilities (less than one) for all premises (Adams, 1966, p. 273f).

Let me now discuss the proof theories of Systems **P**, **P***, and **P**$^+$ and then describe the corresponding semantics for Systems **P** and **P**$^+$. For a more perspicuous treatment of the proof theory of Conditional Logic Systems **P**, **P***, and **P**$^+$, I will first start with an axiomatization of System **P** by Kraus et al. (1990) and Lehmann and Magidor (1992).[13] Here I will formulate System **P** w.r.t. language $\mathcal{L}_{\mathrm{CL}^-}$, which admits only non-conditional formulas or simple conditional formulas (cf. Lemma 7.17):

Definition 3.19. *System* **P** *for language* $\mathcal{L}_{\mathrm{CL}^-}$ *is the smallest logic which is closed under the following rules and axiom schemata (cf. Lehmann & Magidor, 1992, p. 5f):*

LLE	*if* $\alpha \mathrel{\Box\!\!\rightarrow} \gamma$ *and* $\vdash \alpha \leftrightarrow \beta$ *then* $\beta \mathrel{\Box\!\!\rightarrow} \gamma$
RW	*if* $\gamma \mathrel{\Box\!\!\rightarrow} \alpha$ *and* $\vdash \alpha \rightarrow \beta$ *then* $\gamma \mathrel{\Box\!\!\rightarrow} \beta$
*AND**	*if* $\alpha \mathrel{\Box\!\!\rightarrow} \beta$ *and* $\alpha \mathrel{\Box\!\!\rightarrow} \gamma$ *then* $\alpha \mathrel{\Box\!\!\rightarrow} \beta \wedge \gamma$
Refl	$\alpha \mathrel{\Box\!\!\rightarrow} \alpha$
*CM**	*if* $\alpha \mathrel{\Box\!\!\rightarrow} \gamma$ *and* $\alpha \mathrel{\Box\!\!\rightarrow} \beta$ *then* $\alpha \wedge \beta \mathrel{\Box\!\!\rightarrow} \gamma$
*Or**	*if* $\alpha \mathrel{\Box\!\!\rightarrow} \gamma$ *and* $\beta \mathrel{\Box\!\!\rightarrow} \gamma$ *then* $\alpha \vee \beta \mathrel{\Box\!\!\rightarrow} \gamma$

[13] Kraus et al. (1990) and Lehmann and Magidor (1992) describe System **P** effectively in language $\mathcal{L}_{\mathrm{rCL}^*}$, which is as expressive as $\mathcal{L}_{\mathrm{rCL}}$, where in $\mathcal{L}_{\mathrm{rCL}^*}$ only simple or negated conditional formulas are allowed (see Section 3.5.3).

'LLE', 'RW', 'Refl', and 'CM' abbreviate 'Left Logical Equivalence', 'Right Weakening', 'Reflexivity', and 'Cautious Monotonicity', respectively. The asterices in 'AND*', 'CM*' and 'Or*' indicate that the respective principles are formulated as rules rather than an axiom schemata, in contrast to the axiom schema versions for the full language $\mathcal{L}_{\text{CL}}$, which I will discuss in Chapters 4–7. Observe that my use of the term 'logic' presupposes that the respective proof-theoretic system is closed under p.c. consequence (including the rule of substitution).

In addition to the principles of System **P**, System **P*** also draws on Principles Det* ("Detachment) and Cond* ("Conditionalization"), which are rule versions of the Principles Det and Cond, respectively, as discussed in Section 3.4.2:

Det* if $\top \mathrel{\Box\!\!\rightarrow} \alpha$ then α
Cond* if α then $\top \mathrel{\Box\!\!\rightarrow} \alpha$

Principles Det* and Cond* are by Theorem 7.63 logically equivalent to the following two principles on the basis of System **P**, where MP* and CS* are rule versions of Principles MP ("Modus Ponens") and CS ("Conjunctive Sufficiency"), respectively (see also Sections 3.2 and 3.4.2):

MP* if $\alpha \mathrel{\Box\!\!\rightarrow} \beta$ and α then β
CS* if $\alpha \wedge \beta$ then $\alpha \mathrel{\Box\!\!\rightarrow} \beta$

Let me now define the proof theory of System **P*** (cf. Lemma 7.69):

Definition 3.20. *System* **P*** *for language* $\mathcal{L}_{\text{CL}^-}$ *is the smallest logic, which contains System* **P** *and Cond* and Det*.*

Adams (1966, Theorems 7.1 and 7.2, p. 292) proved that System **P*** is sound and complete for the language $\mathcal{L}_{\text{CL}^-}$ w.r.t. the probabilistic validity criterion in Definition 3.18. This completeness result, however, only holds if one limits the consequence relation from formula sets Γ to formulas α (short: $\Gamma \models_p \alpha$) to finite sets of formulas Γ.

System **P⁺** (Adams, 1986) rather than System **P*** results from the validity criterion in Definition 3.18, when one uses the language $\mathcal{L}_{\text{rCL}^*}$ instead of language $\mathcal{L}_{\mathrm{CL}^-}$. Language $\mathcal{L}_{\mathrm{rCL}}$ allows for simple conditional formulas and disjunctions of conditional formulas but not for non-conditional formulas. In Adams' (1986) original proposal, the language contains a further restriction, insofar as disjunctive conditionals are only allowed as conclusions of inferences but not as premises (see Section 3.5.3).

Adams' (1986) System $\mathbf{P}^+$ contains, then, the following rule in addition to the rules and axiom schemata of System **P**:

dRM* if $\alpha \mathrel{\Box\!\!\rightarrow} \gamma$ then $(\alpha \mathrel{\Box\!\!\rightarrow} \neg\beta) \vee (\alpha \wedge \beta \mathrel{\Box\!\!\rightarrow} \gamma)$

The expression 'dRM*' stands for 'Rule Version of a disjunctive Formulation of Rational Monotonicity' (cf. Schurz, 1998, p. 84). In the full conditional logic language $\mathcal{L}_{\mathrm{CL}}$, Principle dRM* is p.c. equivalent to the following rule:

RM* if $\alpha \mathrel{\Box\!\!\rightarrow} \gamma$ and $\neg(\alpha \mathrel{\Box\!\!\rightarrow} \beta)$ then $(\alpha \wedge \beta \mathrel{\Box\!\!\rightarrow} \gamma)$

'RM*' abbreviates 'Rule Version of Rational Monotonicity'. Principle RM* is also expressible in language $\mathcal{L}_{\mathrm{rCL}^*}$ which allows for negations of conditional formulas.

System $\mathbf{P}^+$ can, then, defined the following way:

Definition 3.21. *System* $\mathbf{P}^+$ *for language* $\mathcal{L}_{\mathrm{rCL}}$ *is the smallest logic which contains* **P** *and dRM*.*

Adams (1986, p. 261) effectively specifies the probability of conditionals $\alpha \mathrel{\Box\!\!\rightarrow} \beta$ on the basis of Definition 3.18 for language $\mathcal{L}_{\mathrm{CL}}$ rather than language $\mathcal{L}_{\mathrm{CL}^-}$. I ignore here that Adams (1986, p. 268f) defines empty disjunctions (disjunctions with zero disjuncts) only as "logical falsehoods" and presuppose that all formulas of the language are simple conditional formulas or disjunctions of conditional formulas. If one were to admit non-conditional formulas in one's language (such as $\bot$), his system would become incomplete w.r.t. his semantics, since Adams does not have rules or axiom schemata such as P-Cons (see Section 3.2) which are probabilistically valid (given a sufficiently expressive language) and which make sure that if all conditional formulas are implied by formula set Γ, then also all non-conditional formulas of the language are implied (see below).

Schurz (1998) uses a variant of System $\mathbf{P}^+$ which effectively includes the following principle in addition to rules and axiom schemata of System $\mathbf{P}^+$:

TR* if $\neg\alpha \mathrel{\Box\!\!\rightarrow} \alpha$ then α

'TR*' abbreviates "Rule Version of Total Reflexivity".[14] The expression $\neg\alpha \mathrel{\Box\!\!\rightarrow} \alpha$ is also used as definiens for the definition of $\Box\alpha$, which is in turn interpreted as 'it is necessary that α'. I will postpone the discussion whether such a terminology is justified to Section 7.2.3.

Schurz (1998) escapes the difficulties of Adams (1986) sketched above, since Principle TR implies P-Cons on the basis of the rules and axiom schemata of System **P** as the following lemma shows:

Lemma 3.22. *LLE+TR ⇒ P-Cons*

Proof.

1.	$\top \mathrel{\Box\!\!\rightarrow} \bot$	assumption ind. proof
2.	$\neg\bot \mathrel{\Box\!\!\rightarrow} \bot$	1, LLE
3.	$\bot$	2, TR
4.	$\neg(\top \mathrel{\Box\!\!\rightarrow} \bot)$	1-3, 3, 3, ind. proof

□

Let me now discuss the Central **P** Theorem (Schurz, 2005b), which connects the proof-theoretic side of System **P** with probabilistic semantics and possible worlds semantics. For that purpose I will restrict myself to language $\mathcal{L}_{\text{rrCL}}$, in which every formula is a simple conditional (cf. Section 3.5.3). The Central **P** Theorem of Schurz (2005b) can, then, be described as follows, where the uncertainty $u_P(\cdot)$ for a probability function P is defined as $1 - P(\cdot)$:

Theorem 3.23. *(Central* **P** *Theorem, cf. Schurz, 2005b, p. 43)*
Let α and Γ be a formula of $\mathcal{L}_{\text{rrCL}}$ and a finite set of formulas of language $\mathcal{L}_{\text{rrCL}}$. Then, the following four conditions are equivalent:
(1) (Calculus) α is derivable from Γ in System **P** *(short: $\Gamma \vdash_{\mathbf{P}} \alpha$).*
(2) (Normal Worlds semantics) Γ implies α w.r.t. all finite, ranked Lewis models.

[14] Schurz (1998, p. 84) uses in fact a variant of TR called 'Poss' ("Possibility").

(3) (Infinitesimal Probability Semantics) α is a p-consequence of Γ.
(4) (Non-Infinitesimal Probability Semantics) For all probability functions P holds that $\sum_{\beta\in\Gamma} u_P(\beta) \leq u_P(\alpha)$. (Uncertainty Sum Rule)

By finite ranked models I mean relational Lewis models as described by Definitions 3.4–3.6 with the additional constraint that the set of worlds in these models is finite. The non-infinitesimal probability semantics endorsed by the uncertainty sum rule refers to the uncertainty of formulas for a probability function P as described above. The non-infinitesimal probability semantics allows one – in contrast to the p validity criterion – to determine valid inferences in System **P** in terms of a finite number of steps.

One can also state alternative versions of the Central **P** Theorem for different languages. If one uses language $\mathcal{L}_{\text{CL}^-}$ rather than language $\mathcal{L}_{\text{rrCL}}$, then the derivability relation in point (1) of Theorem 3.23 has to refer to System **P*** rather than System **P** (Adams, 1966, Theorems 7.1 and 7.2, p. 292). In addition one has to change (2) in such a way that it employs finite ranked models with centering axiom schemata (see Section 3.2) rather than the class of all finite ranked models (Adams, 1977, p. 188).

Let me now focus on a version of the Central **P** Theorem for language $\mathcal{L}_{\text{rCL}^*}$ with the additional restriction that disjunctions (with more than one disjunct) are allowed only in the conclusion of inferences. Then, the derivability relation in condition (1) of Theorem 3.23 has to refer to System **P**+ (Adams, 1986, Meta-Metatheorem 1, p. 261; cf. Schurz, 1998, Theorem 3, p. 96) and one has to modify conditions (2) and (4). Adams (1986, Meta-Metatheorem 4, p. 269) shows that it suffices to limit (2) to finite ranked models for which C_{MP} (see Section 3.2) holds. I conjecture here that $C_{\text{P-Cons}}$ (see Section 3.2) suffices for that purpose. In addition, point (4) has to be changed the following way: The sum of the uncertainties of the premises $\sum_{\beta\in\Gamma} u_P(\beta)$ has to be less or equal to the product of the uncertainties of the disjuncts $\prod_{1\leq i\leq n} u_P(\beta_i)$, $n \in \mathbb{N}$, for disjunctive conclusions $\beta_1 \vee \ldots \vee \beta_n$ (cf. Adams, 1986, p. 261; cf. Schurz, 1998, p. 86).

Finally, I conjecture that for language $\mathcal{L}_{\text{rCL}}$ – which contains only simple conditional formulas and negated conditional formulas – one has to change the proof-theoretic derivability relation to **P**+P-Cons in (1). In that case also (2) has to be modified to refer to finite ranked Lewis models for which condition $C_{\text{P-Cons}}$ holds (see Section 3.2).

3.6.2 Threshold Semantics

Adams' probabilistic validity criteria are not the only ones discussed in the literature. Hawthorne and Makinson (2007, p. 250), for example, employ a threshold validity criterion instead. They define meta-language conditional assertions of sort $\alpha \mid\!\sim_{P,r} \beta$ – to be read "if α then β given probability function P and threshold r" (cf. Section 2.2.6) – to hold iff either $P(\alpha) = 0$ or $P(\beta \mid \alpha) \geq r$. Here Hawthorne and Makinson (2007) put a conditional probability interpretation into the definition of $\alpha \mid\!\sim_{P,r} \beta$. This criterion carries, then, through to their second-order consequence relation pertaining to the meta-language conditional assertion $\alpha \mid\!\sim_{P,r} \beta$. In contrast Adams describes conditionals as object formulas of type $\alpha \mathrel{\Box\!\!\rightarrow} \beta$ and defines the probability of conditionals as the respective conditional probability by means of the Stalnaker thesis. In addition he quantifies in his probabilistic validity criterion over probability functions and specific thresholds (see Definition 3.18).

Hence, the second-order consequence relation of Hawthorne and Makinson (2007) and the consequence relation of Adams (1965, 1966, 1977) differ. This can also be seen by the following observation: In the Adams approach the inference from $\alpha \mathrel{\Box\!\!\rightarrow} \beta$ and $\alpha \mathrel{\Box\!\!\rightarrow} \gamma$ to $\alpha \mathrel{\Box\!\!\rightarrow} \beta \wedge \gamma$ (AND*) is p-valid (see previous section). Given the threshold-criterion of Hawthorne and Makinson (2007) this inference is, however, not warranted (see Hawthorne & Makinson, 2007, p. 252).

3.6.3 Adams' System $\mathbf{P}_\epsilon$ and Schurz's Modification

The intuitive starting point for Adams' (1975) modified account lies in the following two observations, one being syntactically and the other semantically motivated. Before I describe Adams' (1975) semantics and Schurz's (1997b) modification I will first focus on the syntactic deviation.

A Syntactic Deviation

For the first point, consider the following natural language statements:

E26 If Adam goes shopping, he is happy.
E27 If Adam goes shopping, he is not happy.
E28 It is not the case that if Adam goes shopping he is happy.

Adams (1965, p. 181) observed that we use statements such as E27 rather than statements of sort E28 to negate statements such as E26. Example E28 is not directly representable in the language $\mathcal{L}_{\mathrm{CL}^-}$ of Adams (1965, 1966, 1977, 1975), since negations of conditional formulas are not formulas of language $\mathcal{L}_{\mathrm{CL}^-}$. On the other hand assertions E26 and E27 can be formalized by $\alpha \mathrel{\Box\!\!\rightarrow} \beta$ and $\alpha \mathrel{\Box\!\!\rightarrow} \neg\beta$, respectively, and represented in language $\mathcal{L}_{\mathrm{CL}^-}$. Adams' observation suggests that $\alpha \mathrel{\Box\!\!\rightarrow} \beta$ and $\alpha \mathrel{\Box\!\!\rightarrow} \neg\beta$ are conjointly inconsistent and that, hence, a variant of the following axiom schema is valid (see Section 3.3.6), where 'CNC' abbreviates 'Conditional Non-Contradiction' (cf. Adams, 1975, p. 46 and p. 56; see also Section 3.3.6):

CNC $\neg((\alpha \mathrel{\Box\!\!\rightarrow} \beta) \wedge (\alpha \mathrel{\Box\!\!\rightarrow} \neg\beta))$

Note that Principle CNC is not expressible in the languages used for Adams' System $\mathbf{P}_\epsilon$, since it draws on Boolean combinations of conditional formulas. One can, however, use the following variant of CNC in Adams' system (Adams, 1975, p. 61):

CNC* if $\alpha \mathrel{\Box\!\!\rightarrow} \beta$ and $\alpha \mathrel{\Box\!\!\rightarrow} \neg\beta$ then γ

Although principles such as CNC are not expressible in the language of Adams' **P** systems, I will nevertheless use the full language $\mathcal{L}_{\mathrm{CL}}$ in the following for the following reason: In my opinion the specifics of these principles are easier to understand in the less restricted language. Moreover, one can more easily compare different systems if one uses the same representation format. For that purpose a full language version seems most appropriate. I will refer to translations into $\mathcal{L}_{\mathrm{CL}^-}$ for each principle discussed in this section, if any exist.

Let me now focus on the Principle CNC. This formula is p.c. equivalent to the following Axiom Schema D (cf. Section 3.3.6):

D $(\alpha \mathrel{\Box\!\!\rightarrow} \beta) \rightarrow (\alpha \mathrel{\Diamond\!\!\rightarrow} \beta)$.

Principle D draws on $\text{Def}_{\Diamond\rightarrow}$, which gives us that $\alpha \mathrel{\Diamond\!\!\rightarrow} \beta$ is nothing but $\neg(\alpha \mathrel{\Box\!\!\rightarrow} \neg\beta)$ (see Section 4.2.1). The name D derives from the fact that this principle represents a generalization from the standard Axiom D in Kripke semantics (cf. Hughes & Cresswell, 1996/2003, p. 43f). [15]

As we saw in Section 3.3.6 Principle CNC is, furthermore, logically equivalent to the following principle, given the basic Conditional Logic Principles RW and AND (or AND*; see Sections 3.3.6 and 3.6.1):

CNC′ $\neg(\alpha \mathrel{\Box\!\!\rightarrow} \bot)$ [16]

For virtually all standard conditional logic systems in the literature both AND (or AND*) and RW are theorems, including Adams' Systems $\mathbf{P}^*$, $\mathbf{P}^+$, and $\mathbf{P}_\epsilon$ and the majority of conditional logic systems investigated in Chapters 4–7. [17]

I will now show that CNC and D are logically equivalent, given $\text{Def}_{\Diamond\rightarrow}$ holds:

Theorem 3.24. *$\text{Def}_{\Diamond\rightarrow} \Rightarrow (CNC \Leftrightarrow D)$*

Proof. We obtain Theorem 3.24 by Lemmata 3.25 and 3.26. □

Lemma 3.25. *$\text{Def}_{\Diamond\rightarrow} + CNC \Rightarrow D$*

[15] This can be seen more clearly if one interprets the conditional operator $\alpha \mathrel{\Box\!\!\rightarrow} \beta$ as $[\alpha]\beta$, where $[\alpha]$ describes a type of unary modal necessity operator (see Chellas, 1975, p. 138).

[16] In the language $\mathcal{L}_{CL}$- Principle CNC* can be represented by the following rule: if $\alpha \mathrel{\Box\!\!\rightarrow} \bot$ then β.

[17] The threshold semantics approach of Hawthorne and Makinson (2007, see Section 3.6.2) is one conditional logic approach which makes RW valid but not AND*.

Proof.

1.	$\neg((\alpha \mathrel{\Box\!\!\rightarrow} \beta) \wedge (\alpha \mathrel{\Box\!\!\rightarrow} \neg\beta))$	CNC
2.	$(\alpha \mathrel{\Box\!\!\rightarrow} \beta) \rightarrow \neg(\alpha \mathrel{\Box\!\!\rightarrow} \neg\beta)$	1, p.c.
3.	$(\alpha \mathrel{\Box\!\!\rightarrow} \beta) \rightarrow (\alpha \mathrel{\Diamond\!\!\rightarrow} \beta)$	2, $\text{Def}_{\Diamond\!\rightarrow}$

□

Lemma 3.26. *$Def_{\Diamond\!\rightarrow}$+D ⇒ CNC*

Proof.

1.	$(\alpha \mathrel{\Box\!\!\rightarrow} \beta) \rightarrow (\alpha \mathrel{\Diamond\!\!\rightarrow} \beta)$	D
2.	$(\alpha \mathrel{\Box\!\!\rightarrow} \beta) \rightarrow \neg(\alpha \mathrel{\Box\!\!\rightarrow} \neg\beta)$	1, $\text{Def}_{\Diamond\!\rightarrow}$
3.	$\neg((\alpha \mathrel{\Box\!\!\rightarrow} \beta) \wedge (\alpha \mathrel{\Box\!\!\rightarrow} \neg\beta))$	2, p.c.

□

Note that Principles D, CNC, CNC′ and their translations into the language $\mathcal{L}_{\text{CL}^-}$ do not hold for the Adams (1965, 1966, 1977, 1975, 1986) Systems $\mathbf{P}^*$, $\mathbf{P}_\epsilon$, and $\mathbf{P}^+$ and for the majority of conditional logic systems discussed in this and the following chapters. This is due to another conditional logic axiom schema which goes as follows:

Refl $\alpha \mathrel{\Box\!\!\rightarrow} \alpha$.

Principle Refl states that if α is the case then α is the case. Although there are some interpretations for specific types of conditionals, such as conditional obligation $\alpha \mathrel{\bigcirc\!\!\rightarrow} \beta$ ("if α then β ought to be the case"), for which this principle might not hold (e.g., Spohn, 1975, pp. 248–250; Bonati, 2005, p. 74f), Refl is valid in the great majority of standard conditional logic systems discussed here.[18]

The problem is that given the very basic conditional logic Principle RW (see Section 3.6.1) – endorsed by virtually all conditional logics, such as the **P** systems, D. Lewis' (1973b) systems, etc. – both Refl and CNC are inconsistent:[19]

[18] The only conditional logic systems described in this book for which Refl does not hold are the proof theoretic system for CS semantics described in Chapter 5.

[19] An analogous proof can be given for the formulation of CNC in the restricted language $\mathcal{L}_{\text{CL}^-}$ of Adams (1965, 1966, 1977, 1975), as described in Footnote 16.

Theorem 3.27. *Given RW, Principles CNC and Refl are inconsistent.*

Proof.

1.	$\bot \,\Box\!\!\rightarrow \bot$	Refl
2.	$\neg((\bot \,\Box\!\!\rightarrow \alpha) \wedge (\bot \,\Box\!\!\rightarrow \neg\alpha))$	CNC
3.	$\bot \,\Box\!\!\rightarrow \alpha$	1, RW
4.	$\bot \,\Box\!\!\rightarrow \neg\alpha$	1, RW
5.	$\bot$	3, 4, 2 p.c.

□

The inconsistency is, however, easier avoided than it might first seem. The only problematic case is the one in which a logically false formula, such as $\bot$, is the antecedent of a conditional formula. To exclude this case for the **P** systems, one can, hence, use the following rule:

RCNC if $\nvdash_{p.c.} \neg\alpha$ then $\neg((\alpha \,\Box\!\!\rightarrow \beta) \wedge \neg(\alpha \,\Box\!\!\rightarrow \neg\beta))$.

$\nvdash_{p.c.} \alpha$ says α is not derivable in p.c. Hence, $\nvdash_{p.c.} \alpha$ ensures that the antecedent of a formula is p.c. consistent (see Sections 2.2.3 and 2.2.4). P.c. consistency rather than consistency w.r.t. Adams' System $\mathbf{P}_\epsilon$ suffices, as Adams (1975, Theorem 2, p. 56) proves. [20]

As RCNC is a non-trivially non-monotonic rule, any proof-theoretic system endorsing this principle results – provided sufficiently reasonable conditions – in a default logic (see Section 2.2.3). Adams (1975, p. 61, Rule R7) uses a version of this rule in the specification of his proof theory (see Schurz, 1998, p. 84) [21] so that his system is effectively a default-logic. Since default

[20] The following rule RCNC′ is a translation of RCNC into $\mathcal{L}_{\mathrm{CL}^-}$: if $\alpha \,\Box\!\!\rightarrow \beta, \alpha \,\Box\!\!\rightarrow \neg\beta$ and $\nvdash_S \neg\alpha$, then γ, where $\nvdash_S \alpha$ abbreviates non-derivability of α in system S. In Adams' language $\mathcal{L}_{\mathrm{CL}^-}$ the formula α is not allowed to contain an instance of a conditional operator. Otherwise α could neither be an antecedent formula nor could its negation be a formula of that language. So, it suffices to require α being consistent according to p.c.

[21] Strictly speaking Adams (1975, p. 61) uses following rule RCNC″: if $\alpha \,\Box\!\!\rightarrow \beta, \nvdash \neg\alpha$, $\vdash \neg(\alpha \wedge \beta)$ then γ. Principle RCNC″ is equivalent to RCNC′ (see Footnote 20) given RW, AND*, and Refl. Note that Principles RW, AND*, and Refl are all valid in Adams' (1975) System $\mathbf{P}_\epsilon$.

logic approaches are not unproblematic with respect to their axiomatization (see Section 2.2.4), also the Adams (1975) system is prone to the same types of drawbacks. Despite this fact, this issue of Adams' (1975) System $\mathbf{P}_\epsilon$ is seldomly addressed in the conditional logic literature.

In addition to a version of RCNC, Adams (1975) imposes following further requirement: He accepts conditional formulas of language $\mathcal{L}_{\text{CL}^-}$ for his language $\mathcal{L}^{\text{cons}}_{\text{CL}^-}$ (see also next section) only if the antecedent of the conditional formula is p.c. consistent (p. 46, see my Footnote 20). Due to that fact he has to adopt the rules of his proof theory in such a way that he allows for conditional formulas only in case they have a p.c. consistent antecedent.[22] This is not the only possible approach. I will sketch an alternative approach by Schurz (1997b) below.

Adams in his earlier and later approaches (Adams, 1965, 1966, 1977, 1986) neither employs the non-monotonic rule RCNC nor variants of it (see Schurz, 1998, p. 84). Since Adams (1965, 1966, 1977, 1986) uses only monotonic rules (see Adams, 1965, p. 189, Adams, 1966, p. 277, Adams, 1986, p. 257f), his 1965/1966 System $\mathbf{P}^*$ and his 1986 System $\mathbf{P}^+$ are not default logics, and, hence, do not suffer from the drawbacks of default logic systems (see Section 2.2.4). Most importantly the second-order inference relation is still monotonic, although the conditional operator $\Box\!\!\rightarrow$ is not (cf. Sections 2.2.6 and 3.6.2). The proof-theoretic extensions of System **CK** described in Section 4.2.6, the D. Lewis (1973b), Stalnaker's (1968) systems, and the systems described by Kraus et al. (1990) and partially by Lehmann and Magidor (1992) share the property of having a (second-order) monotonic inference relation.

The Semantic Modification

In this section I will first describe the probabilistic validity criterion of Adams (1975) and then discuss Schurz's (1997b) modification.

Let me first focus on Adams' (1975) validity criterion. The first impor-

[22] This seems, moreover, to be the motivation for the addition of a conditional consistency conditions for his other rules (p. 60f).

tant difference between the semantics of Adams (1975) and his semantics in Adams (1965, 1966, 1977, 1986) described in Section 3.6.1 is the fact that Adams (1975) modifies the definition of the probability of conditionals. We saw in Section 3.6.1 that Adams deviated from Kolmogorov's account in specifying the conditional probability $P(\beta\,|\,\alpha)$ as one in case $P(\alpha) = 0$ (see Definition 3.17.b.iv). Adams (1965, p. 176) points out that he endorsed this approach on behalf of technical completeness and simplicity only.

Although Adams' argumentation suggests that the decision to assign a probability value of one to $P(\beta\,|\,\alpha)$ given $P(\alpha) = 0$ is rather arbitrary, there is a certain intuition guiding it, as specified by the probability function in Definition 3.17: Since p.c. inconsistent statements, such as $\bot$, have to be assigned a zero-probability, it follows that all conditional formulas with p.c. inconsistent antecedents have probability one according to the above definition. In particular both $P(\alpha\,|\,\bot) = 1$ and $P(\neg\alpha\,|\,\bot) = 1$ hold. However, since for any formula α it is the case that α and $\neg\alpha$ are p.c. consequences of $\bot$, this specification seems rather adequate.

Adams (1975, p. 48f and p. 50), then, uses the following modified version of Definition 3.17, where formulas of $\mathcal{L}^{\text{cons}}_{\text{CL}^-}$ are formulas of language $\mathcal{L}_{\text{CL}^-}$ with the additional requirement that the antecedent of any conditional formula has to be p.c. consistent (see above):

Definition 3.28. *P is a probability function over* $\mathcal{L}^{\text{cons}}_{\text{CL}^-}$ *iff*
(a) P is a partial function from $\mathcal{L}^{\text{cons}}_{\text{CL}^-}$ *to* $\mathbb{R}$ *so that*
(b) for all formulas $\alpha, \beta \in \mathcal{L}_{\text{prop}} \subseteq \mathcal{L}^{\text{cons}}_{\text{CL}^-}$ *holds:*

(i) $0 \leq P(\alpha) \leq 1$ *and* $P(\top) = 1$,
(ii) if $\models \alpha \rightarrow \beta$, *then* $P(\alpha) \leq P(\beta)$,
(iii) if $\models \neg(\alpha \wedge \beta)$, *then* $P(\alpha \vee \beta) = P(\alpha) + P(\beta)$, *and*
(iv) if $P(\alpha) > 0$, *then* $P(\alpha \mathbin{\Box\!\!\rightarrow} \beta) = P(\alpha \wedge \beta) / P(\alpha)$.

The expression $\mathcal{L}_{\text{prop}}$ refers here to the set of formulas of $\mathcal{L}^{\text{cons}}_{\text{CL}^-}$ which do not contain an instance of a conditional operator $\Box\!\!\rightarrow$. P is only a partial function from the whole set $\mathcal{L}^{\text{cons}}_{\text{CL}^-}$, since for $P(\alpha) = 0$ the probability $P(\alpha \mathbin{\Box\!\!\rightarrow} \beta)$ is undefined. Definition 3.28 differs from Definition 3.17 in two ways: (a)

The probability function is defined over language $\mathcal{L}^{\text{cons}}_{\text{CL}^-}$ rather than language $\mathcal{L}_{\text{CL}^-}$ and (b) $P(\alpha \mathbin{\Box\!\!\rightarrow} \beta)$ is defined only if $P(\alpha) > 0$.

Adams (1975, p. 49f), then, proceeds to define the following two notions:

Definition 3.29. *Suppose that $\alpha, \beta \in \mathcal{L}_{\text{prop}} \subseteq \mathcal{L}^{\text{cons}}_{\text{CL}^-}$ and that P is a probability function according to Definition 3.28 over language $\mathcal{L}^{\text{cons}}_{\text{CL}^-}$. Then, P is proper for $\alpha \mathbin{\Box\!\!\rightarrow} \beta$ iff $P(\alpha) > 0$.*

Definition 3.30. *For all formula sets $\Gamma \subseteq \mathcal{L}^{\text{cons}}_{\text{CL}^-}$ and probability functions P over $\mathcal{L}^{\text{cons}}_{\text{CL}^-}$ as specified in Definition 3.28 holds: P is proper for Γ iff P is proper for all $\alpha \in \Gamma$.*

Definition 3.29 gives us that a probability function P is proper for a conditional formula $\alpha \mathbin{\Box\!\!\rightarrow} \beta$ iff $P(\alpha) > 0$. Definition 3.30 extends this definition to sets of formulas.

Adams (1975, p. 57), then, employs the notion of properness to specify a modified criterion of valid inference:

Definition 3.31. *(P^* Validity)*
Let $\alpha \in \mathcal{L}^{\text{cons}}_{\text{CL}^-}$, $\Gamma \subseteq \mathcal{L}^{\text{cons}}_{\text{CL}^-}$, ϵ, $\delta \in \mathbb{R}$, and let P be a probability function over language $\mathcal{L}^{\text{cons}}_{\text{CL}^-}$. Then, α is probabilistic consequence of Γ (short: $\Gamma \models_{p^*} \alpha$) iff $\forall \epsilon > 0, \exists \delta > 0, \forall P$, which are proper for Γ and α, the following is the case: If $\forall \beta \in \Gamma$ holds that $P(\beta) \geq 1 - \delta$, then $P(\alpha) \geq 1 - \epsilon$.*

In Definition 3.31 the notion of properness is used to restrict probability functions. According to Definition 3.31 no probability is assigned to a conditional formula $\alpha \mathbin{\Box\!\!\rightarrow} \beta$ if $P(\alpha) = 0$ and Adams' reference to his notion of properness excludes those cases: In his p^*-criterion (Definition 3.31) the properness of α and Γ guarantees that neither the antecedent of α (if it is a conditional formula) nor any antecedent of a conditional formula in Γ receives a zero-probability assignment. In addition only conditional formulas with a consistent antecedent are permitted in Definition 3.31 (due to the restrictions of $\mathcal{L}^{\text{cons}}_{\text{CL}^-}$). Adams (1975, Theorem 4.2, p. 60f), then, proves that System $\mathbf{P}_\epsilon$ is sound and complete w.r.t. all p^*-consequence $\Gamma \models_{p^*} \alpha$ if we restrict Γ to a finite set of formulas of $\mathcal{L}^{\text{cons}}_{\text{CL}^-}$, including a version of the uncertainty sum rule described in the Central **P** Theorem (his Theorem 3.1, p. 57; see also Theorem 3.23 herein).

Let me now describe Schurz's (1997b) alternative characterization of Adams' (1975) probabilistic semantics. Schurz (1997b) uses Definition 3.17 rather than Definition 3.28 as a basis and accordingly employs language $\mathcal{L}_{\mathrm{CL^-}}$ for that purpose.

Schurz (1997b, p. 539), then, defines the notion of properness the following way:

Definition 3.32. *Suppose that $\alpha, \beta \in \mathcal{L}_{\mathrm{prop}} \subseteq \mathcal{L}_{\mathrm{CL^-}}$ and that P is a probability function according to Definition 3.17 over language $\mathcal{L}_{\mathrm{CL^-}}$. Then, P is proper for $\alpha \mathrel{\Box\!\!\rightarrow} \beta$ iff the following holds: either $\vdash_{p.c.} \neg\alpha$ or $P(\alpha) > 0$.*

Definition 3.33. *For all formula sets $\Gamma \subseteq \mathcal{L}_{\mathrm{CL^-}}$ and probability functions P over language $\mathcal{L}_{\mathrm{CL^-}}$ as specified in Definition 3.17 holds: P is proper for Γ iff P is proper for all $\alpha \in \Gamma$.*

Schurz (1997b) modifies Adams' (1975) notion of properness in such a way that all probability functions over language $\mathcal{L}_{\mathrm{CL^-}}$ are proper for all conditionals with inconsistent antecedent formulas. This move allows Schurz to extend his system to arbitrary formulas of language $\mathcal{L}_{\mathrm{CL^-}}$.

Let me now describe Schurz's (1997b, p. 539) alternative version of the notion of p* validity (which I will call 'p′ validity'):

Definition 3.34. *(P′ Validity)*
Let $\alpha \in \mathcal{L}_{\mathrm{CL^-}}$, $\Gamma \subseteq \mathcal{L}_{\mathrm{CL^-}}$, $\epsilon, \delta \in \mathbb{R}$, and let P be a probability function over language $\mathcal{L}_{\mathrm{CL^-}}$. Then, α is probabilistic consequence′ of Γ (short: $\Gamma \models_{p'} \alpha$) iff $\forall\epsilon > 0, \exists\delta > 0, \forall P$, which are proper for Γ and α, the following is the case: If $\forall\beta \in \Gamma$ holds that $P(\beta) \geq 1 - \delta$, then $P(\alpha) \geq 1 - \epsilon$.

Schurz's modification of the notions of properness and p* validity admits all conditional formulas with inconsistent antecedent formulas such as $\bot \mathrel{\Box\!\!\rightarrow} \alpha$ and makes them valid. In all other cases Adams' (1975) p* validity criterion and System $\mathbf{P}_\epsilon$ are not changed. Most notably, in both versions, Adams' (1975) account and Schurz's (1997b) modification, the inference RCNC is valid.

One might argue that in Adams' (1975) approach CNC holds rather than RCNC for Adams' (1975) semantics, since conditionals with p.c. inconsistent antecedents are not expressible in the corresponding language $\mathcal{L}^{\text{cons}}_{\text{CL}^-}$. It is true that CNC is valid in Adams' (1975) semantics for System $\mathbf{P}_\epsilon$ due to its restriction to language $\mathcal{L}^{\text{cons}}_{\text{CL}^-}$. However, in order to apply Adams' (1975) system to conditionals in natural language, one effectively has to apply RCNC and cannot restrict oneself to CNC: To draw inferences in the Adams (1975) System $\mathbf{P}_\epsilon$ for natural language conditionals of the form $\alpha \mathrel{\Box\!\!\rightarrow} \beta$. First, one has to determine whether its antecedent α is p.c. consistent. Otherwise one is not allowed to represent the conditional in Adams' (1975) language $\mathcal{L}^{\text{cons}}_{\text{CL}^-}$. This p.c. consistency check is equivalent to the non-derivability condition in RCNC (see previous section). Moreover, the discussion of Schurz's alternative p′ validity criterion shows that the restriction of Adams (1975) to language $\mathcal{L}^{\text{cons}}_{\text{CL}}$ is in no way essential.

3.6.4 Possible Worlds Semantics for (Indicative) Conditionals and Quasi Truth Value Assignments in Adams' Probabilistic Semantics

In this section I will show where and in which way possible worlds semantics for conditionals, such as CS semantics (see Chapters 4–5; see also Section 3.1), differs from probabilistic semantics for Adams' **P** systems. The relation between both types of semantics is important, since one might argue that the probabilistic semantics for Adams's **P** systems are more general than any truth value accounts and that, hence, truth value accounts can adequately be described within Adams' probabilistic approaches. This point is important, since if the foregoing thesis were true, then D. Lewis' (1976) triviality result would also hold for possible worlds semantics. I will, however, postpone a discussion of the implications concerning D. Lewis' (1976) triviality result to Section 3.7.

To compare the probabilistic semantics as described in Sections 3.6.1 and 3.6.3 on the one hand and possible worlds semantics, such as CS semantics, on the other (i) I will first review the role of the material implication in Adams' **P** systems (1965, 1966, 1977, 1975, 1986) semantics. (ii) I will then discuss in which way we can account for the material implication and possible worlds semantics for (indicative) conditionals in a probabilistic semantics. For that purpose I will use probability-based quasi truth value assignments, that is probability assignments which draw only on the values one and zero (corresponding to 'true' and 'false', respectively). We will see that possible worlds semantics for (indicative) conditionals cannot be appropriately accounted for within this framework, whereas we can do so for the material implication. Finally, (iii) I will discuss an argument by Adams (1975, pp. 5–7) against truth value approaches for indicative conditionals which relies on the assumption that we can describe truth value approaches – including possible worlds semantics for conditionals – by means of quasi truth values within Adams' probabilistic framework.

Let us now focus on (i). In general, the following result holds for the material implication in Adams' **P** systems semantics, where Lemma 3.35 draws on Lemma 3.36:

Lemma 3.35. *Let function P be defined as in Definition 3.17. Then, the following holds:* $P(\alpha \,\Box\!\!\rightarrow \beta) = 1$ *iff* $P(\alpha \rightarrow \beta) = 1$

Proof. "$\Rightarrow$": This direction is immediately given by Lemma 3.36.

"$\Leftarrow$": Let $P(\alpha \rightarrow \beta) = 1$. I treat the cases (i) $P(\alpha) = 0$ and (ii) $P(\alpha) > 0$ separately. Suppose that (i) is the case. Then, by Definition 3.17.b.iv it follows trivially that $P(\alpha \,\Box\!\!\rightarrow \beta) = 1$. Suppose that (ii) holds. Then, by Definition 3.17.b.iv we have (1) $P(\alpha \,\Box\!\!\rightarrow \beta) = \frac{P(\alpha \wedge \beta)}{P(\alpha)}$. We get by Definition 3.17.b.i–iii that (2) $P(\alpha \,\Box\!\!\rightarrow \beta) = \frac{P(\alpha \wedge \beta)}{P(\alpha \wedge \beta) + P(\alpha \wedge \neg\beta)}$. Since by assumption $P(\alpha \rightarrow \beta) = 1$ we obtain $P(\alpha \wedge \neg\beta) = 0$. Hence, $P(\alpha \,\Box\!\!\rightarrow \beta) = \frac{P(\alpha \wedge \beta)}{P(\alpha \wedge \beta)} = 1$. □

Lemma 3.36. *Let function P be defined as in Definition 3.17. Then, the following holds:* $P(\alpha \rightarrow \beta) \geq P(\alpha \,\Box\!\!\rightarrow \beta)$.

Proof. I distinguish two cases here: (i) $P(\alpha \wedge \neg\beta) > 0$ and (ii) $P(\alpha \wedge \neg\beta) = 0$.

(A) Suppose (i) and assume for an indirect proof that (1) $P(\alpha \rightarrow \beta) < P(\alpha \,\Box\!\!\rightarrow \beta)$. Since by (i) $P(\alpha \wedge \neg\beta) > 0$, it follows from (1) by Definition

3.17.b.i–iv that (2) $1 - P(\alpha \wedge \neg\beta) < \frac{P(\alpha\wedge\beta)}{P(\alpha)}$. By Definition 3.17.b.ii and iii we infer that (3) $1 < \frac{P(\alpha\wedge\beta)}{P(\alpha\wedge\beta)+P(\alpha\wedge\neg\beta)} + P(\alpha \wedge \neg\beta)$. I abbreviate $P(\alpha \wedge \beta)$ and $P(\alpha \wedge \neg\beta)$ by X and Y, respectively. Then, (3) is equivalent to (4) $1 < \frac{X}{X+Y} + Y$. It follows from (4) that (5) $X + Y < X + Y(X + Y)$ and, hence, (6) $0 < -Y + XY + Y^2$. Since by assumption (i) holds that $Y > 0$, this implies that $0 > -1 + X + Y$ and thus, (7) that $X + Y > 1$. So, one gets (8) $P(\alpha \wedge \beta) + P(\alpha \wedge \neg\beta) > 1$ which contradicts points b.i–iii of Definition 3.17. Hence, it follows that $P(\alpha \rightarrow \beta) \geq P(\alpha \,\Box\!\!\rightarrow \beta)$.

(B) Suppose that (ii). Then, by Definition 3.17.b.i–iii $P(\alpha \rightarrow \beta) = 1$. Since $P(\alpha \,\Box\!\!\rightarrow \beta)$ is a probability assignment proper in the sense of Definition 3.17, it follows by point b.i of Definition 3.17 that $P(\alpha \rightarrow \beta) \geq P(\alpha \,\Box\!\!\rightarrow \beta)$. □

Hence, we can prove that for probability value one, both the probability of the material implication and the probability of conditionals conincide. To be able to compare the probability of the material implication (a non-conditional formula) and the probability of the corresponding conditional formula we have to presuppose a conditional language, such as $\mathcal{L}_{\text{CL}^-}$ or $\mathcal{L}^{\text{cons}}_{\text{CL}^-}$, which allows us to represent both types of formulas.

We can prove a further interesting result concerning the material conditional in Adams' probabilistic semantics given the following definition of a strict *p*-consequence criterion (Adams, 1965, p. 185; Adams, 1966, p. 274):

Definition 3.37. *Let $\alpha \in \mathcal{L}_{\text{CL}^-}$ and $\Gamma \subseteq \mathcal{L}_{\text{CL}^-}$, and let P be a probability function over language $\mathcal{L}_{\text{CL}^-}$. Then, α is a strict p-consequence of Γ (short: $\Gamma \models^s_p \alpha$) iff $\forall P$ the following is the case: If $\forall\beta \in \Gamma$ holds that $P(\beta) = 1$, then $P(\alpha) = 1$.*

Adams (1966), then, shows that his notion of strict *p*-consequence and p.c. entailment coincide (p. 274f, Theorem 1.1).

A consequence of both results is the following. We are well justified to assume that the probabilistic semantics of Adams is more general than the extensional semantics of the material implication and that the semantics for the material implication is a limiting case of Adams' (1965, 1966, 1977) **P** systems semantics.

Table 3.1
Quasi Truth Values for Conditionals in (a) Adams (1965, 1966, 1977, 1986) and Schurz (1997b) and (b) Adams (1975) and (c) Truth Values in Lewis models

$V_{1\text{-}3}(\alpha)$	$V_{1\text{-}3}(\beta)$	$V_{1\text{-}3}(\alpha \rightarrow \beta)$	$V_1(\alpha \,\Box\!\!\rightarrow \beta)$	$V_2(\alpha \,\Box\!\!\rightarrow \beta)$	$V_3(\alpha \,\Box\!\!\rightarrow \beta)$
1	1	1	1	1	0/1
1	0	0	0	0	0/1
0	1	1	1	undef.	0/1
0	0	1	1	undef.	0/1

V_1 and V_2 are quasi truth values in Adams (1965, 1966, 1977, 1986) and Schurz (1997b) on the one hand and Adams (1975) on the other, respectively. Quasi truth values result from restricting Definitions 3.17 and 3.28 to the values zero and one. V_3 describes truth values in Lewis models (Definitions 3.1–3.3), where Definitions 3.1–3.3 do not make any bridge principles valid. The expressions 'undef.' and '0/1' indicate that the quasi truth value is undefined and may be zero or one, respectively.

Let me now discuss point (ii). To compare Adams' probabilistic semantics with possible worlds semantics for (indicative) conditionals, such as CS semantics, I will use quasi truth values – probability values one (truth) and zero (falsity) (cf. Adams, 1966, p. 297, Definition 10.1; Adams, 1975, p. 47). In this approach atomic formulas of the respective language are assigned the quasi truth values zero and one, and quasi truth values for conditional formulas and Boolean combinations of formulas are calculated by Definitions 3.17 or 3.28.

In Table 3.1 I summarize quasi truth values for conditional formulas in (a) Adams (1965, 1966, 1977, 1986) and Schurz (1997b) and (b) Adams (1975), and (c) truth values for conditional formulas in Lewis models (see Definitions 3.1 3.3). As we can see, both the material implication and the representation of truth values for (a) concur. Furthermore, the quasi truth values are not defined ('undef.') for (b), when the antecedent formula is assigned the value zero. In contrast the truth values in (c) for conditional formulas $\alpha \,\Box\!\!\rightarrow \beta$ are undetermined in all four rows of the (quasi) truth table.

D. Lewis' semantics is not the only semantics for which the above result holds. Among these are the basic CS semantics (see Chapters 4–7; see also Section 3.1) and the ordering semantics of Kraus et al. (1990) and Lehmann and Magidor (1992) (cf. Section 3.2). These semantics have in common that they do not make bridge principles valid, viz. principles, which suppose a fixed relationship between conditional and non-conditional formulas (see Section 3.4.2). In the systems described by Kraus et al. (1990) and Lehmann and Magidor (1992) bridge principles cannot even be expressed since the authors consider only conditional assertions or negations thereof (see Section 3.5.3).

Let us now come back to (a) and (b). Note here a particular feature of the probabilistic semantics for conditionals. Suppose that α and β are non-conditional formulas. Then, in Adams' (1965, 1966, 1977, 1986) and Schurz's (1997b) framework the probability value of a conditional formula $\alpha \,\Box\!\!\rightarrow \beta$ is a function of the probability values for the conjunction of the antecedent and the consequent $\alpha \wedge \beta$ and the antecedent α. I will call this property – the property that probability values of $\alpha \wedge \beta$ and α determine the probability value of $\alpha \,\Box\!\!\rightarrow \beta$ for $P(\alpha) > 0$ – 'extended value functionality (of conditional formulas)'. Observe that for non-conditional formulas α and β the probabilities $P(\alpha)$ and $P(\beta)$ determine only $P(\neg\alpha)$ and $P(\neg\beta)$, but not $P(\alpha \,\Box\!\!\rightarrow \beta)$, $P(\alpha \wedge \beta)$, and $P(\alpha \vee \beta)$.

In constrast, in possible worlds semantics without bridge principles, such as Lewis models (see Definitions 3.1–3.3) and basic CS semantics, no non-conditional formula determines the value of conditional formulas. Hence, possible worlds semantics in general do not have the extended value functionality (of conditional formulas). In this type of semantics, however, the truth value of disjuncts, conjuncts, and negated formulas gives one the truth value of the respective disjunctions, conjunctions, and negations, respectively.

There is a range of possible worlds semantics for indicative (and counterfactual) conditionals which make bridge principles valid. Among these are Stalnaker models (1968; Stalnaker & Thomason, 1970; see Definition 3.7) and Lewis models with centering conditions. As we saw in Section 3.2, these centering conditions correspond to Principles MP (i.e., $(\alpha \mathrel{\Box\!\!\rightarrow} \beta) \rightarrow (\alpha \rightarrow \beta)$) and CS (i.e., $\alpha \wedge \beta \rightarrow (\alpha \mathrel{\Box\!\!\rightarrow} \beta)$). Observe that the rule versions of both principles (see also Section 3.6.1) are also valid in Adams' System $\mathbf{P}^*$ (Adams, 1965, 1966, 1977) and the extended Adams (1975) System $\mathbf{P}_\epsilon$ (see Section 3.6.1 and 3.6.3). For possible worlds semantics, however, the Principles MP and CS give us neither value functionality nor extended value functionality as described above: In case α is false, the truth values of $\alpha \wedge \beta$ and α do not determine the truth value of $\alpha \mathrel{\Box\!\!\rightarrow} \beta$.

Let me now describe which impact bridge principles have for (extended) value functionality in a truth value approach. If one adds CS to a conditional logic without bridge principles, then the truth of both the antecedent formula α and the consequent formula β leads to the truth of the conditional formula $\alpha \mathrel{\Box\!\!\rightarrow} \beta$. Likewise, if MP is valid in a truth value semantics, then the truth of the antecedent formula α and the falsehood of the consequent formula β lead to the falsehood of the conditional formula $\alpha \mathrel{\Box\!\!\rightarrow} \beta$. If one adds both CS and MP the resulting truth table for conditionals resembles the one quasi truth values in Adams' (1975) approach.

There is, however, an essential difference between both cases: The quasi truth assignments of conditionals in Adams' (1975) semantics are merely undefined in the third and fourth row, while in the possible worlds accounts the truth value of conditional formulas is not determined by the truth values of antecedent and consequent formulas. So, since rule versions of MP and CS are valid in Adams' probabilistic semantics, it is not the general lack of bridge principles which causes a failure of (extended) value functionality. To see that more clearly, let us focus on Schurz's (1997b) approach. In Schurz's (1997b) semantics, conditionals with inconsistent antecedent formulas are also assigned probability values and these assignments for the values zero and one concur perfectly with the probability assignments of Adams' Systems $\mathbf{P}^*$ and $\mathbf{P}^+$ and, hence, with the material implication analysis of conditional described in Table 3.1.

Let us now discuss in which way we can obtain full truth functionality for possible worlds semantics. To achieve this one can add a frame condition which corresponds to the Bridge Principle EFQ, that is $\neg\alpha \rightarrow (\alpha \,\Box\!\!\rightarrow \beta)$ (see Section 3.4.2). If one does so, then the conditional received the truth value 'true' in the third and fourth row of the truth table. (We will discuss a possible worlds semantics of this sort in Section 7.3.4.) Observe that the Principle EFQ holds in the semantics for Adams' **P** systems only when the strict p validity criterion is applied and not when we use the p validity criterion: In the probabilistic semantics for the **P** systems the probability $P(\neg\alpha)$ does not guarantee an arbitrarily high probability $P(\alpha \,\Box\!\!\rightarrow \beta)$ (smaller than one). Hence, this inference is not p-valid (and analogously for p^* validity).

Let us now draw some conclusions from the discussion in this section. We saw that we can model the extensional semantics of the material implication when we use quasi truth assignments in a probabilistic framework as described above. Furthermore, as we shall see in Section 7.3.4, we can also account for the extensional semantics of the material implication by means of possible worlds semantics for conditionals, such as CS semantics. So, the material implication is in fact a limiting case of both types of semantics.

Both semantics, however, differ from each other in the way they generalize the extensional semantics of the material implication. In a probabilistic semantics the number of possible assignments to p.c. formulas is increased, while extended value functionality for conditionals is retained. In contrast, in possible worlds semantics from p.c. both the bivalent assignment and the value functionality w.r.t. Boolean combinations of non-conditional formulas are kept. Possible worlds semantics for conditionals differs from the extensional semantics of the material implication and probabilistic semantics insofar as both classical truth functionality and extended value functionality of conditional formulas are given up. With that in mind it seems clear that possible worlds semantics for (indicative) conditionals is not a special case of the probabilistic semantics described herein.

Let me now focus on point (iii). There are arguments against truth value semantics for indicative conditionals which use the method of quasi truth value assignments in a probabilistic framework. Adams (1975, pp. 5–7), for example, aims to show that truth value approaches and in particular possible worlds approaches such as Stalnaker (1968; see Section 3.3.3) have absurd consequences when applied to indicative conditionals. For that purpose Adams proves that for any such approach there exist no propositions α, β, and γ, and quasi truth value assignments t_1, t_2, and t_3 such that $t_1(\alpha) \neq t_2(\alpha)$, $t_1(\beta) \neq t_3(\alpha)$ and $t_2(\gamma) \neq t_3(\alpha)$. For his proof to go through, Adams (1975) presupposes – among other things – that truth values of conditional formulas in truth value semantics, including possible worlds semantics, can be adequately described by a variant of the quasi truth value assignments described above. While this is not plausible, as our discussion in this section shows, one might argue that if Adams uses his 1975 System $\mathbf{P}_\epsilon$ then his argument has some force, at least for possible worlds semantics with bridge principles.

I disagree. Undefined values and undetermined values are completely distinct cases. Any argument which is based on one approach might not generalize to the other. Moreover, the restriction of a conditional logic to language $\mathcal{L}^{\text{cons}}_{\text{CL}^-}$ for $\mathbf{P}_\epsilon$ semantics such as in Adams (1975) is inessential, as our discussion of the alternative approach to $\mathbf{P}_\epsilon$ semantics by Schurz (1997b) in Section 3.6.3 shows. Even if Adams were right, one could still escape the partial extended value functionality by rendering the Bridge Principles CS and MP invalid. Moverover, as we saw in Section 3.4.2, there are independent reasons to reject both MP and CS for indicative conditionals.

3.7 D. Lewis' Triviality Result and Logics for Indicative Conditionals

In this section I will describe D. Lewis' (1976) triviality result and discuss its consequences for logics for indicative conditionals. In Section 3.7.1 I will start with a description of both triviality proofs of D. Lewis based on different versions of the Stalnaker thesis. In Section 3.7.2 I will, then, show that in a semantics such as Adams' **P** systems semantics (see Section 3.6), not only the admission of conjunctions of conditionals in the language leads to counter-intuitive consequences – as D. Lewis' triviality result shows – but also the admission of iterations (and nestings) of conditional formulas. In Section 3.7.3 I will address Bennett's argument (a) for his thesis NTV ("No Truth Value") from Section 3.1, that is that indicative conditionals are not propositions and cannot have truth values, where argument (a) is based on D. Lewis' (1976) triviality result. For that purpose I will discuss the implications of D. Lewis' triviality result for probabilistic conditional logic systems and investigate which bearing his result has on possible worlds semantics such as Lewis models (see Definitions 3.1–3.3) and CS semantics (see Chapters 4–7; see also Section 3.1). In that context I will heavily draw on the earlier sections of this chapter.

3.7.1 D. Lewis' Triviality Result

For his triviality proofs, D. Lewis (1976, p. 299) uses the following definition of probability functions:

Definition 3.38. *P is a probability function over $\mathcal{L}_{\mathrm{CL}}$ iff*
(a) P is a partial function from $\mathcal{L}_{\mathrm{CL}}$ to $\mathbb{R}$ so that
(b) for all $\alpha, \beta \in \mathcal{L}_{\mathrm{CL}}$ holds:
(i) $0 \leq P(\alpha) \leq 1$ and $P(\top) = 1$,
(ii) if $\models \alpha \rightarrow \beta$, then $P(\alpha) \leq P(\beta)$,
(iii) if $\models \neg(\alpha \wedge \beta)$, then $P(\alpha \vee \beta) = P(\alpha) + P(\beta)$, and
(iv) if $P(\alpha) > 0$, then $P(\beta | \alpha) = P(\alpha \wedge \beta) / P(\alpha)$.

P is only a partial function from $\mathcal{L}_{\mathrm{CL}}$ to $\mathbb{R}$ since since for $P(\alpha) = 0$ the probability $P(\alpha \mathbin{\Box\!\!\rightarrow} \beta)$ is undefined. Note that the postulates of Definition 3.38 hold also for the specifications of probability functions in the context of Adams' **P** systems semantics, that is Definitions 3.17 and 3.28,. Definition 3.38 differs from Definitions 3.17 and 3.28, insofar as Definition 3.38 employs the language $\mathcal{L}_{\mathrm{CL}}$ rather than the restricted languages $\mathcal{L}_{\mathrm{CL^-}}$ and $\mathcal{L}^{\mathrm{cons}}_{\mathrm{CL^-}}$, respectively. Furthermore, unlike in Definitions 3.17 and 3.28 in condition b.iv of Definition 3.38 no probabilities of conditional formulas (i.e., $\alpha \mathbin{\Box\!\!\rightarrow} \beta$) are defined except conditional probabilities (cf. Section 3.6.1).

Let us now focus on D. Lewis' (1976) triviality result. At the center of the D. Lewis' triviality proofs stands the Stalnaker thesis (cf. Stalnaker, 1970; see also Section 3.5.2). One can distinguish at least the following three versions of the Stalnaker thesis, where $P_\gamma(\cdot)$ and $P_\gamma(\cdot|\cdot)$ are the unconditional and conditional probability functions, respectively, which result from conditionalizing on a formula γ of language $\mathcal{L}_{CL}$ for $P(\gamma) > 0$, viz. the probability function that is determined by setting $P(\gamma)$ to the value one (cf. Section 3.3.7):[23]

Assumption 3.39. *(Unrestricted Stalnaker Thesis)*
For all probability functions P over language $\mathcal{L}_{CL}$ and for all $\alpha,\beta,\gamma \in \mathcal{L}_{CL}$ with $P(\gamma) > 0$ holds: $P_\gamma(\alpha \,\Box\!\!\rightarrow \beta) = P_\gamma(\beta|\alpha)$ if $P_\gamma(\alpha) > 0$.

Assumption 3.40. *(Very Restricted Stalnaker Thesis)*
There is a probability function P over language $\mathcal{L}_{CL}$ such that for all $\alpha, \beta \in \mathcal{L}_{CL}$ and some formula $\gamma \in \mathcal{L}_{CL}$ with $P(\gamma) > 0$ holds: $P_\gamma(\alpha \,\Box\!\!\rightarrow \beta) = P_\gamma(\beta|\alpha)$ if $P_\gamma(\alpha) > 0$.

Assumption 3.40*. *(Restricted Stalnaker Thesis)*
There is a probability function P over language $\mathcal{L}_{CL}$ such that for all $\alpha,\beta,\gamma \in \mathcal{L}_{CL}$ with $P(\gamma) > 0$ holds: $P_\gamma(\alpha \,\Box\!\!\rightarrow \beta) = P_\gamma(\beta|\alpha)$ if $P_\gamma(\alpha) > 0$.

Note in this context that P is also a probability function which results from P by conditionalization, since it holds that $P(\cdot) = P_\top(\cdot)$.

The above versions of the Stalnaker thesis define probabilities of conditionals by means of conditional probabilities. They do so to varying degrees of generality. While the unrestricted Stalnaker thesis (Assumption 3.39) equates the probability $P(\alpha \,\Box\!\!\rightarrow \beta)$ with $P(\beta|\alpha)$, given $P(\alpha) > 0$, for *all* probability functions, the restricted Stalnaker thesis (Assumption 3.40*) states only that there *exists* a probability function P such that for *all* probability functions P' which result from conditionalization of P by some $\mathcal{L}_{CL}$ formula

[23] The Stalnaker thesis is named after Stalnaker's proposal as in Stalnaker (1970; cf. D. Lewis, 1976, p. 308).

it is the case that $P'(\alpha \mathrel{\Box\!\!\rightarrow} \beta) = P'(\beta \mid \alpha)$, provided $P'(\alpha) > 0$. The very restricted Stalnaker thesis (Assumption 3.40) is still weaker than the restricted version, since it only claims that there *exists* a probability function P such that for *some* probability function P' holds, which results from conditionalization, $P'(\alpha \mathrel{\Box\!\!\rightarrow} \beta) = P'(\beta \mid \alpha)$ provided $P'(\alpha) > 0$.

D. Lewis (1976) shows that the unrestricted Stalnaker thesis leads to a triviality result for all probability functions (first triviality theorem). He proves in addition that any probability function for which the description in the restricted Stalnaker thesis (Assumption 3.40*) applies, must itself be trivial (second triviality theorem). Only the very restricted Stalnaker thesis does not lead to any sort of triviality result. It should, however, be clear that the very restricted Stalnaker thesis does not provide one with more than a toy version of a probabilistic conditional logic semantics (cf. Section 3.7.3).

Let me now describe the two triviality theorems of D. Lewis. For this purpose I will prove some lemmata first.

Lemma 3.41. *For any probability function P of $\mathcal{L}_{\mathrm{CL}}$ and any formulas α, $\beta \in \mathcal{L}_{\mathrm{CL}}$ such that $P(\alpha \wedge \beta), P(\alpha \wedge \neg\beta) > 0$, holds: $P(\alpha \mathrel{\Box\!\!\rightarrow} \beta) = P(\alpha \mathrel{\Box\!\!\rightarrow} \beta \mid \beta) \cdot P(\beta) + P(\alpha \mathrel{\Box\!\!\rightarrow} \beta \mid \neg\beta) \cdot P(\neg\beta)$.*

Proof.

1. $P(\alpha \mathrel{\Box\!\!\rightarrow} \beta) = P(((\alpha \mathrel{\Box\!\!\rightarrow} \beta) \wedge \beta) \vee ((\alpha \mathrel{\Box\!\!\rightarrow} \beta) \wedge \neg\beta))$ by Def 3.38.b.ii
2. $= P((\alpha \mathrel{\Box\!\!\rightarrow} \beta) \wedge \beta) + P((\alpha \mathrel{\Box\!\!\rightarrow} \beta) \wedge \neg\beta)$ by Def 3.38.b.iii
3. $= P(\alpha \mathrel{\Box\!\!\rightarrow} \beta \mid \beta) \cdot P(\beta) + P(\alpha \mathrel{\Box\!\!\rightarrow} \beta \mid \neg\beta) \cdot P(\neg\beta)$ by Def 3.38.b.iv since $P(\alpha \wedge \beta), P(\alpha \wedge \neg\beta) > 0$.

□

Observe that Lemma 3.41 explicitly draws on conjunctions involving conditional formulas.

Lemma 3.42. *(Conditionalization Lemma)*
Let P be a probability function over language $\mathcal{L}_{\mathrm{CL}}$ and let α be an arbitrary formula of $\mathcal{L}_{\mathrm{CL}}$. Then, there is a non-conditional probability function P_α over $\mathcal{L}_{\mathrm{CL}}$ such that for any formula $\beta \in \mathcal{L}_{\mathrm{CL}}$ it holds that $P_\alpha(\beta) = P(\beta \mid \alpha)$, provided $P(\alpha) > 0$.

Proof. Let P be a probability function over language $\mathcal{L}_{\mathrm{CL}}$ and let $\alpha \in \mathcal{L}_{\mathrm{CL}}$ be such that $P(\alpha) > 0$. Due to the latter fact, $P(\beta \mid \alpha)$ is defined by Definition 3.38 for all $\beta \in \mathcal{L}_{\mathrm{CL}}$. It is trivially the case for all formulas $\gamma, \delta \in \mathcal{L}_{\mathrm{CL}}$ and all probability assignments P' over $\mathcal{L}_{\mathrm{CL}}$ that $P'(\delta \mid \gamma)$ is a probability function according to Definition 3.38 provided $P'(\gamma) > 0$. Hence, the probabilities assigned by $P(\beta \mid \alpha)$ for all $\beta \in \mathcal{L}_{\mathrm{CL}}$ can be described by a non-conditional probability function P'' over $\mathcal{L}_{\mathrm{CL}}$. Thus, a non-conditional probability function P_α over $\mathcal{L}_{\mathrm{CL}}$ exists, that is P'', such that $P_\alpha(\beta) = P(\beta \mid \alpha)$ for all $\beta \in \mathcal{L}_{\mathrm{CL}}$.

□

Let us now prove the essential step in D. Lewis' triviality proof:

Lemma 3.43. *Let P be an arbitrary probability function over $\mathcal{L}_{\mathrm{CL}}$. Furthermore, suppose that either (i) the unrestricted Stalnaker thesis (Assumption 3.39) holds for probability functions over $\mathcal{L}_{\mathrm{CL}}$ or, alternatively, (ii) P is one of the probability functions which is described by the restricted Stalnaker thesis (Assumption 3.40*). Then, for any α, β, $\gamma \in \mathcal{L}_{\mathrm{CL}}$ it is the case that $P(\beta \mathrel{\Box\!\!\rightarrow} \gamma \mid \alpha) = P(\gamma \mid \alpha \wedge \beta)$ provided $P(\alpha \wedge \beta) > 0$.*

Proof. By assumption $P(\alpha \wedge \beta) > 0$ we can infer by Lemma 3.42 that there exists a probability function P_α such that $P(\beta \mathrel{\Box\!\!\rightarrow} \gamma \mid \alpha) = P_\alpha(\beta \mathrel{\Box\!\!\rightarrow} \gamma)$. Moreover, since $P(\alpha \wedge \beta) > 0$ implies $P(\alpha) > 0$, we get $P(\beta \mid \alpha) = P(\beta \wedge \alpha)/P(\alpha)$. As $P(\alpha \wedge \beta) > 0$ and $P(\alpha) > 0$, it follows that $P(\beta \wedge \alpha)/P(\alpha) > 0$ and, hence, $P(\beta \mid \alpha) > 0$. This implies by definition of P_α that $P_\alpha(\beta) > 0$. Thus $P_\alpha(\gamma \mid \beta)$ is defined as well. By (i) and by (ii) we obtain $P_\alpha(\beta \mathrel{\Box\!\!\rightarrow} \gamma) = P_\alpha(\gamma \mid \beta)$. Thus, $P(\beta \mathrel{\Box\!\!\rightarrow} \gamma \mid \alpha) = P_\alpha(\gamma \mid \beta)$.

Let us now show that $P_\alpha(\gamma \mid \beta) = P(\gamma \mid \alpha \wedge \beta)$ for $P(\alpha \wedge \beta) > 0$. Due to $P_\alpha(\beta) > 0$ and Definition 3.38.b.iv we have $P_\alpha(\gamma \mid \beta) = P_\alpha(\gamma \wedge \beta)/P_\alpha(\beta)$. As $P(\alpha) > 0$, we obtain by the definition of P_α that $P_\alpha(\gamma \wedge \beta) = P(\gamma \wedge \beta \mid \alpha)$ and $P_\alpha(\beta) = P(\beta \mid \alpha)$. Hence, it follows that $P_\alpha(\gamma \mid \beta) = P(\gamma \wedge \beta \mid \alpha)/P(\beta \mid \alpha)$. Since $P(\alpha) > 0$, on the basis of Definition 3.38.b.iv the latter term is $(P(\gamma \wedge \beta \wedge \alpha) \cdot P(\alpha))/((P(\alpha) \cdot P(\beta \wedge \alpha))$ which equals $P(\gamma \wedge \beta \wedge \alpha)/P(\beta \wedge \alpha)$. Hence, $P_\alpha(\gamma \mid \beta) = P(\gamma \wedge \beta \wedge \alpha)/P(\beta \wedge \alpha)$. Since $P(\alpha \wedge \beta) > 0$, by points 3.38.b.ii and 3.38.b.iv it follows that $P_\alpha(\gamma \mid \beta) = P(\gamma \mid \alpha \wedge \beta)$ and thus we obtain $P(\beta \mathrel{\Box\!\!\rightarrow}$

$\gamma|\alpha) = P(\gamma|\alpha \wedge \beta)$.

□

Observe that conditions (i) and (ii) in Lemma 3.43 draw on the unrestricted and the restricted Stalnaker assumption, respectively.

Let us finally prove both versions of D. Lewis' triviality result (Theorem 3.44) where Theorem 3.44 draws on Lemmata 3.41 and 3.43.

Theorem 3.44. *(Lewis' Triviality Result)*
Let P be a probability function over $\mathcal{L}_{\mathrm{CL}}$. Furthermore, suppose that either (i) the unrestricted Stalnaker thesis (Assumption 3.39) holds for probability functions over $\mathcal{L}_{\mathrm{CL}}$ or, alternatively, (ii) P is one of the probability functions which is described by the restricted Stalnaker thesis (Assumption 3.40). Then, for all formulas $\alpha, \beta \in \mathcal{L}_{\mathrm{CL}}$ such that $P(\alpha \wedge \beta), P(\alpha \wedge \neg\beta) > 0$ it is the case that $P(\alpha \mathrel{\Box\!\!\rightarrow} \beta) = P(\beta)$.*

Proof. By assumption it holds that $P(\alpha \wedge \beta) > 0$ and $P(\alpha \wedge \neg\beta) > 0$. Thus, by Lemma 3.41 we obtain $P(\alpha \mathrel{\Box\!\!\rightarrow} \beta) = P(\alpha \mathrel{\Box\!\!\rightarrow} \beta|\beta) \cdot P(\beta) + P(\alpha \mathrel{\Box\!\!\rightarrow} \beta|\neg\beta) \cdot P(\neg\beta)$. Due to (i) or (ii) it follows by Lemma 3.43 that $P(\alpha \mathrel{\Box\!\!\rightarrow} \beta|\beta) = P(\beta|\alpha \wedge \beta)$ and that $P(\alpha \mathrel{\Box\!\!\rightarrow} \beta|\neg\beta) = P(\beta|\alpha \wedge \neg\beta)$. Hence, (a) $P(\alpha \mathrel{\Box\!\!\rightarrow} \beta) = P(\beta|\alpha \wedge \beta) \cdot P(\beta) + P(\beta|\alpha \wedge \neg\beta) \cdot P(\neg\beta)$.

By assumption $P(\alpha \wedge \beta) > 0$ we obtain $P(\beta|\alpha \wedge \beta) = P(\beta \wedge \alpha \wedge \beta)/P(\alpha \wedge \beta)$ based on Definition 3.38.b.vi. Hence, $P(\beta \wedge \alpha \wedge \beta)/P(\alpha \wedge \beta) = 1$ and thus, $P(\beta|\alpha \wedge \beta) = 1$. Moreover, as $P(\alpha \wedge \neg\beta) > 0$, by Definition 3.38.b.iv we get $P(\beta|\alpha \wedge \neg\beta) = P(\beta \wedge \alpha \wedge \neg\beta)/P(\alpha \wedge \neg\beta)$. It follows that $P(\beta \wedge \alpha \wedge \neg\beta) = 0$. Thus, we obtain $P(\beta|\alpha \wedge \neg\beta) = 0$. As $P(\beta|\alpha \wedge \beta) = 1$ and $P(\beta|\alpha \wedge \neg\beta) = 0$ we get by (a) that $P(\alpha \mathrel{\Box\!\!\rightarrow} \beta) = P(\beta|\alpha \wedge \beta) \cdot P(\beta) + P(\beta|\alpha \wedge \neg\beta) \cdot P(\neg\beta) = 1 \cdot P(\beta) + 0 \cdot P(\neg\beta) = P(\beta)$. □

D. Lewis' (1976) first triviality theorem is described by condition (i) in Theorem 3.44 and Lewis' second triviality theorem is specified by condition (ii) in the same theorem.

3.7.2 Triviality due to Iterations (and Nestings) of Conditional Formulas

The aim of the present section is to show that not only the admission of conjunctions of conditional formulas in the language leads to counter-intuitive consequences for a probabilistic semantics – as we saw in the previous section – but also the admission of iterated conditional formulas (e.g., $\alpha \,\Box\!\!\rightarrow (\beta \,\Box\!\!\rightarrow \gamma)$). For that purpose I will demonstrate that given fairly uncontroversial assumptions – as provided in Adams' **P** systems semantics (see Sections 3.6.1 and 3.6.3) – a probabilistic semantics collapses with the extensional semantics of the material implication, insofar as it licenses for (indicative) conditionals all inferences of the latter sort. Observe that since iterated conditionals are a special case of nested conditionals (e.g., $\alpha \,\Box\!\!\rightarrow (\beta \wedge (\gamma \,\Box\!\!\rightarrow \delta))$; see Section 4.2.1) the present result also applies to languages which allow for the latter type of formulas.

To prove the present result (i) I assume that the unrestricted Stalnaker thesis (Definition 3.39) holds – also given in Definitions 3.17 and 3.28 for Adams' **P** systems semantics – with the modification that the languages for which these probability functions are defined allow also for iterated conditionals. Furthermore, (ii) I presuppose that as in Adams' Systems $\mathbf{P}^*$ and $\mathbf{P}_\epsilon$ semantics (see Sections 3.6.1 and 3.6.3) the following (probabilistic) standard principles are valid, where MP* is a rule version of principle MP (see Section 3.4.2): [24]

Refl $\alpha \,\Box\!\!\rightarrow \alpha$
RW if $\vdash \alpha \rightarrow \beta$ and $\gamma \,\Box\!\!\rightarrow \alpha$ then $\gamma \,\Box\!\!\rightarrow \beta$
MP* if $\alpha \,\Box\!\!\rightarrow \beta$ then $\alpha \rightarrow \beta$

Principles RW and Refl are considered cornerstones of the great majority of standard conditional logic semantics in the literature, including Adams' $\mathbf{P}^*$ and $\mathbf{P}_\epsilon$ semantics (see Sections 3.6.1 and 3.6.3) and the probabilistic thresh-

[24] I use Principle MP* rather than Principle MP, since MP* is a formula of the language used in this section whereas MP is not.

old semantics of Hawthorne and Makinson (2007).[25]

Let us now consider the following two inference schemata which involve iterated conditional formulas:

Ex* if $\alpha \wedge \beta \boxright \gamma$ then $\alpha \boxright (\beta \boxright \gamma)$
Im* $\alpha \boxright (\beta \boxright \gamma)$ then $\alpha \wedge \beta \boxright \gamma$

The expressions 'Ex' and 'Im' abbreviate 'Exportation' and 'Importation', respectively. The asterices indicate that the respective principles are described as rules rather than in terms of axiom schemata. (See Table 5.2 for axiom schemata version of both principles.)[26]

Suppose that the unrestricted Stalnaker thesis holds. Then, by Lemma 3.42 it is the case that $P(\beta \boxright \gamma|\alpha) = P(\gamma|\alpha \wedge \beta)$ provided $P(\alpha \wedge \beta) > 0$. Since we explicitly allow for iterations of conditionals, it follows that $P(\beta \boxright \gamma|\alpha) = P(\alpha \boxright (\beta \boxright \gamma))$ and $P(\gamma|\alpha \wedge \beta) = P(\alpha \wedge \beta \boxright \gamma)$. Hence, for $P(\alpha \wedge \beta) > 0$ we obtain $P(\alpha \boxright (\beta \boxright \gamma)) = P(\alpha \wedge \beta \boxright \gamma)$. Furthermore, given that $P(\alpha \boxright (\beta \boxright \gamma)) = P(\alpha \wedge \beta \boxright \gamma)$ holds for $P(\alpha \wedge \beta) > 0$, this implies that Ex* and Im* are valid according to Adams' validity criteria discussed in Section 3.6.

Moreover, RW, Refl, MP*, and Ex* imply the following equivalence, where $\rightarrow$ describes the material implication:

Triv′ $\alpha \boxright \beta$ iff $\alpha \rightarrow \beta$.

So, the admission of iterated conditionals in one's language trivializes Adams' probabilistic validity criteria.

Let me now prove the latter result, where Triv′$^{\Rightarrow}$ and Triv′$^{\Leftarrow}$ describe the left-to-right and right-to-left direction of Triv′, respectively.

[25] Principles RW and Refl are also valid in Stalnaker models (Definition 3.7), Lewis models (Definitions 3.1–3.3), and the ordering semantics of Kraus et al. (1990) and Lehmann and Magidor (1992; see Section 3.2). Moreover, MP* is also valid in a range of possible worlds semantics for conditionals, such as Stalnaker models and Lewis models with centering conditions.

[26] I use here rule versions of both principles since otherwise they cannot be formulated in our language.

Theorem 3.45. *RW+Refl+MP*+Ex ⇒ Triv′*

Proof. MP is nothing but Triv′$^{\Rightarrow}$. Moreover, Lemma 3.46 gives us Triv′$^{\Leftarrow}$ on the basis of RW, Refl, MP*, and Ex*. □

Lemma 3.46. *RW+Refl+MP*+Ex* ⇒ Triv′$^{\Rightarrow}$.*

Proof.

1.	$\alpha \rightarrow \beta$	given
2.	$(\alpha \rightarrow \beta) \wedge \alpha \mathrel{\Box\!\!\rightarrow} (\alpha \rightarrow \beta) \wedge \alpha$	Refl
3.	$(\alpha \rightarrow \beta) \wedge \alpha \mathrel{\Box\!\!\rightarrow} \beta$	2, RW
4.	$(\alpha \rightarrow \beta) \mathrel{\Box\!\!\rightarrow} (\alpha \mathrel{\Box\!\!\rightarrow} \beta)$	3, Ex*
5.	$\alpha \mathrel{\Box\!\!\rightarrow} \beta$	4, 1, MP*

□

Observe that the proof of Lemma 3.46 is analogous to a proof by McGee (1985, p. 466).

In sum, Adams' **P** systems semantics, as endorsed in Systems **P*** and $\mathbf{P}_{\epsilon}$, becomes trivial if one allows for iterated conditional formulas in one's language. Since it is the main objective of conditional logics that they provide an alternative and less counter-intuitive analysis of (indicative) conditionals compared to the material implication analysis (see Chapter 1), such a result is highly undesirable.

3.7.3 D. Lewis' Triviality Result, Restrictions of Languages, and Truth Value Accounts of Indicative Conditionals

In this section I will first discuss (i) D. Lewis' (1976) triviality theorems and my result from the previous section regarding iterated conditional formulas for probabilistic semantics. Then, (ii) I will criticize Bennett's (2003) argument for NTV based on D. Lewis' (1976) triviality result (argument (a) from Section 3.1), in particular for possible worlds semantics for (indicative) conditionals. For that purpose I will heavily draw on the earlier sections of this chapter.

Let me focus on point (i). D. Lewis' (1976) first triviality result implies that any probabilistic conditional logic semantics which endorses Definition

3.38 and accepts the unrestricted Stalnaker thesis runs into the triviality result, viz. that for $P(\alpha) > 0$, it holds that $P(\alpha \mathrel{\Box\!\!\rightarrow} \beta) = P(\beta)$. Furthermore, D. Lewis' (1976) second triviality result shows that any probability function P based on Definition 3.38 and for which the restricted Stalnaker thesis holds, faces the same problem. Only when the very restricted Stalnaker thesis is employed, does D. Lewis' (1976) triviality result ensue. However, the very restricted Stalnaker thesis, that is Definition 3.40, is too weak, insofar as it only makes sure that for a given probability function P and some probability function P_γ such that $P(\gamma) > 0$ and $P_\gamma(\alpha) > 0$ holds that $P_\gamma(\alpha \mathrel{\Box\!\!\rightarrow} \beta) = P_\gamma(\beta|\alpha)$. So, the very weak Stalnaker thesis does not imply that for a given probability function the probability of conditionals is specified in a systematic manner. The latter fact, can, however, be regarded as a minimum requirement for any adequate probabilistic conditional logic semantics (cf. Section 3.7.1).

Let us now apply D. Lewis' (1976) triviality result to Adams' **P** systems semantics discussed in Section 3.6. D. Lewis' (1976) triviality result implies that, given one accepts (a) the unrestricted Stalnaker thesis – as given Adams' definitions of probability functions in Definitions 3.17 and 3.28 – or (b) the restricted Stalnaker thesis, one cannot allow for arbitrary Boolean combinations involving conditional formulas. Since Adams' **P** systems semantics endorse a definition of probability semantics in line with (strong) Stalnaker thesis (Definition 3.39), these semantics would be prone to D. Lewis' triviality result if one were to extend the language of these semantics to the full language $\mathcal{L}_{\mathrm{CL}}$. The probabilistic Conditional Logic Systems $\mathbf{P}, \mathbf{P}^*, \mathbf{P}^+$, and $\mathbf{P}_\epsilon$ escape the triviality result by using the restricted languages $\mathcal{L}_{\mathrm{rrCL}}, \mathcal{L}_{\mathrm{CL}^-}, \mathcal{L}_{\mathrm{rCL}}$, and $\mathcal{L}^{\mathrm{cons}}_{\mathrm{CL}^-}$, respectively, rather than the full language $\mathcal{L}_{\mathrm{CL}}$ (see Sections 3.6.1 and 3.6.3). Languages $\mathcal{L}_{\mathrm{rrCL}}, \mathcal{L}_{\mathrm{CL}^-}, \mathcal{L}_{\mathrm{rCL}}$, and $\mathcal{L}^{\mathrm{cons}}_{\mathrm{CL}^-}$ have in common that they do not allow for conjunctions involving conditional formulas. Note that it is exactly this feature that blocks D. Lewis' (1976) triviality theorems (D. Lewis, 1976, p. 304). This is due to the fact that Theorem 3.41 cannot be formulated in one's probabilistic language since this theorem explicitly refers to conjunctions involving conditional formulas.

Furthermore, the discussion in Section 3.7.2 shows that not just Boolean combinations involving conditional formulas are problematic, but also nestings and iterations of conditional formulas, given one applies in addition validity criteria as in Adams's **P** systems semantics. Interestingly, Adams avoids also the latter type of problem, insofar he neither allows for nestings nor for iterations of conditional formulas.

Let me now discuss point (ii), namely Bennett's argument that D. Lewis' (1976) triviality result shows that his thesis NTV ("No Truth Value") is true, that is that indicative conditionals are not propositions and cannot have truth values. Bennett (2003) argues that D. Lewis' (1976) triviality result counts against truth value approaches to indicative conditionals, since his NTV thesis has the "unique power to protect the Equation [the (strong) Stalnaker thesis] from the 'triviality' proofs of [D.] Lewis and others" (Bennett, 2003, p. 103).

There are two main points which speak against Bennett's argument for his NTV thesis. First, as we saw earlier in this section, D. Lewis' (1976) triviality result is formulated in a probabilistic framework and is, therefore, first and foremost a problem of probabilistic approaches to indicative conditionals. Furthermore, we noted in Section 3.6.4 that one *might* argue that possible world semantics for (indicative) conditionals can be accounted for in a probabilistic framework and that for that reason a possible worlds semantics for (indicative) conditionals is also prone to D. Lewis' (1976) triviality result. However, as we saw in Section 3.6.4, possible worlds semantics for (indicative) conditionals, such as CS semantics (see Chapters 4–7) and Lewis models (Definitions 3.1–3.3), cannot adequately be described in a probabilistic framework. Hence, D. Lewis' (1976) triviality proof has no direct bearing on possible worlds semantics for (indicative) conditionals. Moreover, we observed in Section 3.6.4 that it is not so much the notion of truth that is problematic but rather in line with Schurz the notion of probability combined with the (strong) Stalnaker thesis. Note that Schurz (1997b, 1998, 2005b) escapes D. Lewis' (1976) triviality result since he only associates conditionals with their conditional probabilities (see Section 3.5.2). In my view it seems sensible to give up the (strong) Stalnaker thesis rather than to restrict one's language (cf. Section 3.5.5). One reason Bennett (2003) might still prefer to stick with the (strong) Stalnaker thesis is that he thinks that it is essential for a Ramsey test interpretation (of a probabilistic semantics) (cf. Section 3.5.2). As we saw in Section 3.5.2 this does not seem to be true.

Second, it is not a specific property of probabilistic systems that they allow for restricted languages. As we saw in Section 3.5.5 truth value semantics for indicative conditionals also exist, such as in Kraus et al. (1990) and Lehmann and Magidor (1992), which use a restricted conditional logic language (see Section 3.5.3). Furthermore, we saw in Sections 3.5.3 and 3.5.4 that D. Lewis' (1976) triviality proof also seems to be the main motivation for the restriction of conditional logic languages for probabilistic semantics. Although Adams (1965) argues for a restriction of one's probabilistic language on different grounds, he does so in terms of a (subjective) probabilistic framework (see Section 3.5.4) which – as we saw earlier – does not have a direct bearing in possible worlds semantics for (indicative) conditionals.

3.7.4 Conclusion

The present discussion shows that there seems to be no reason to assume that D. Lewis' (1976) triviality result counts against truth value approaches, such as CS semantics (Chellas, 1975; Segerberg, 1989). On the contrary, D. Lewis' (1976) triviality result seems to be a specific problem of probabilistic semantics for indicative conditional logics.

3.8 Bennett's Gibbardian Stand-Off Argument

Bennett (2003, Chapters 6 and 7) provides an extended argument against (i) truth value approaches and (ii) objective probabilistic approaches to indicative conditionals based on an argument by Gibbard (1980, p. 231f), the so-called Gibbardian stand-off argument (Bennett's argument (b) for NTV from Section 3.1). In Section 3.8.1 I will describe Bennett's version and extension of Gibbard's stand-off argument. In Section 3.8.2 I will, then, inquire in which way Bennett's Gibbardian stand-off argument is decisive against approaches of type (i) including possible worlds semantics for indicative conditionals such as CS semantics (see Chapters 4–7; see also Section 3.1) and approaches of sort (ii). I will thereby heavily draw on earlier sections of this chapter.

3.8.1 Bennett's Extended Gibbardian Stand-Off Argument

In this section I will first describe Bennett's extended Gibbardian stand-off argument. Let me first describe Gibbardian stand-off situations. Gibbardian stand-offs are situations that support belief in pairs of indicative conditionals of the following form:

(a) $\alpha \mathrel{\Box\!\!\rightarrow} \beta$
(b) $\alpha \mathrel{\Box\!\!\rightarrow} \neg\beta$

The situation that gives rise to a Gibbardian stand-off makes sure that the belief in (a) and the belief in (b) are equally well justified and that both beliefs are justified rather than not. Since Bennett (2003, p. 84) assumes that (a) and (b) cannot be both true at the same time, (i) truth value approaches and (ii) objective probabilistic accounts of Gibbardian stand-offs seem to have a problem, as we will see below.

Let me at this point describe Bennett's (2003) Gibbardian stand-off case and then discuss in which way Gibbardian stand-offs might threaten both approaches (i) and (ii). Bennett's Gibbardian stand-off case goes as follows:

> "Top Gate holds back water in a lake behind a dam; a channel running down from it splits into two distributaries, one (blockable by East Gate) running eastwards and the other (blockable by West Gate) running westwards. The gates are connected as follows: if east lever is down, opening Top Gate will open East Gate so that the water will run eastwards; and if west lever is down, opening Top Gate will open West Gate so that the water will run westwards. On the rare occasions when both levers are down, Top Gate cannot be opened because the machinery cannot move three Gates at once.
>
> Just after the lever-pulling specialist has stopped work, Wesla knows that west lever is down, and thinks 'If Top Gates opens, all the water will run west-

wards'; Esther knows that east lever is down, and thinks 'If Top Gate opens, all the water will run eastwards'." (Bennett, 2003, p. 85)

The following two indicative conditionals are the conditionals of sort (a) and (b) in Bennett's Gibbardian stand-off case:

E29 If Top Gates opens, then the water will only run westwards.
E30 If Top Gate opens, then the water will only run eastwards.

The word 'only' is essential here for the formulation of E29 and E30, for otherwise the consequents E29 and E30 would not contradict each other (assuming that westwards and eastwards are different directions).[27]

For indicative conditionals of sort E29 and E30, then, only the following four possible cases exist (Bennett, 2003, p. 94), where '1' and '0' either stand for 'true' and 'false' or 'supported' and 'unsupported', respectively:

	E29	E30
(1)	1	1
(2)	1	0
(3)	0	1
(4)	0	0

Let me now explain how Gibbardian stand-offs threaten approaches (i) and (ii). As we saw earlier, it is one of the core assumptions of Gibbardian stand-offs that pairs of indicative conditionals, such as E29 and E30, are equally well supported by the Gibbardian stand-off situation. Therefore, it cannot be the case that either one is believed and the other is not. Thus, cases (2) and (3) are excluded. It is, moreover, assumed for the Gibbardian stand-off that

[27] There seems to be a slight problem with the assumption that Gibbardian stand-offs are an empirically significant phenomenon (cf. Bennett, 2003, p. 86f) due to the assumption that the Gibbardian stand-off situations support beliefs in conditionals of sort E29 and E30 equally well. As Bennett (2003, p. 85) admits, Gibbard's (1980) original stand-off case does not in fact support beliefs in conditionals of sort E29 and E30 in a perfectly analogous way. Furthermore, Bennett's own example is quite artificial. These facts, hence, rather count against the empirical significance of Gibbardian stand-offs.

conditionals of sort E29 and E30 are supported rather than not and, therefore, case (4) is ruled out as well. Finally, Bennett (2003) assumes – as we saw earlier – that conditionals of sort E29 and E30 cannot both be true either. Hence, since all possible truth assignments to both conditionals are excluded, truth value approaches to conditionals fail and due to the above argument conditionals cannot in general have truth values. Moreover, on perfectly analogous grounds for objective probabilistic semantics neither cases (1)–(4) are possible in terms of (evidential) support of E29 and E30. So, approach (ii) seems to fail as well.

Bennett (2003), then, extends Gibbard's (1980) stand-off argument insofar as he aims to show why a subjective probabilistic approach escapes the Gibbardian stand-off argument while subjective truth value accounts (approach (i)) and objective probabilistic accounts (approach (ii)) do not. According to Bennett's terminology, objective approaches in general take the whole (truth of the) situation into account (as I did above), while subjective approaches only consider an agent's beliefs and most typically do *not* use the whole truth (see Bennett, 2003, p. 80f).[28] Bennett divides his argument, then, into two parts. In the first part he argues against objective approaches in general (approach (ii) above in particular, see pp. 80–88), in the second part he addresses subjective truth value accounts (approach (i), see pp. 88–93).

Before I continue with Bennett's (2003) extension of Gibbard's (Gibbard, 1980) stand-off argument, let me make two points here. First, the exclusion of case (4) is actually based on nothing but an application of Principle CEM″, a variant of the Principle CEM ("Conditional Excluded Middle") from Section 3.3.6. Second, since Bennett (2003) explicitly excludes case (1) on principle grounds, he effectively uses Principle CNC (see also Bennett, 2003, p. 84; cf. Sections 3.3.6 and 3.6.3). Let me repeat both principles.

[28] It is a matter of debate whether an objective approach in Bennett's terminology is applicable at all, since it is not clear what it means to take the whole truth into account concerning any state of affairs.

CEM″ $\neg(\neg(\alpha \,\Box\!\!\rightarrow \beta) \wedge \neg(\alpha \,\Box\!\!\rightarrow \neg\beta))$
CNC $\neg((\alpha \,\Box\!\!\rightarrow \beta) \wedge (\alpha \,\Box\!\!\rightarrow \neg\beta))$

The Gibbardian stand-off argument is constructed in such a way that Bennett (2003, p. 84) only needs to presuppose CNC, since (4) is already given by the Gibbardian stand-off situation. Note that CNC represents nothing but the strong conditional consistency criterion discussed in Sections 3.1 and 3.3.6.

Let me now come back to Bennett's extension of Gibbard's argumentation, focus on approach (ii) and consider for that purpose Wesla and Esther's beliefs. If one takes all information into account – as would be required in order to have an objective belief in Bennett's sense – one would be justified to endorse both conditionals E29 and E30, which would, then, contradict CNC. Since Bennett (2003, pp. 83–88) presupposes that CNC holds unversally, he concludes that an objective approach has to fail for Gibbardian stand-offs.

I shall now describe Bennett's line of reasoning against approach (i). According to Bennett's argument a subjective truth value account for indicative conditionals runs into the following problem: It has to relativize its truth value to a belief system, otherwise it is not clear to which subjective position it refers to (p. 88f). Bennett (2003, pp. 89–91) argues that one can either use (A) a self-descriptive expression (i.e., indicating that it is "my belief system") or alternatively (B) a fixed-reference term (i.e., Wesla's belief system). The subjective truth value approach (i) fails, since (indicative) conditionals in natural language are most typically not used that way: In a majority of cases, assertions of conditionals do not serve as reports on the agent's beliefs (which the self-description approach predicts). Instead they are often used as claims that the conditionals in fact hold (Bennett, 2003, p. 90). (Note that Bennett phrases his argumentation here in terms of probabilities rather than truth.)

According to Bennett (2003) approach (B) fails for case (i) as well. In that approach any conditional Wesla asserts has a fixed frame of reference which is Wesla's belief system. If someone asks 'If Top Gate opens, where does the water go?', Wesla and Esther would answer 'west' and 'east', respectively. Both only refer to their own fixed belief system. S, if this were the case, then, communication between Esther and Wesla would be impossible, since Wesla and Esther always refer to distinct belief systems when they assert conditionals of sort (a) and (b), respectively (Bennett, 2003, p. 91).

So according to Bennett, in both cases (A) and (B) we cannot pick out the same belief system for Wesla and Esther, since this would result again in a belief system that would contain E29 and E30 and that would therefore contradict CNC. The subjective approach, however, escapes this difficulty, since Wesla's and Esther's belief systems are considered separately and, thus E29 and E30 might be believed (or accepted), but not both.

3.8.2 A Criticism of Bennett's Gibbardian Stand-Off Argument

Let me critically review Bennett's (2003) Gibbardian stand-off argument in this section and thereby also defend possible worlds semantics for indicative conditionals. I shall for that purpose (I) focus on Bennett's argument against truth value approaches (including possible worlds semantics), (II) discuss whether his argument has implications for objective frequency-based approaches described in Section 3.5.1 and, finally, (III) inquire whether the Gibbardian stand-off argument has also some force against subjective probabilistic approaches, such as in Adams' (1975) System $\mathbf{P}_\epsilon$ semantics, which Bennett explicitly endorses (cf. Bennett, 2003, pp. 129).

Let me now focus on (I). Bennett presupposes throughout his argument that CNC, that is $\neg((\alpha \mathrel{\Box\!\!\rightarrow} \beta) \wedge (\alpha \mathrel{\Box\!\!\rightarrow} \neg\beta))$, is universally valid with the only exception of classical validity – validity in p.c. and f.o.l. semantics where in CNC $\Box\!\!\rightarrow$ is replaced by the material implication $\rightarrow$. Bennett argues in particular that CNC is valid for possible world semantics for (indicative) conditionals and draws on Davis (1979). According to Bennett (2003, p. 84) possible worlds semantics for conditionals tie the truth of a conditional $\alpha \mathrel{\Box\!\!\rightarrow} \beta$ to the truth of β at a certain world. CNC would then be valid in possible worlds semantics for (indicative) conditionals, since β and $\neg\beta$ cannot both be true at a single possible world (p. 84). The problem with Bennett's argument is that it is based on a false premise: In his account of indicative conditionals

Davis (1979, p. 544) explicitly refers to Stalnaker (1968) semantics (see Section 3.3.3). As we saw in Section 3.3.6, Stalnaker's (1968) semantics does not make a strong conditional consistency criterion such as CNC valid, although he argues for such a criterion on the basis of the Ramsey test (see Section 3.3.6). It can be seen that CNC is not valid in Stalnaker's conditional logic system from the fact that Stalnaker uses an absurd world λ also for his semantics, at which all formulas are true, including both β and $\neg\beta$ (see Definition 3.8).

In my view it is quite surprising that Bennett (2003, p. 84) and also Gibbard (1980, p. 231) assume that CNC is a universally valid principle in the first place. In the great majority of standard conditional logic semantics as described in this book, neither CNC nor its restricted version RCNC is valid (see Sections 3.3.6 and 3.6.3). This includes Lewis' (1973b) semantics (see Section 3.2), Stalnaker's (1968) semantics (see Section 3.3.3), set selection semantics (see Section 3.3.4), probabilistic threshold semantics of Hawthorne and Makinson (2007; see Section 3.6.2), and the probabilistic semantics of Systems $\mathbf{P}^*$ (Adams, 1965, 1966, 1977) and $\mathbf{P}^+$ (Adams, 1986; Schurz, 1998; see Section 3.6.1). Also the great majority of systems described by CS semantics in Chapters 4–7 makes neither Principle CNC nor RCNC valid. In fact, a range of the above semantics cannot account for Principle CNC, since in these semantics Principle Refl is valid which contradicts CNC given minimal assumptions (see Section 3.6.3). That leaves the Principle RCNC, which restricts Principle CNC to cases in which the antecedent is consistent and thereby allows one to add this principle to conditional logic systems conjointly with Refl, without making these systems inconsistent (see Section 3.6.3). The semantics of Adams' (1975) System $\mathbf{P}_\epsilon$ and Schurz's (1997b) modified version are the only conditional logic semantics reviewed in this chapter which make RCNC valid (see Section 3.6.3). Both approaches, however, deviate from standard approaches to conditionals due to Principle RCNC insofar as they are default logics in the sense of Section 2.2.

Let me now focus on point (II), that is let us inquire into Bennett's (2003) argument against objective approaches to indicative conditionals as described in Section 3.5.1. I will first review some specific points he makes against objective probabilistic approaches and then describe whether his line of reasoning also applies to the objective frequency-based approach described earlier.

To argue against objective probabilistic approaches, Bennett uses an objective Ramsey test interpretation (pp. 78–80). This Ramsey test interpretation takes the whole truth into account but is otherwise analogous to Bennett's subjective version of the Ramsey test (see Section 3.3.7). When one applies Bennett's objective Ramsey test – which makes a consistency adjustment if necessary (Bennett, 2003, p. 80) – both conditional probabilities $P(\beta|\alpha)$ and $P(\neg\beta|\alpha)$ cannot be greater than .5 (cf. Definition 3.17). Therefore, both corresponding indicative conditionals $\alpha \mathrel{\Box\!\!\rightarrow} \beta$ and $\alpha \mathrel{\Box\!\!\rightarrow} \neg\beta$ cannot be rationally accepted by a single agent at a single occasion. Hence, CNC would hold in an objective probabilistic semantics (Bennett, 2003, p. 84) and Bennett's Gibbardian stand-off argument would go through.

Let me now apply the latter argument to the objective frequency-based approach of Schurz (1997b, 2001b, 2005b) I described in Section 3.5.1. In Schurz's (1997b, 2001b, 2005b) approach, conditional probabilities are not interpreted in terms of the Ramsey test and do not lend themselves to a Ramsey test interpretation, not even the objective version of Bennett. Furthermore, this approach needs not draw on a version of the Ramsey test, since this semantics is already rationally justified by its reference to frequencies (see Section 3.5.1). Hence, Bennett's latter argument is not applicable for this type of semantics.

Let me finally inquire into point (III), viz. which implications Bennett's Gibbardian stand-off argument has for subjective probabilistic approaches. Observe that Bennett does not exempt subjective probabilistic interpretations of conditionals from the list of systems for which CNC is valid (Bennett, 2003, p. 84). Bennett (2003) even includes a type of consistency criterion in his Ramsey test interpretation (see Section 3.3.6). Furthermore, Bennett sees Adams' (1975) system as a plausible candidate for an appropriate subjective probabilistic account of conditionals (cf. Bennett, 2003, p. 129). Since the

Principle RCNC is valid in Adams' (1975; see also Schurz, 1997b) System $\mathbf{P}_\epsilon$ (see Section 3.6.3), in that framework any set of formulas containing conditionals of the form (a) and (b) is inconsistent, where the antecedents of (a) and (b) are themselves p.c. consistent.[29] Since System $\mathbf{P}_\epsilon$ is specifically interpreted in a subjective way (see Sections 3.5.1 and 3.6.3), it is possible to take only an agent's belief state into account and *not* the whole truth. This way subjective approaches can obey both the Principles CNC or RCNC in Gibbardian stand-offs.

Such a subjective probabilistic approach falls, however prey to the same criticism as the subjective truth value approaches in the previous section. In order to succeed at communication, the subjective probabilistic approach also has to refer to the reasoner's belief system. Otherwise it is not clear whose beliefs it refers to. It cannot refer to the whole truth, since otherwise it would contradict CNC and RCNC. Analogously to part (B) of Bennett's argument described above, the self-reference can be done via self-description (approach (A); i.e., indicating that it is my belief system) or a fixed reference account (approach (B); i.e., Wesla's belief system). Both, however, run into the same problem as the subjective truth value approach in the previous section.

3.8.3 Summary

In this section I discussed Bennett's (2003) Gibbardian stand-off argument against objective probabilistic semantics and truth value based semantics for indicative conditionals. In order for Bennett's argument to go through, he has to refer to the Principle CNC or at least some restricted version of it, such as RCNC. Unfortunately, neither of these principles is valid for the majority of (indicative) standard conditional logic semantics reviewed here, including Lewis' (1973b) semantics (Section 3.2), Stalnaker's (1968) semantics (see Section 3.3.3), set selection semantics (see Section 3.3.4), the probabilistic

[29] A set of formulas is *p*-consistent in Adams (1975) if one can assign arbitrarily high probabilities smaller than one to all members of that set such that the probability assignment is proper (Adams 1975, p. 51; see Section 3.6.3).

threshold semantics of Hawthorne and Makinson (2007; see Section 3.6.2), and the semantics of Systems **P** (Adams, 1965, 1966, 1977) and $\mathbf{P}^{+}$ (Adams, 1986; Schurz, 1998). In addition the majority of CS semantics investigated in Chapters 4–7 do not render either principle valid. Finally, Bennett's Gibbardian stand-off argument does not address the objective frequency-based conditional logic approach of Schurz (2001b, 2005b). Interestingly the only conditional logic systems reviewed here for which the argument seems to have some force, are the subjective probabilistic approach of Adams (1975), which also Bennett (2003) endorses, and Schurz' (1997b) account.

3.9 Conclusion

In this chapter I (i) gave an overview of standard possible worlds semantics for conditionals, (ii) discussed Ramsey test interpretations of conditionals, (iii) inquired into the difference between indicative and counterfactual conditionals, and (iv) described probabilistic conditional logics and some of their fundamental assumptions and problems. (v) Last but not least I defended truth value accounts for indicative conditionals such as the CS semantics. For that purpose I argued that there are no principal grounds to assume that conditionals were no propositions and addressed two arguments against truth value accounts for indicative conditionals by Bennett (2003): (a) D. Lewis' (1976) triviality result and (b) Bennett's Gibbardian stand-off argument (Gibbard, 1980). My discussion showed that both arguments against truth value analyses of indicative conditionals are in no way decisive against possible worlds semantics for indicative conditionals such as CS semantics.

II Formal Results for Chellas-Segerberg Semantics

Chapter 4

Formal Framework

4.1 Why Chellas-Segerberg Semantics?

In this and the following chapters I investigate and discuss Chellas-Segerberg (CS) semantics (Chellas, 1975; Segerberg, 1989) from a formal and a philosophical perspective. I have so far not fully answered why I have picked CS semantics rather than an alternative possible worlds semantics for conditionals, such as Lewis models (Definitions 3.1–3.3), the relational semantics of Kraus et al. (1990) and Lehmann and Magidor (1992) (see Section 3.2) or the propositional neighborhood semantics also described by Chellas (1975, pp. 144–147). Hence, let me provide reasons for investigating of CS semantics.

First, the proof-theoretic system corresponding to basic CS semantics is by far weaker than the minimal systems investigated by D. Lewis (1973b), Kraus et al. (1990), and Lehmann and Magidor (1992). In particular, CS semantics allows one to describe weaker systems than those accounted for by these alternative semantics. In addition, it is possible to describe strong conditional logic systems within CS semantics by superimposing additional frame conditions. In this book I shall even describe conditional logics which make conditionals as strong as the material implication (see Section 7.3.4).

A second reason for choosing CS semantics is that this semantics allows – as we shall see – for a simple, but intuitive interpretation of conditionals in terms of both a modified Ramsey test and an objective alethic interpretation (cf. Section 7.1). These interpretations can then be used to account for both indicative and counterfactual conditionals (cf. Section 3.4). In contrast, semantics such as in D. Lewis (1973b), Kraus et al. (1990), Lehmann and Magidor (1992), and Stalnaker (1968), employ an ordering semantics approach and, thereby, do not lend themselves easily into a Ramsey test interpretation (see Section 3.3.5). Moreover, I do not see that the propositional neighborhood semantics (Chellas, 1975, pp. 144–147) also allows for a natural interpretation in terms of the Ramsey test.

Finally, CS semantics is also interesting from a formal point of view, since it deviates in essential ways from standard Kripke semantics and its multi-modal extensions (see Sections 4.3.1–4.3.3). Moreover, the general formal properties of CS semantics are, except for Chellas (1975) and Segerberg (1989), virtually unexplored. Hence, before I focus on CS semantics, let me first describe some proof-theoretic notions.

4.2 Proof-Theoretic Notions

4.2.1 Languages $\mathcal{L}_{\mathrm{CL}}$, $\mathcal{L}_{\mathrm{CL}^-}$, $\mathcal{L}_{\mathrm{rCL}}$, $\mathcal{L}_{\mathrm{rCL}^*}$, and $\mathcal{L}_{\mathrm{rrCL}}$

In this section I define several conditional logic languages. I start with the full language $\mathcal{L}_{\mathrm{CL}}$ and then describe the more restricted languages $\mathcal{L}_{\mathrm{CL}^-}$, $\mathcal{L}_{\mathrm{rCL}}$, $\mathcal{L}_{\mathrm{rCL}^*}$, and $\mathcal{L}_{\mathrm{rrCL}}$. Here CL, CL$^-$, rCL, rCL*, and rrCL stand for 'conditional logic', 'qualified conditional logic', 'restricted conditional logic', 'variant of the restricted conditional logic', and 'more restricted conditional logic', respectively. All languages are based on the same propositional vocabulary, but differ with respect to (w.r.t.) the type of expressions, which count as formulas.

Let me specify the following meta-language expressions for the above languages: $\neg$ ("negation"), $\wedge$ ("conjunction"), $\vee$ ("disjunction"), $\Rightarrow$ ("material implication"), $\Leftrightarrow$ ("material equivalence"), $\forall$ ("universal quantification"), $\exists$ ("existential quantification"), $=$ ("identity"), $=_{\mathrm{df}}$ ("identity by definition") and $\neq$ ("negation of identity"). In addition, the meta-language symbols w,

$w', w'', \ldots, w_0, w_1, \ldots$ (individual variables), and $X, Y, Z, X_0, X_1, \ldots, Y_0, Y_1,$ ... (variables ranging over sets of individuals) are used. All meta-language expressions serve as abbreviations for respective natural language phrases.

Let us now focus on the vocabulary of our object languages: The propositional variables $p, q, \ldots, p_1, p_2, \ldots$ denote atomic propositional variables (I restrict myself to formal languages with countably but infinitely many atomic propositions). The expressions $\mathcal{AV}$ and $\mathcal{L}$ denotes the set of all atomic propositional variables and the set of all formulas in a given language $\mathcal{L}$, respectively; our propositional logical constants are: $\neg$ ("negation") and $\vee$ ("disjunction"). The symbol $\Box\!\!\rightarrow$ is a two-place modal operator, also called 'conditional operator'.

The expressions $\alpha, \beta, \ldots \alpha_0, \alpha_1, \alpha_2, \ldots$ stand for arbitrary object language formulas of the respective language. I also use these expressions to refer to formula schemata rather than formulas. For the sake of brevity I shall often refer to these expressions indiscrimately as formulas. In addition, the auxiliary symbols '(' and ')' are used. (I employ parentheses also in the meta-language.) The expressions $\wedge$ ("conjunction"), $\rightarrow$ ("material implication" or "subjunction"), $\leftrightarrow$ ("bisubjunction") $\top$ ("verum"), $\bot$ ("falsum"), and $\Diamond\!\!\rightarrow$ ("conditional possibility operator") are meta-language abbreviations for formulas in the object language. This way I can restrict myself to the discussion of a small range of primitive symbols, but use the meta-language abbreviations to make the discussion and proofs more perspicuous.

In particular, I use the following definitions: Def$_\wedge$: $\alpha \wedge \beta =_{\mathrm{df}} \neg(\neg\alpha \vee \neg\beta)$; Def$_\rightarrow$: $\alpha \rightarrow \beta =_{\mathrm{df}} \neg\alpha \vee \beta$; Def$_\leftrightarrow$: $\alpha \leftrightarrow \beta =_{\mathrm{df}} (\alpha \rightarrow \beta) \wedge (\beta \rightarrow \alpha)$; Def$_\top$: $\top =_{\mathrm{df}} p \vee \neg p$; Def$_\bot$: $\bot =_{\mathrm{df}} p \wedge \neg p$; Def$_{\Diamond\!\rightarrow}$: $\alpha \Diamond\!\!\rightarrow \beta =_{\mathrm{df}} \neg(\alpha \Box\!\!\rightarrow \neg\beta)$. The definitions apply to formulas in a language $\mathcal{L}$, if the expressions by which they are defined can occur in formulas of $\mathcal{L}$.

I will sometimes use the concept of Boolean combinations of formulas in a language $\mathcal{L}$. Boolean combinations of formulas are combinations of formulas by means of the connectives $\neg$, $\vee$, $\wedge$, $\rightarrow$, and $\leftrightarrow$ (where formulas with the connectives $\wedge$, $\rightarrow$, and $\leftrightarrow$ are again meta-language abbreviations of corresponding object language formulas).

Let us now describe the set of formulas of the languages $\mathcal{L}_{\mathrm{CL}}$, $\mathcal{L}_{\mathrm{CL}^-}$, $\mathcal{L}_{\mathrm{rCL}}$, $\mathcal{L}_{\mathrm{rCL}}$, and $\mathcal{L}_{\mathrm{rrCL}}$. The full language $\mathcal{L}_{\mathrm{CL}}$ resembles p.c. insofar as it allows the conditional operator $\Box\!\!\rightarrow$ to combine in formulas in any way the material implication $\rightarrow$ does in p.c. In particular, any type of Boolean combination of conditionals (e.g., $(\alpha \Box\!\!\rightarrow \beta) \wedge \gamma)$ and nestings of conditionals (e.g., $\alpha \Box\!\!\rightarrow (\beta \wedge (\gamma \Box\!\!\rightarrow \delta)))$ are formulas of $\mathcal{L}_{\mathrm{CL}}$. The set of formulas of the language $\mathcal{L}_{\mathrm{CL}}$ can be formally described the following way:

(a) $p, q, \ldots, p_1, p_2, \ldots$ are formulas of $\mathcal{L}_{\mathrm{CL}}$.
(b) If α and β are formulas of $\mathcal{L}_{\mathrm{CL}}$, then $\neg\alpha$ and $(\alpha \vee \beta)$ are formulas of $\mathcal{L}_{\mathrm{CL}}$.
(c) if α and β are formulas of $\mathcal{L}_{\mathrm{CL}}$, then $(\alpha \Box\!\!\rightarrow \beta)$ is a formula of $\mathcal{L}_{\mathrm{CL}}$.
(d) No other expression is a formula of $\mathcal{L}_{\mathrm{CL}}$.

The languages $\mathcal{L}_{\mathrm{CL}^-}$, $\mathcal{L}_{\mathrm{rCL}}$, $\mathcal{L}_{\mathrm{rCL}^*}$, and $\mathcal{L}_{\mathrm{rrCL}}$ differ from $\mathcal{L}_{\mathrm{CL}}$ insofar as they do not allow one to represent nested conditionals: Any conditional formula $\alpha \Box\!\!\rightarrow \beta$ in languages $\mathcal{L}_{\mathrm{CL}^-}$, $\mathcal{L}_{\mathrm{rCL}}$, $\mathcal{L}_{\mathrm{rCL}^*}$, and $\mathcal{L}_{\mathrm{rrCL}}$ requires α and β not to contain any instance of the conditional operator $\Box\!\!\rightarrow$. The language $\mathcal{L}_{\mathrm{CL}^-}$ differs from the languages $\mathcal{L}_{\mathrm{rCL}}$, $\mathcal{L}_{\mathrm{rCL}^*}$, and $\mathcal{L}_{\mathrm{rrCL}}$ insofar as $\mathcal{L}_{\mathrm{CL}^-}$ allows for both conditional and non-conditional formulas, while the set of formulas of $\mathcal{L}_{\mathrm{rCL}}$, $\mathcal{L}_{\mathrm{rCL}^*}$, and $\mathcal{L}_{\mathrm{rrCL}}$ do not contain formulas without instances of the conditional operator $\Box\!\!\rightarrow$. Hence, for example, $p \wedge q$ is a formula of the language $\mathcal{L}_{\mathrm{CL}^-}$, but not of the languages $\mathcal{L}_{\mathrm{rCL}}$, $\mathcal{L}_{\mathrm{rCL}^*}$, and $\mathcal{L}_{\mathrm{rrCL}}$. Language $\mathcal{L}_{\mathrm{CL}^-}$ neither allows for Boolean combinations of conditional formulas on the one hand nor conditional and non-conditional formulas on the other. The language $\mathcal{L}_{\mathrm{CL}^-}$ is, then, formally described the following way:

(a^-) $p, q, \ldots, p_1, p_2, \ldots$ are formulas of $\mathcal{L}_{\mathrm{prop}}$.
(b^-) If α and β are formulas of $\mathcal{L}_{\mathrm{prop}}$, then $\neg\alpha$ and $(\alpha \vee \beta)$ are formulas of $\mathcal{L}_{\mathrm{prop}}$.
(c^-) No other expression is a formula of $\mathcal{L}_{\mathrm{prop}}$.
(d^-) If α is a formula of $\mathcal{L}_{\mathrm{prop}}$, then α is a formula of $\mathcal{L}_{\mathrm{CL}^-}$.
(e^-) If α and β are formula of $\mathcal{L}_{\mathrm{prop}}$, then $(\alpha \Box\!\!\rightarrow \beta)$ is a formula of $\mathcal{L}_{\mathrm{CL}^-}$.
(f^-) No other expression is a formula of $\mathcal{L}_{\mathrm{CL}^-}$.

One could specify languages which also include Boolean combinations of conditional formulas and Boolean combinations of conditional and non-conditional formulas. Since I will not employ these languages here, I shall not discuss them any further.

The languages $\mathcal{L}_{\mathrm{rCL}}$, $\mathcal{L}_{\mathrm{rCL}^*}$, and $\mathcal{L}_{\mathrm{rrCL}}$ differ from each other to the extent they allow for Boolean combinations, where $\mathcal{L}_{\mathrm{rCL}}$, $\mathcal{L}_{\mathrm{rCL}^*}$, and $\mathcal{L}_{\mathrm{rrCL}}$ allow only for disjunctions of conditional formulas, negations of conditional formulas, and no Boolean combinations of conditional formulas, respectively. The set of formulas of $\mathcal{L}_{\mathrm{rCL}}$ can accordingly be determined by the following set of clauses:

(a_1)–(c_1) Conditions (a_1)–(c_1) are conditions (a^-)–(c^-), respectively.
(d_1) If α and β are formulas of $\mathcal{L}_{\mathrm{prop}}$, then $(\alpha \mathrel{\Box\!\!\rightarrow} \beta)$ is a formula of $\mathcal{L}_{\mathrm{rCL}}$.
(e_1) If α and β are formulas of $\mathcal{L}_{\mathrm{rCL}}$, then $(\alpha \vee \beta)$ is a formula of $\mathcal{L}_{\mathrm{rCL}}$.
(f_1) Nothing else is a formula of $\mathcal{L}_{\mathrm{rCL}}$.

For the specification of the set of formulas of language $\mathcal{L}_{\mathrm{rCL}^*}$ one formulates (d_1) w.r.t language $\mathcal{L}_{\mathrm{rCL}^*}$ and replaces conditions (e_1) and (f_1) in the definition of formulas of $\mathcal{L}_{\mathrm{rCL}}$ by the following clauses (e_2) and (f_2), respectively:

(e_2) If α is a formula of $\mathcal{L}_{\mathrm{rCL}^*}$, then $\neg\alpha$ is a formula of $\mathcal{L}_{\mathrm{rCL}^*}$.
(f_2) Nothing else is a formula of $\mathcal{L}_{\mathrm{rCL}^*}$.

For the definition of formulas of language $\mathcal{L}_{\mathrm{rrCL}}$ one applies (d_1) to language $\mathcal{L}_{\mathrm{rrCL}}$, deletes condition ($f_1$) and replaces ($e_1$) in the definition of the set of formulas of $\mathcal{L}_{\mathrm{rCL}}$ by the following clause (e_3), respectively:

(e_3) No other expression is a formula of $\mathcal{L}_{\mathrm{rrCL}}$.

Let me now discuss which systems in the conditional logic literature draw on the languages $\mathcal{L}_{\mathrm{CL}}$, $\mathcal{L}_{\mathrm{CL}^-}$, $\mathcal{L}_{\mathrm{rCL}}$, $\mathcal{L}_{\mathrm{rCL}^*}$, and $\mathcal{L}_{\mathrm{rrCL}}$: D. Lewis' (1973b, p. 118) counterfactual logic and the indicative and counterfactual conditional logic of Stalnaker (1968, p. 105) and Stalnaker and Thomason (1970, p. 24f) are formulated w.r.t. the full language $\mathcal{L}_{\mathrm{CL}}$, whereas Adams' (1965, p. 184;

1966, p. 270; 1977, p. 186f) conditional logic system is based on language $\mathcal{L}_{\mathrm{CL}^-}$. Adams (1975, p. 45f) uses a modified version of language $\mathcal{L}_{\mathrm{CL}^-}$, namely language $\mathcal{L}^{\mathrm{cons}}_{\mathrm{CL}^-}$, in which all conditional formulas are required to have p.c. consistent antecedents. In contrast, Adams (1986) effectively employs language $\mathcal{L}_{\mathrm{rCL}^*}$: Adams allows only for inferences of type $\{\alpha_1 \wedge \ldots \wedge \alpha_m\} \vdash \beta_1 \vee \ldots \vee \beta_n$ (p. 259), where $\alpha_1,\ldots,\alpha_m,\beta_1,\ldots,\beta_n$ are simple conditionals and it is the case that $m, n \in \mathbb{N}_0$ (p. 256f).[1]

Schurz (1998, p. 84f) shows that in a probabilistic system S the above inference schemata can also be expressed in restricted conditional logic language $\mathcal{L}_{\mathrm{rCL}}$ (which allows for disjunctions of simple conditionals). For that purpose Schurz (1998) has to assume that for S (a) the Deduction Theorem holds, which goes as follows: $\Gamma \vdash_S \alpha$ holds iff $\exists \Gamma_f \subseteq \Gamma$: $\vdash_S \bigwedge \Gamma_f \rightarrow \alpha$, where Γ refers to an arbitrary set of formulas, Γ_f denotes a finite set of formulas and $\bigwedge X$ stands for the conjunction of all elements in the set of formulas X in a pre-specified order (see also Schurz, 2002/2005a, p. 453). (The Deduction Theorem makes sure that the relevant formal properties of the material implication are transfered to the consequence relation in system S.) Moreover, Schurz (1998) has to presuppose (b) that all p.c. valid inferences are valid in system S (short: S-valid).

According to Schurz (1998) inferences of type $\{\alpha_1,\ldots,\alpha_m\} \vdash \beta_1 \vee \ldots \vee \beta_n$ are, then, S-valid exactly if all inferences in $\{\{\alpha_1,\ldots,\alpha_m\} \vdash \beta_1, \ldots, \{\alpha_1,\ldots, \alpha_m\} \vdash \beta_n\}$ are S-valid. Moreover, Schurz's (1998, p. 84f) proof also gives us that all inferences in $\mathcal{L}_{\mathrm{rCL}^*}$ in a probabilistic system S can be expressed by (simple) inferences in $\mathcal{L}_{\mathrm{rCL}}$: Any inference in $\mathcal{L}_{\mathrm{rCL}}$ (with a finite set of premises) is valid iff in language $\mathcal{L}_{\mathrm{rCL}^*}$ the inference $\{\alpha_1,\ldots,\alpha_m,\neg\beta_1,\ldots,\neg\beta_n\} \vdash \gamma$ with $m,n \in \mathbb{N}_0$ and specific formulas $\alpha_1,\ldots,\alpha_m,\beta_1,\ldots,\alpha_n,\gamma$ is valid iff $\{\alpha_1,\ldots,\alpha_m\} \vdash \beta_1 \vee \ldots \vee \beta_n \vee \gamma$ is valid. This equivalence result presupposes again that (a) and (b) hold.

Let me now focus on Schurz's (1998) proof. W.r.t. the probabilistic system

[1] In case of $m = 0$ the conjunctive premise is interpreted as $\top \mathrel{\Box\!\!\rightarrow} \top$ and in case of $n = 0$ the disjunctive conclusion is interpreted as $\top \mathrel{\Box\!\!\rightarrow} \bot$. I will ignore this complication here (see also Section 3.5.3).

S points (a) and (b) are assumed to hold and one has in addition to presuppose (c) in the probabilistic system S inferences are restricted to a finite number of premises, since systems such as system $\mathbf{P}^+$ are non-compact (Schurz, 1998, p. 84). Finally I require that (d) $\top \Box\!\!\rightarrow \top$ is valid in system S. On the basis of our assumptions regarding inferences of a probabilistic system S in the language $\mathcal{L}_{\text{rCL}}$ take in general the form $\Gamma_f \vdash \beta$, where Γ_f represents a finite set of formulas in $\mathcal{L}_{\text{rCL}}$. Hence, any inference in such a system can be described by $\alpha \vdash_{\mathbf{L}} \beta$, where α is the conjunction of all elements in Γ (formally: $\alpha = \bigwedge \Gamma$). **L** is here also assumed to be closed under p.c. consequences, including the rule of substitution (see Section 4.2.5). Moreover, since language $\mathcal{L}_{\text{rCL}}$ allows only for Boolean combinations of conditional formulas, one can transform α and β into disjunctive and conjunctive normal forms $\bigvee_i \bigwedge_j \alpha_{ij}$, and $\bigwedge_k \bigvee_l \beta_{kl}$, respectively, where $i \in \{1,\ldots,m\}$, $j \in \{1,\ldots,n\}$, $k \in \{1,\ldots,o\}$, and $l \in \{1,\ldots,p\}$ so that α_{ij} and β_{kl} represent simple or negated conditional formulas.[2] This observation implies that any inference in a probabilistic system S can be represented in the following equivalent form: $\bigvee_i \bigwedge_j \alpha_{ij} \vdash_{\mathbf{L}} \bigwedge_k \bigvee_l \beta_{kl}$. This inference is valid in a probabilistic logic **L** iff $\{\bigwedge_j \alpha_{1j} \vdash_{\mathbf{L}} \bigvee_l \beta_{1l}, \ldots, \bigwedge_j \alpha_{mj} \vdash_{\mathbf{L}} \bigvee_l \beta_{ol}\}$ is valid in **L** iff $\{(\bigwedge_j \alpha_{1j}) \wedge (\bigwedge_l \neg \beta_{1l}) \vdash_{\mathbf{L}} \neg(\top \Box\!\!\rightarrow \top), \ldots, (\bigwedge_j \alpha_{mj}) \wedge (\bigwedge_l \neg \beta_{ol}) \vdash_{\mathbf{L}} \neg(\top \Box\!\!\rightarrow \top)\}$. Hence for all probabilistic systems for which assumptions (a)–(d) hold (e.g., System $\mathbf{P}^+$), languages $\mathcal{L}_{\text{rCL}}$ and $\mathcal{L}_{\text{rCL}^*}$ are equally expressive. Schurz (1998) neither uses language $\mathcal{L}_{\text{rCL}}$ nor $\mathcal{L}_{\text{rCL}^*}$, since he allows for non-conditional formulas in his language (p. 85). Moreover, since in Schurz's (1998) language inferences can involve disjunctions of conditionals (p. 85), his language is in fact less restrictive than language $\mathcal{L}_{\text{CL}^-}$ (which does not allow for arbitrary Boolean combinations of condi-

[2] Our representation of disjunctive [conjunctive] normal forms presupposes that all conjunctions [disjunctions] of simple and negated formulas in the disjunctive [conjunctive] normal form have the same length m [p]. The classical description of conjunctive and disjunctive normal forms does not assume. In order to transform the classical versions of conjunctive and disjunctive normal forms into our version, one can fill up "empty" slots in disjunctions and conjunctions with the formulas $\neg(\top \Box\!\!\rightarrow \top)$ and $\top \Box\!\!\rightarrow \top$, respectively.

tional formulas).[3]

The language $\mathcal{L}_{rCL^*}$ is effectively used by Lehmann and Magidor (1992, p. 5; see Section 2.2.6). Hence, Lehmann and Magidor (1992) can also represent inferences of probabilistic systems w.r.t. language $\mathcal{L}_{rCL}$ or inferences described by Adams (1986). In contrast, Kraus et al. (1990) do not explicitly indicate whether they effectively employ the language $\mathcal{L}_{rCL^*}$ or the language $\mathcal{L}_{rrCL}$. In particular they do not state whether they regard negations of conditionals also as formulas of the object language or only as a meta-language abbreviation (see Section 2.2.6).

The use of the restricted languages $\mathcal{L}_{CL^-}$, $\mathcal{L}_{rCL}$, $\mathcal{L}_{rCL^*}$, and $\mathcal{L}_{rrCL}$ has a profound impact on a system's soundness and completeness properties (see Section 3.5.3 and 3.6.1). Since the use of the full language has several advantages over a more restricted version (see Sections 3.5.5 and 3.7.3), I formulate the model-theory and proof-theory investigated in Chapters 5–7 in the full language $\mathcal{L}_{CL}$.

Furthermore, for the sake of perspicuity, I shall omit outer bracket and other brackets in all languages according to following rules: The opterator $\neg$ binds strongest. The operator $\Box\!\!\rightarrow$ is less strong, but binds stronger than both $\wedge$ and $\vee$, which in turn bind stronger than both connectives $\Box\!\!\rightarrow$ and $\rightarrow$.

Let me now describe some terminology used throughout this book. I start with the definition of scopes of conditional operators. The antecedent scope of a conditional operator $\Box\!\!\rightarrow$ in language $\mathcal{L}$ is defined as the formula α of $\mathcal{L}$, which precedes $\Box\!\!\rightarrow$. The consequent scope of a conditional operator $\Box\!\!\rightarrow$ in a language $\mathcal{L}$ is defined as the formula α of $\mathcal{L}$, which is preceded by $\Box\!\!\rightarrow$. An antecedent scope and a consequent scope must always exist for any conditional operator occurring in formulas of the languages $\mathcal{L}_{CL}$, $\mathcal{L}_{CL^-}$, $\mathcal{L}_{rCL}$, $\mathcal{L}_{rCL^*}$, and $\mathcal{L}_{rrCL}$. I often call the antecedent scope and the consequent scope of a conditional operator its 'antecedent formula' and its 'consequent formula', respectively. Let us now focus on the concepts of iteration and nestedness for $\mathcal{L}_{CL}$ (note that both iterated and nested formulas are not allowed for

[3] Schurz (1997b, 1997b, 2005b) escapes Lewis' (1976) triviality result by using a weaker version of the Stalnaker thesis (see Section 3.5.2).

the languages $\mathcal{L}_{\text{CL}^-}$, $\mathcal{L}_{\text{rCL}}$, $\mathcal{L}_{\text{rCL}^*}$, and $\mathcal{L}_{\text{rrCL}}$): A formula α in language $\mathcal{L}_{\text{CL}}$ is iterated iff there exist formulas β, γ, δ of $\mathcal{L}_{\text{CL}}$ such that $\alpha = (\beta \mathrel{\Box\!\!\rightarrow} (\gamma \mathrel{\Box\!\!\rightarrow} \delta))$ or $\alpha = ((\beta \mathrel{\Box\!\!\rightarrow} \gamma) \mathrel{\Box\!\!\rightarrow} \delta)$. A formula α in language $\mathcal{L}_{\text{CL}}$ is nested, if there exist formulas β and γ of $\mathcal{L}_{\text{CL}}$ such that $\alpha = (\beta \mathrel{\Box\!\!\rightarrow} \gamma)$ and either β or γ contains an instance of $\mathrel{\Box\!\!\rightarrow}$. From both definitions follows immediately that any formula of $\mathcal{L}_{\text{CL}}$ which is iterated is also nested. The converse does not hold, since, for example, the formula $p \mathrel{\Box\!\!\rightarrow} \neg(p \mathrel{\Box\!\!\rightarrow} r)$ of $\mathcal{L}_{\text{CL}}$ is nested but not iterated. Furthermore, left-nestedness and right-nestedness are often distinguished in the literature. One can make those notions precise as follows: A formula α in $\mathcal{L}_{\text{CL}}$ is left-nested iff there are formulas β and γ of $\mathcal{L}_{\text{CL}}$ such that $\alpha = (\beta \mathrel{\Box\!\!\rightarrow} \gamma)$ and β contains an instance of $\mathrel{\Box\!\!\rightarrow}$. A formula α in $\mathcal{L}_{\text{CL}}$ is right-nested iff there are formulas β and γ of $\mathcal{L}_{\text{CL}}$ such that $\alpha = (\beta \mathrel{\Box\!\!\rightarrow} \gamma)$ and γ contains an instance of $\mathrel{\Box\!\!\rightarrow}$.

In this book I often use the term 'bridge principle'. This notion can be made more precise the following way: A formula (schema) α is a bridge principle in language $\mathcal{L}_{\text{CL}}$ iff it contains the conditional operator $\mathrel{\Box\!\!\rightarrow}$ and there exists a formula β of $\mathcal{L}_{\text{CL}}$ for which the following holds: (a) β is contained in α, (b) β does not contain any occurrence of $\mathrel{\Box\!\!\rightarrow}$, (c) β lies outside the antecedent scope and the consequent scope of any conditional operator, and (d) α is not logically equivalent to a conjunction of formulas which do not satisfy (a)–(c). Typical examples for bridge principles are the formulas $(\alpha \mathrel{\Box\!\!\rightarrow} \beta) \rightarrow (\alpha \rightarrow \beta)$ (MP, "Modus Ponens") and $\alpha \wedge \beta \rightarrow (\alpha \mathrel{\Box\!\!\rightarrow} \beta)$ (CS; "Conjunctive Sufficiency"; see Table 5.1). I include condition (d) to make sure that no principles of, for example, the form $(\alpha \mathrel{\Box\!\!\rightarrow} \beta) \wedge \gamma$ are regarded as bridge-principles (cf. Schurz, 1997a, p. 91f). The addition of (d) is essential, since formulas such as $(\alpha \mathrel{\Box\!\!\rightarrow} \beta) \wedge \gamma$, are p.c. equivalent to sets of formulas (e.g., $\{\alpha \mathrel{\Box\!\!\rightarrow} \beta, \gamma\}$), which do not contain bridge principles (cf. Schurz, 1997a, p. 92).

One cannot express any type of bridge principle (as defined here) in the languages $\mathcal{L}_{\text{CL}^-}$, $\mathcal{L}_{\text{rCL}}$, $\mathcal{L}_{\text{rCL}^*}$, and $\mathcal{L}_{\text{rrCL}}$ directly. However, one can use rules of inference in the language $\mathcal{L}_{\text{CL}^-}$ which correspond to bridge principles in the full language $\mathcal{L}_{\text{CL}}$ (cf. Section 3.6.1). The inferences 'if $\alpha \mathrel{\Box\!\!\rightarrow} \beta$ then

$\alpha \rightarrow \beta$' and 'if $\alpha \wedge \beta$ then $\alpha \mathrel{\Box\!\!\rightarrow} \beta$' are, for example, rules in language $\mathcal{L}_{CL}$-which correspond to MP and E, respectively. I do not distinguish between both types of bridge principles (formulas on the one hand and rules on the other) in the discussion, but use the labels 'MP', 'E', etc. and 'bridge principle' indiscriminately. It is, furthermore, interesting to note that no bridge principles (in that wide sense) can be expressed in languages $\mathcal{L}_{rCL}$, $\mathcal{L}_{rCL^*}$, and $\mathcal{L}_{rrCL}$, since all formulas of these language have to contain at least one conditional operator.

4.2.2 Logics

In this section I define a number of syntactic notions. These syntactic notions are not restricted to a particular system, but rather apply to all systems discussed in this book (except if specified otherwise). This enables me to communicate many concepts more efficiently.

First, I use the notion of a logic **L** in two ways: On the one hand I refer by 'logic' to a *logical systems* with specific axioms and rules of derivation (see below); on the other hand, I associate with this term a *set of formulas* which are valid in such systems. In the latter sense I always presuppose that a logic **L** is closed under propositional consequence and substitution of non-logical symbols (cf. Schurz, 2002/2005a, pp. 447–449; see also the next section). My use of that term in general should not lead to ambiguities. I specify my use of the term accordingly, when there might be an ambiguity. All logical systems considered here – if not specified otherwise – refer to the full conditional language $\mathcal{L}_{CL}$.

Let $A, B, \ldots$ denote sequences of formulas and Γ, Δ, and Σ sets of formulas of language $\mathcal{L}$. A proof in a system **L** is, then, a finite sequence of formulas, for which every formula in that sequence is either an instance of an axiom schema of **L** or follows by the rules of **L** from preceding formulas. Both the set of axioms and rules are specified in such a way that they are effectively enumerable. A formula α is derivable in system **L** (short: $\vdash_{\mathbf{L}} \alpha$) iff there exists a proof for α in **L**. $\Gamma \vdash_{\mathbf{L}} \alpha$ (α is derivable from Γ in **L**) iff the following holds: if $\vdash_{\mathbf{L}} \beta$ for all $\beta \in \Gamma$, then $\vdash_{\mathbf{L}} \alpha$.

Let us now take a closer look at the notions of axioms and rules. A rule can be described as an ordered pair $\langle P,C \rangle$, where P is a set of derivability or non-derivability claims (cf. Schurz, 1996, p. 205) and C is a derivability claim. Derivability claims and non-derivability claims have the form $\vdash \alpha$ and $\nvdash \alpha$, respectively (see also Section 2.2). A rule might be read intuitively in the following way: If $\alpha_1, \alpha_2, \ldots$ are derivable and $\beta_1, \beta_2, \ldots$ are not derivable, then γ is derivable. A rule $\langle P,C \rangle$ is monotonic iff P contains only derivability claims; a rule $\langle P,C \rangle$ is non-monotonic iff P contains at least one non-derivability claim. Please note that our notion of rule is somewhat broader than the usual notion, which is restricted to monotonic rules.

4.2.3 Non-Monotonicity

A monotonic logic is, then, a logic which can be described by monotonic rules only. The basic idea behind the restriction to monotonic rules is the fact that if I add monotonic rules to monotonic systems, then the set of theorems cannot decrease. More formally, for a monotonic system **L** the following holds: if $\Gamma \vdash_{\mathbf{L}} \beta$, then $\Gamma \cup \{\alpha\} \vdash_{\mathbf{L}} \beta$. (I call the derivability relation of a system **L** monotonic iff **L** is monotonic.) The upshot of this is that in monotonic systems no rule prevents another rule from applying. For non-monotonic systems this is not the case. Since the conclusion of a non-monotonic rule depends on non-derivability claims, the addition of axioms can result in the retraction of a conclusion. The non-derivability conditions of non-monotonic rules are to blame here. In monotonic systems this can, hence, not be the case (see Section 2.2.3). Please note that the addition of axioms to a monotonic logic system cannot make the monotonic system non-monotonic. This is due to the fact that axioms are also rules, namely rules of type $\langle P,C \rangle$ with P being the empty set. This means, in other words, that axioms are rules which are applied unconditionally. Hence, axioms do not have non-derivability conditions and, hence, adding them to monotonic systems does not make those systems non-monotonic.

Although I investigate non-monotonic properties of conditionals in this book, the formal investigation in Chapters 5–7 is limited to monotonic logics in the above sense. I do this by specifying the logical properties of the conditional operator $\Box\!\!\rightarrow$ in such a way that $\Box\!\!\rightarrow$ is non-monotonic, but the inference relation $\vdash$ remains monotonic. That our inference relation for all systems described in Chapters 5 and 6 is monotonic (in the above sense) can be seem from the fact that I only use axioms and monotonic rules. Hence, these systems are not default logics in the sense specified in Section 2.2.3. Due to these facts, these systems do not suffer from the difficulties of default logic approaches (see Section 2.2.5).

4.2.4 Consistency and Maximality

Let me at this point define the notions of consistency and maximality: A set of formulas Γ is **L**-consistent iff $\Gamma \nvdash_{\mathbf{L}} \bot$, where $\nvdash_{\mathbf{L}}$ denotes non-derivability in logic **L**. A set of formulas Γ is **L**-maximal iff there is no set of formulas Δ such that $\Gamma \subset \Delta$ and Δ is **L**-consistent. A **L**-maximal set of formulas needs not be consistent. Finally, a formula set Γ is maximally **L**-consistent iff it is **L**-consistent and **L**-maximal.

4.2.5 A Propositional Basis for Conditional Logics

The lattice of conditional logics investigated in the following chapters are all closed under propositional consequence. The set of propositional tautologies can be axiomatized by the following set of axioms plus the additional rules *modus ponens* and *substitution* (specified below):

1. $p \rightarrow (q \rightarrow p)$
2. $(p \rightarrow (q \rightarrow r)) \rightarrow ((p \rightarrow q) \rightarrow (p \rightarrow r))$
3. $(\neg p \rightarrow \neg q) \rightarrow (q \rightarrow p)$

Let me now describe the propositional rules of inference. Modus ponens tells us that if $\alpha \rightarrow \beta$ and α are the case, then β holds. The rule of substitution allows for the uniform substitution of any atomic proposition by an arbitrary formula in any derivable formula, formally, if $\vdash \alpha$ then $\vdash s(\alpha)$, where $s(\alpha)$ refers to the simultaneous and uniform substitution of atomic propositions by some arbitrary formulas in the formula α (cf. Schurz, 2002/2005a,

Table 4.1
Rules and Axioms of System CK

LLE	if $\vdash \alpha \leftrightarrow \beta$ and $\alpha \mathrel{\Box\!\!\to} \gamma$, then $\beta \mathrel{\Box\!\!\to} \gamma$
RW	if $\vdash \alpha \rightarrow \beta$ and $\gamma \mathrel{\Box\!\!\to} \alpha$, then $\gamma \mathrel{\Box\!\!\to} \beta$
AND	$(\alpha \mathrel{\Box\!\!\to} \beta) \wedge (\alpha \mathrel{\Box\!\!\to} \gamma) \rightarrow (\alpha \mathrel{\Box\!\!\to} \beta \wedge \gamma)$
LT	$\alpha \mathrel{\Box\!\!\to} \top$

Note. 'LLE', 'RW', and 'LT' stand for 'Left Logical Equivalence', 'Right Weakening', and 'Logical Truth', respectively.

p. 448f). The rule of substitution is not truth-preserving but only validity-preserving: It does not suffice for the application of the rule of substitution to presuppose that α is true, but I have to assume that α is in fact valid or derivable (cf. Schurz, 2002/2005a, p. 448f). Observe, moreover, that the rule of substitution does not hold for default logics in general (see Section 2.2.4).

Moreover, since I do not use defined expressions in the object language (see Section 4.2.1), I do not need an additional rule of replacement, which guarantees that these definitions hold. Furthermore, I use axiom schemata rather than axioms in Chapters 5–7. Hence, I do not need to employ the substitution rule explicitly. Moreover, as is standard in the conditional logical literature, I will also not refer to p.c. rules and p.c. axioms explicitly.

4.2.6 System CK

System **CK** is syntactically characterized the following way:

Definition 4.1. *System* **CK** *is the smallest logic containing LLE+RW+AND +LT (see Table 4.1).*

My characterization of System **CK** (and henceforth all definitions of conditional logics) presupposes closure under p.c. rules and p.c. axioms. So, System **CK** is LLE+RW+AND+LT plus axioms of propositional logic and the rule modus ponens. The Deduction Theorem holds for System **CK** and the stronger systems discussed in Chapters 5–7 (see Section 4.2.1).[4] The Deduction Theorem allows us, for example, to conclude from LLE the following: If $\alpha \leftrightarrow \beta$ is derivable in S, then $(\alpha \,\Box\!\!\rightarrow \gamma) \rightarrow (\beta \,\Box\!\!\rightarrow \gamma)$ is derivable in S. One can, furthermore, by the same theorem infer from AND that if $(\alpha \,\Box\!\!\rightarrow \beta) \wedge (\alpha \,\Box\!\!\rightarrow \gamma)$ is derivable in S, then $\alpha \,\Box\!\!\rightarrow \beta \wedge \gamma$ is derivable in S. I shall neither prove the Deduction Theorem for System **CK** nor for the stronger systems discussed in this book.

From time to time I will also provide object language proofs of lemmata and theorems associated with System **CK**. For the sake of perspicuity I employ natural deduction proofs rather than axiomatic proofs. The natural deduction elements refers to the p.c. part of the respective systems only. Premises and presuppositions of conditional proofs are indicated by 'given'. Moreover, assumptions for proofs by cases and indirect proofs are marked and numbered by expressions such as 'assumption 1, proof by cases' and 'assumption, ind. proof', respectively. Furthermore, less obvious steps referring to p.c. are indicated by 'p.c.'. Examples of object language proofs can be found below.

Observe that our axiomatization of System **CK** represents a variation of the axiomatization by Segerberg (1989, p. 158). There are two main differences: First, my axiomatization is more in line with Kraus et al. (1990). In particular I use RW instead of the rule RLE ("Right Logical Equivalence") plus Axiom Schema CW ("Consequent Weakening"), as Segerberg (1989) does:

RLE if $\vdash \beta \leftrightarrow \gamma$ and $\alpha \,\Box\!\!\rightarrow \beta$ then $\alpha \,\Box\!\!\rightarrow \gamma$
CW $(\alpha \,\Box\!\!\rightarrow \beta \wedge \gamma) \rightarrow (\alpha \,\Box\!\!\rightarrow \beta) \wedge (\alpha \,\Box\!\!\rightarrow \gamma)$

Since Lemmata 4.3–4.5 give us that RLE+CW are logically equivalent to RW, we obtain the following lemma:

Lemma 4.2. *System* **CK** *can be axiomatized as LLE+RLE+CW+LT.*

Proof. By Lemmata 4.3–4.5, we get Lemma 4.2. □

[4] The deduction theorem, however, does not hold for System $\mathbf{P}_\epsilon$ of Adams (1975).

Lemma 4.3. *RLE+CW* $\Rightarrow$ *RW*

Proof.

1.	$\vdash \alpha \rightarrow \beta$	given
2.	$\gamma \,\Box\!\!\rightarrow \alpha$	given
3.	$\vdash \alpha \leftrightarrow \alpha \wedge \beta$	1, p.c.
4.	$\gamma \,\Box\!\!\rightarrow \alpha \wedge \beta$	3, 2, RLE
5.	$(\gamma \,\Box\!\!\rightarrow \alpha) \wedge (\gamma \,\Box\!\!\rightarrow \beta)$	4, CW
6.	$(\gamma \,\Box\!\!\rightarrow \beta)$	5, p.c.

□

Lemma 4.4. *RW* $\Rightarrow$ *RLE*

Proof.

1.	$\vdash \beta \leftrightarrow \gamma$	given
2.	$\alpha \,\Box\!\!\rightarrow \beta$	given
3.	$\vdash \beta \rightarrow \gamma$	1, p.c.
4.	$\alpha \,\Box\!\!\rightarrow \gamma$	3, 2, RW

□

Lemma 4.5. *RW* $\Rightarrow$ *CW*

Proof.

1.	$\alpha \,\Box\!\!\rightarrow \beta \wedge \gamma$	given
2.	$\alpha \,\Box\!\!\rightarrow \beta$	1, RW
3.	$\alpha \,\Box\!\!\rightarrow \gamma$	1, RW
4.	$(\alpha \,\Box\!\!\rightarrow \beta) \wedge (\alpha \,\Box\!\!\rightarrow \gamma)$	2, 3, p.c.

□

The second difference between Segerberg's (1989) axiomatization and my axiomatization of conditional logic systems lies in the fact that I formulate principles in terms of axioms rather than rules whenever possible. This works for many principles from Tables 5.1 and 5.2, because I use the full language $\mathcal{L}_{\mathrm{CL}}$ and employ an inference relations $\vdash$, which is monotonic (see Section 4.2.3). This treatment of logical principles allows for an easier application in terms of correspondence and completeness proofs.

4.2.7 A Further Axiomatization of System CK

Let us now turn to a further axiomatization of System **CK**. Originally, Chellas (1975, p. 137f) defined System **CK** in terms of the propositional part plus LLE plus the following rule RCK:

RCK if $\vdash \beta_1 \wedge \cdots \wedge \beta_n \rightarrow \gamma$, then $(\alpha \Box\!\!\rightarrow \beta_1) \wedge \cdots \wedge (\alpha \Box\!\!\rightarrow \beta_n) \rightarrow (\alpha \Box\!\!\rightarrow \gamma)$, $n \geq 0$

For $n = 0$ the material implication $\beta_1 \wedge \cdots \wedge \beta_n \rightarrow \gamma$ is identified with its consequent formula γ (see Chellas, 1975, Footnote 6). One can easily see that Chellas' axiomatization and the axiomatization described in Table 4.1 agree. Hence, we have the following theorem:

Theorem 4.6. *System* **CK** *can be axiomatized by RCK+LLE (cf. Chellas, 1975, Footnote 8).*

Proof. By Lemmata 4.7–4.10 we obtain Theorem 4.6. □

Lemma 4.7. *RCK ⇒ RW (see Nejdl, 1992, Theorem 2.1)*

Proof.

1.	$\alpha \Box\!\!\rightarrow \beta$	given
2.	$\vdash \beta \rightarrow \gamma$	given
3.	$\alpha \Box\!\!\rightarrow \gamma$	2, 1, RCK

□

Lemma 4.8. *RCK ⇒ AND (see Nejdl, 1992, Theorem 2.4)*

Proof.

1.	$\alpha \Box\!\!\rightarrow \beta$	given
2.	$\alpha \Box\!\!\rightarrow \gamma$	given
3.	$\vdash \beta \wedge \gamma \rightarrow \beta \wedge \gamma$	p.c.
4.	$\alpha \Box\!\!\rightarrow \beta \wedge \gamma$	3, 1, 2, RCK

□

By Lemma 4.9 one can derive LT from RCK focusing on the case $n = 0$:

Lemma 4.9. *RCK ⇒ LT*

Proof.

1.	$\vdash \top$	p.c.
2.	$\alpha \mathrel{\Box\!\!\rightarrow} \top$	1, RCK ($n = 0$)

□

Furthermore, the following lemma holds:

Lemma 4.10. *RW+AND+LT ⇒ RCK*

Proof.

for $n = 0$:

1.	$\vdash \gamma$	given
2.	$\alpha \mathrel{\Box\!\!\rightarrow} \top$	LT
3.	$\vdash \top \rightarrow \gamma$	1, p.c.
4.	$\alpha \mathrel{\Box\!\!\rightarrow} \gamma$	3, 2, RW

for $n > 0$:

1.	$\vdash \beta_1 \wedge \cdots \wedge \beta_n \rightarrow \gamma$	given
2.	$(\alpha \mathrel{\Box\!\!\rightarrow} \beta_1) \wedge \cdots \wedge (\alpha \mathrel{\Box\!\!\rightarrow} \beta_n)$	given
3.	$\alpha \mathrel{\Box\!\!\rightarrow} \beta_1 \wedge \cdots \wedge \beta_n$	2, $n-1$ times AND
4.	$\alpha \mathrel{\Box\!\!\rightarrow} \gamma$	1,3, RW

□

4.3 Model-Theoretic Notions

Let us now describe the basic Chellas-Segerberg (CS) semantics. CS semantics is a possible worlds semantics and represents a generalization of standard Kripke semantics as, for example, described in Hughes and Cresswell (1996/2003, p. 38f) and Blackburn, de Rijke, and Venema (2001, pp. 16–18). In CS semantics not a single accessibility relation among possible worlds in W is used as in standard Kripke semantics, but rather a multitude of such accessibility relations.

The basic idea of this semantics is that for each antecedent α of a conditional formula $\alpha \boxright \beta$ one has an accessibility relation R which is relativized to α. In this context let 'R_α' denote an accessibility relation R relativized to α. The formula $\alpha \boxright \beta$ is, then, true at a possible world w in W if and only if at all possible worlds w', which are accessible to w by R_α the consequent β is true.

In CS semantics the accessibility relation R is not directly relativized to formulas (i.e., α), but instead to subsets of W at which the formulas (i.e., α) are true. I denote the sets of possible worlds at which α is true by '$||\alpha||$'. In my terminology the set $||\alpha||$ is the proposition which is expressed by α. Propositions are, hence, subsets of the set of possible worlds W. Moreover, I write '$R_{||\alpha||}$' and 'R_X' for an accessibility relation relativized to a proposition $||\alpha||$ or a proposition X, respectively. In this approach I can, then, define frames in the tradition of Kripke frames (Hughes & Cresswell, 1996/2003, p. 38f), which results in ordered pairs of a non-empty set of possible worlds W and an accessibility relation R defined on members and subsets of W. This approach avoids direct reference to formulas on the level of frames and seems, hence, more natural. This definition of frames, moreover, parallels the notion of Kripke frames, for which no such relativization to propositions is needed (cf. Hughes & Cresswell, 1996/2003, p. 38f). Moreover, the use of sets of possible worlds makes naturally LLE (see Table 4.1) valid. Due to the specification of valuations V on possible worlds (see below), any two formulas being logically equivalent describe the same set of possible worlds.

Let me now turn to the notions of frames and models investigated in Chapters 5 and 6. I distinguish between two types of frames and models. The most general notions of frames and models are Chellas frames and models. Less general, but useful notions are Segerberg frames and Segerberg models. Both types of frames and models are based on the ideas described above. For the sake of perspicuity, I will, first, introduce the notions of Chellas frames and models and then discuss in which way Segerberg frames and models might be more appropriate notions regarding soundness and completeness.

4.3.1 Chellas Frames and Chellas Models

The basic units of CS semantics investigated in this book are Chellas frames and Chellas models. These are generalizations of Kripke frames and Kripke models (Hughes & Cresswell, 1996/2003, p. 38f). To my knowledge they were first formally described by Chellas (1975, p. 134f) and can be defined as follows:

Definition 4.11. *$\mathcal{F}_C = \langle W, R\rangle$ is a Chellas frame iff*
(a) W is a non-empty set of possible worlds and
(b) R is a relation on $W \times W \times Pow(W)$.

Definition 4.12. *Let $\mathcal{F}_C = \langle W, R\rangle$ be a Chellas frame. Then, $\mathcal{M}_C = \langle W, R, V\rangle$ is a Chellas model iff V is a valuation function from $\mathcal{AV} \times W$ to $\{0, 1\}$.*

If $\mathcal{M}_C = \langle W, R, V\rangle$ and $\mathcal{M}'_C = \langle W, R, V'\rangle$ are such that $V = V'$, I shall call $\mathcal{M}_C$ and $\mathcal{M}'_C$ 'elementary equivalent'.

Definition 4.13. *Let $\mathcal{M}_C = \langle W, R, V\rangle$ be a Chellas model. Then V^* is an extension of V to arbitrary formulas iff*
(a) $\forall p \in \mathcal{AV}\,\forall w \in W$: $V^*(p, w) = 1$ iff $V(p, w) = 1$ and
(b) for all formulas α, β and $w \in W$ holds:

$$\begin{array}{lll} \models^{\mathcal{M}_C}_w \neg\alpha & \textit{iff } \not\models^{\mathcal{M}_C}_w \alpha, & (V_\neg) \\ \models^{\mathcal{M}_C}_w \alpha \vee \beta & \textit{iff } \models^{\mathcal{M}_C}_w \alpha \vee \models^{\mathcal{M}_C}_w \beta, \textit{ and} & (V_\vee) \\ \models^{\mathcal{M}_C}_w \alpha \Box\!\!\rightarrow \beta & \textit{iff } \forall w'(wR_{||\alpha||}w' \Rightarrow \models^{\mathcal{M}_C}_{w'} \beta). & (V_{\Box\!\rightarrow}) \end{array}$$

The expressions $\models^{\mathcal{M}}_w \alpha$ and $\not\models^{\mathcal{M}}_w \alpha$ stand for $V^*(\alpha, w) = 1$ and $V^*(\alpha, w) = 0$ for V as specified in $\mathcal{M}$, respectively. Intuitively, $\models^{\mathcal{M}}_w \alpha$ and $\not\models^{\mathcal{M}}_w \alpha$ are to be interpreted as α being true at world w in model $\mathcal{M}$ and being false at world w in model $\mathcal{M}$, respectively. When I refer to a world w in a model $\mathcal{M}$, I mean in fact a world $w \in W$ such that $\mathcal{M} = \langle W, R, V\rangle$. The expression $||\alpha||^{\mathcal{M}}$ is defined for a CS model $\mathcal{M} = \langle W, R, V\rangle$ the following way: $||\alpha||^{\mathcal{M}} =_{\text{df}} \{w \in W \mid \models^{\mathcal{M}}_w \alpha\}$ ($\text{Def}_{||.||}$).

Strictly speaking only the extension of V, namely V^*, generalizes to arbitrary formulas. Since V determines V^* in a unique way, I will often use V

and V^* indiscriminately. Moreover, I sometimes write $||\alpha||$ when the context makes it clear to which model $||\alpha||$ refers.

The function V is a valuation function, which assigns truth values to atomic formulas of the language $\mathcal{L}_{\mathrm{CL}}$ relative to possible worlds. Conditions $\mathrm{V}_\neg$ and $\mathrm{V}_\vee$ give us truth conditions for Boolean combinations of formulas. For truth conditions of conditional formulas, Chellas models refer to the three-place accessibility relation R. This accessibility relation in Chellas models is a relation R between possible worlds relativized to sets of possible worlds. To indicate that a relation R holds between $w, w' \in W$ and $X \subseteq W$ I sometimes write wR_Xw'.

4.3.2 Chellas Models and Frames and Kripke Semantics

Chellas frames [models] differ from Kripke frames [models] (e.g., Hughes & Cresswell, 1996/2003, p. 38f; Blackburn et al., 2001, pp. 16–18) and their multi-modal extensions (e.g., Blackburn et al., 2001, p. 20) in the following three ways: (a) For any Kripke model $\mathcal{M} = \langle W, R, V \rangle$ the cardinality of the set of accessibility relations R is always greater than the cardinality of W. This is due to the fact that for any subset $X \subseteq W$, there exists an accessibility relation R_X between possible worlds in W. Any subset X of W is an element of the power set of W. The power set Pow(W) has higher cardinality than W, except if $W = \varnothing$ holds. The latter fact is, however, excluded by condition (a) of Definition 4.11. In contrast, Kripke semantics for normal modal logics – where only a single accessibility relation is used – does not have this property (Hughes & Cresswell, 1996/2003, p. 38f; Blackburn et al., 2001, pp. 16–18). Even in multi-modal explorations of modal logic this is not the case. Either a constant finite number of n-ary accessibility relations is investigated (cf. Blackburn et al., 2001, p. 20) or a constant but potentially infinite number of accessibility relations (cf. Schurz, 1997a, p. 166). These investigations do not require that the number of accessibility relations in a frame is strictly greater than the number of possible worlds in that frame.

(b) The second difference lies in the use of the accessibility relation. From a proof-theoretic perspective I can regard $\alpha \mathrel{\Box\!\!\rightarrow} \beta$ as a type of modality where

$\alpha \Box\!\!\rightarrow \cdot$ expresses the modal operator. For this purpose Chellas (1975, p. 138f) introduces the inofficial terminology $[\alpha]\beta$. In CS semantics the formal relation between R_X and a modal operator $[\alpha]$ is only determined on the basis of the valuation function V. Hence, for Chellas frames $\mathcal{F}_C = \langle W, R \rangle$ it is not specified which accessibility relation R_X is associated with the modalities $[\alpha]$, $[\beta]$, etc. Again, this differs from Kripke frames and their multi-modal extensions (cf. Blackburn et al., 2001, p. 20; Schurz, 1997a, p. 166; Gabbay, Kurucz, Wolter, & Zakharyaschev, 2003, p. 21). For the latter types of frames there is a predetermined connection between modal operators and accessibility relations.[5]

(c) The third difference lies in the fact that the relativization of the accessibility relation R in Chellas frames $\mathcal{F}_C = \langle W, R \rangle$ is stronger than what is actually needed for the truth conditions in Chellas models. The relativization is done for every subset X of W. So, there exists an accessibility relation for each subset X of W. The truth conditions for Chellas models can, however, refer only to subsets of W in $V_{\Box\!\rightarrow}$ for which a formula α exists such that $||\alpha||$. I call this type of subsets of the Chellas model $\mathcal{M}_C$ 'syntactically representable sets in $\mathcal{M}_C$' and give the following formal definition:

Definition 4.14. *Let $\mathcal{M}_C = \langle W, R, V \rangle$ be a Chellas model. Then, a subset $X \subseteq W$ is syntactically representable in $\mathcal{M}_C$ (by formulas of $\mathcal{L}_{\mathrm{CL}}$) iff there exists a formula α (of $\mathcal{L}_{\mathrm{CL}}$) such that $X = ||\alpha||^{\mathcal{M}_C}$.*

One can distinguish the following two aspects for point (c), both of which I shall discuss separately: First, (c1) the truth conditions of a Chellas model $\mathcal{M}_C = \langle W, R, V \rangle$ (Definition 4.12) do not refer to syntactically non-representable subsets in $\mathcal{M}_C$. Second, (c2) there may exist syntactically non-representable subsets X in $\mathcal{M}_C$ for which the accessibility relation R_X is defined.

Let me first focus on (c2). In general not all subsets X of W of a Chellas model $\mathcal{M}_C = \langle W, R, V \rangle$ are syntactically representable. For a subset X in $\mathcal{M}_C$

[5] CS semantics is related to standard semantics for dynamic logic (Gabbay et al., 2003, p. 63) where the accessibility relation is relativized to actions. In that framework actions are, however, *not* represented by subsets of possible worlds (see above).

to be syntactically representable there has to exist a single formula which is true at all and only the worlds contained in X. If ones finds a finite collection of formulas Γ which are true at all and only those worlds in X, then X is syntactically representable, since there exists such a formula, namely the conjunction of the finite collection of formulas Γ. Since any formula of $\mathcal{L}_{\mathrm{CL}}$ can be described in terms of a conjunctive normal form by atomic propositional variables and conditional formulas, the search for syntactically representable formulas boils down to finding finite sets of formulas which conjointly represent the respective set. It should be clear that such a finite collection of formulas might not exist for all subsets of W if W is infinite.

To make this more plausible, let us take a look at the proof-theoretic analogue of semantic representability. This is (propositional) finite axiomatizability of theories, namely axiomatizability by a finite set of eigenaxioms. A set of possible worlds X might be associated with the set of formulas which are true and only true at the possible worlds in X. Due to the definition of Chellas models, sets of formulas of this kind are closed under propositional consequence and, hence, represent propositional theories. It should be clear that not all theories of this type are axiomatizable by a finite set of eigenaxioms.

Let us now focus on (c1). We can prove accordingly the following lemma:

Lemma 4.15. *Let $\mathcal{M}_C = \langle W, R, V\rangle$ and $\mathcal{M}'_C = \langle W, R', V'\rangle$ be Chellas models. Suppose (a) that $V = V'$ and suppose (b) that for all syntactically representable sets $X \subseteq W$ in $\mathcal{M}_C$ and all w, $w' \in W$ holds that wR_Xw' iff wR'_Xw'. Then, for arbitrary formulas α of $\mathcal{L}_{\mathrm{CL}}$ and worlds $w \in W$ holds: $\models_w^{\mathcal{M}_C} \alpha$ iff $\models_w^{\mathcal{M}'_C} \alpha$.*

Proof. The proof is by induction on the construction of formulas. Condition (a) gives us that $\mathcal{M}_C$ and $\mathcal{M}'_C$ are elementary equivalent. Hence, it holds for all for all atomic propositions $p \in \mathcal{AV}$ and all worlds $w \in W$: $\models_w^{\mathcal{M}_C} p$ iff $\models_w^{\mathcal{M}'_C} p$.

Non-atomic propositional case: By the induction hypothesis $V(\alpha, w) = V'(\alpha, w)$ and $V(\beta, w) = V'(\beta, w)$ for arbitrary worlds $w \in W$ and given formulas α and β. By Definition 4.12 it follows that for any Boolean combination γ of α and β it holds that $V(\gamma,w) = V'(\gamma,w)$.

Modal case: By the induction hypothesis $\models_w^{\mathcal{M}_C} \alpha$ iff $\models_w^{\mathcal{M}'_C} \alpha$, and $\models_w^{\mathcal{M}_C} \beta$ iff $\models_w^{\mathcal{M}'_C} \beta$ for arbitrary worlds $w \in W$ and given formulas α and β. Suppose for an arbitrary $w \in W$ that $\models_w^{\mathcal{M}_C} \alpha \Box\!\!\rightarrow \beta$. Then, by $V_{\Box\!\rightarrow}$ it follows that (i) $\forall w' \in W(wR_{||\alpha||^{\mathcal{M}_C}} w' \Rightarrow \models_{w'}^{\mathcal{M}_C} \beta)$. By the induction hypothesis holds that $||\alpha||^{\mathcal{M}_C} = ||\alpha||^{\mathcal{M}'_C}$ and that for all w it is the case that $\models_w^{\mathcal{M}_C} \beta$ iff $\models_w^{\mathcal{M}'_C} \beta$. Hence, it follows by (b) that (ii) $\forall w' \in W(wR_{||\alpha||^{\mathcal{M}'_C}} w' \Rightarrow \models_{w'}^{\mathcal{M}'_C} \beta)$ iff (i). By $V_{\Box\!\rightarrow}$ condition (ii) holds iff $\models_w^{\mathcal{M}'_C} \alpha \Box\!\!\rightarrow \beta$ is the case. □

Lemma 4.15 shows that any two Chellas models which are elementary equivalent concur w.r.t. the extension of their valuation functions. This lemma, hence, gives us that any accessibility relations R_X, where X is a syntactically non-representable subset in a Chellas model, does not have an impact on the truth of formulas at any possible world in that model. Hence, any of these R_X are irrelevant in the sense that they do not contribute to the determination of the truth of formulas at any world in that Chellas model. Despite this fact, all those R_X are defined on the basis of Chellas models and Chellas frames as we saw from the discussion of (c2).

Let me summarize the result of the discussion of this section. We saw in this section that Chellas frames $\mathcal{F}_C = \langle W,R\rangle$ and Chellas models $\mathcal{M}_C = \langle W,R,V\rangle$ deviate from Kripke frames and Kripke models (and their multimodal extensions) in three ways: (a) For any Chellas frame [model] there are strictly more accessibility relations R_X than possible worlds in W. (b) The association between modal operators $[\alpha]$ and the accessibility relation is not fixed on the level of frame, but only determined on the basis of Chellas models. (c1) The truth of formulas depends only on accessibility relations R which are relativized to syntactically representable sets in the Chellas model. (c2) Accessibility relations w.r.t. syntactically non-representable subsets in Chellas models are also defined.

4.3.3 Segerberg Frames and Segerberg Models

We saw in the previous section that in Chellas models $\mathcal{M}_C = \langle W, R, V \rangle$ only syntactically representable sets (in that model) determine the truth value of formulas at a world, whereas syntactically non-representable sets in such a model are not relevant in that context (Lemma 4.15, point (c1)). The set of syntactically representable sets is essentially determined by the valuation function V and cannot be specified on the level of frames. Since no restriction on the valuation function exists for Chellas models for any subset of X of W the accessibility relation R_X is defined and is potentially syntactically representable. This is not the case for Segerberg models and frames due to the additional parameter P.

A Segerberg frame is a triple $\mathcal{F}_S = \langle W, R, P \rangle$ and a Segerberg model is a quadruple $\mathcal{M}_S = \langle W, R, P, V \rangle$. The parameter P is, then, defined as a subset of the power set Pow(W) of W and is assumed to be closed under the logical operations (negation, disjunction, and conditional operator). Segerberg frames and Segerberg models, hence, are general frames and general frame-based models (cf. Hughes & Cresswell, 1996/2003, p. 167; Blackburn et al., 2001, p. 28f) rather than Kripke frames. They owe their name to Segerberg, since they were to my knowledge first defined in Segerberg (1989, p.160f). Models on general frames are restricted in such a way that all syntactically representable sets are in that set P. Segerberg models share this property. They differ, however, from general frames in one important respect: The additional parameter P is used not only to restrict the range of admissible valuations, but also the accessibility relation R_X. The accessibility relation R is not defined on $W \times W \times \text{Pow}(W)$, but rather on $W \times W \times P$. In Segerberg models [frames] the accessibility relation, hence, is relativized to a (possibly proper) subset of Pow(W). In contrast, in Kripke frames [models] with n-ary modal operator (cf. Blackburn et al., 2001, p. 20) and their multi-modal extensions (cf. Schurz, 1997a, p. 166; Gabbay et al., 2003, p. 21) there is no restriction of the accessibility relation R to a parameter P.

Let me now define the notions of Segerberg frames and Segerberg models formally. I shall specify the notion of Segerberg models w.r.t. the set of admissible valuations on Segerberg frames (Definition 4.17).

Definition 4.16. *$\mathcal{F}_S = \langle W,R,P\rangle$ is a Segerberg frame iff*
(a) W is a non-empty set of possible worlds,
(b) R is a relation on $W \times W \times P$, and
(c) $P \subseteq Pow(W)$ is such that for all $X, Y \subseteq W$ holds:

$\varnothing \in P$, *(Def$_{P_\varnothing}$)*
if $X \in P$ *then* $-X \in P$, *(Def$_{P_-}$)*
if $X, Y \in P$ *then* $X \cup Y \in P$, *and* *(Def$_{P_\cup}$)*
if $X, Y \in P$ *then* $\{w \in W \mid \forall w' \in W(wR_Xw' \Rightarrow w' \in Y)\} \in P$. *(Def$_{P_{Mod}}$)*

The conditions on (c) in Definition 4.16 are used to provide closure conditions for the logical operators $\neg$, $\vee$, $\wedge$, $\rightarrow$, $\Box\!\!\rightarrow$, and $\Diamond\!\!\rightarrow$ regarding the set P. These closure conditions are needed to ensure that all propositions in a Segerberg model are in P. Thereby, (c) ensures that validity in a Segerberg frame is closed under the rule of substitution. Let us now focus on the notion of Segerberg models:

Definition 4.17. *Let $\mathcal{F}_S = \langle W,R,P\rangle$ be a Segerberg frame and let the valuation function V and its extension V^* be defined w.r.t. W and R as in Definitions 4.12 and 4.13, respectively, such that $\mathcal{M}_C = \langle W,R,V\rangle$. The valuation V is, then, admissible in $\mathcal{F}_S$ iff for all formulas α and the extension V^* of V holds: $||\alpha||^{\mathcal{M}_C} = \{w \in W \mid V^*(\alpha,w) = 1\} \in P$.*

Definition 4.18. *$\mathcal{M}_S = \langle W,R,P,V\rangle$ is a Segerberg model iff*
(a) $\langle W,R,P\rangle$ is a Segerberg frame, as described in Definition 4.16,
(b) V is a valuation function w.r.t. W and R as in Definition 4.12, and
(c) V is admissible in $\langle W,R,P\rangle$.

Let us now prove the following theorem:

Theorem 4.19. *Let $\mathcal{F}_S = \langle W,R,P\rangle$ be a Segerberg frame and let the valuation function V be such that $\mathcal{M}_C = \langle W,R',V\rangle$ is a Chellas model, where R'_X is an expansion of R_X to arbitrary subsets X of W. Then, V is admissible in $\mathcal{F}_S$ iff for all $p \in \mathcal{AV}$ holds: $||p||^{\mathcal{M}_C} = \{w \in W \mid V(p,w) = 1\} \in P$.*

Proof. "⇒": The left-to-right direction is trivial and follows directly from Definitions 4.16, 4.17, 4.12, and 4.13. So, the right-to-left direction remains to be shown.

"⇐": The proof is by induction on the construction of formulas. Suppose that $||p||^{\mathcal{M}_C} \in P$. In order for V to be admissible, $||\alpha||^{\mathcal{M}_C} \in P$ has to be the case for all formulas α and the extension V^* of V (see Definition 4.13). (I shall henceforth use V and V^* indiscriminately.) For atomic propositions this holds by assumption. Hence, the non-atomic propositional case and the modal case remain to be proven.

$\alpha = \neg\beta$: By induction hypothesis it holds that $||\beta||^{\mathcal{M}_C} \in P$. Def_{P_-} of Definition 4.16 gives us that $W - ||\beta||^{\mathcal{M}_C} \in P$. Thus, due to $\mathrm{V}_\neg$ and $\mathrm{Def}_{||\cdot||}$ it holds that $||\neg\beta||^{\mathcal{M}_C} \in P$.

$\alpha = \beta \vee \gamma$: By induction hypothesis $||\beta||^{\mathcal{M}_C}, ||\gamma||^{\mathcal{M}_C} \in P$ is the case. Due to $\mathrm{Def}_{P_\cup}$ of Definition 4.16 and $\mathrm{Def}_{||\cdot||}$ we obtain $||\beta \vee \gamma||^{\mathcal{M}_C} \in P$.

$\alpha = \beta \,\Box\!\!\rightarrow \gamma$: By induction hypothesis, it holds that $||\beta||^{\mathcal{M}_C}, ||\gamma||^{\mathcal{M}_C} \in P$. Due to $\mathrm{Def}_{P_{Mod}}$ of Definition 4.16, it follows that $\{w \in W \mid \forall w' \in W(wR_{||\beta||^{\mathcal{M}_C}}w' \Rightarrow w' \in ||\gamma||^{\mathcal{M}_C})\} \in P$. By $V_{\Box\!\rightarrow}$ and $\mathrm{Def}_{||\cdot||}$ this implies that $||\beta \,\Box\!\!\rightarrow \gamma||^{\mathcal{M}_C} \in P$. □

Segerberg frames [models] are more general than Chellas frames [models] in the following sense: For any Chellas frame $\mathcal{F}_C = \langle W, R\rangle$ there exists a Segerberg frame $\mathcal{F}_S = \langle W, R', P\rangle$ such that $P = \mathrm{Pow}(W)$ and $R = R'$. Then, the valuation function V on $\mathcal{F}_C$ is also admissible in $\mathcal{F}_S$ and vice versa. For any valuation function V admissible in a Segerberg frame $\mathcal{F}_S = \langle W, R, P\rangle$ the Segerberg model $\mathcal{M}_S = \langle W, R, P, V\rangle$ has the following property: All syntactically representable subsets of W in the corresponding Chellas model are in P, and R_X is defined for all syntactically representable sets X in that model. The parameters V and P might, then, be defined in such a way that only syntactically representable subsets of W are in P. In that case R_X is restricted to the set of syntactically representable subsets X. Due to these facts one can construct a Segerberg model $\mathcal{M}_S = \langle W, R', P, V\rangle$ for each Chellas model $\mathcal{M}_C = \langle W, R, V\rangle$ such that R' is restricted to the subsets which are syntactically representable in $\mathcal{M}_C$. By these means one can remedy points (a) and (c2) discussed in Section 4.3.2.

4.3.4 Validity, Logical Consequence, and Satisfiability

In this section I define different notions of validity, logical consequence, and satisfiability w.r.t. Chellas and Segerberg frames [models]. I indicate different versions of these definitions by means of square brackets. I abbreviate Chellas frames [models] and Segerberg frames [models] by $\mathcal{F}_C$, $\mathcal{F}'_C$, $\mathcal{F}''_C$, ... [$\mathcal{M}_C$, $\mathcal{M}'_C$, $\mathcal{M}''_C$, ...], and $\mathcal{F}_S$, $\mathcal{F}'_S$, $\mathcal{F}''_S$, ... [$\mathcal{M}_S$, $\mathcal{M}'_S$, $\mathcal{M}''_S$, ...], respectively. Classes of Chellas frames [models] and Segerberg frames [models] are denoted by $\mathbf{F}_C$, $\mathbf{F}'_C$, $\mathbf{F}''_C$, ...[$\mathbf{M}_C$, $\mathbf{M}'_C$, $\mathbf{M}''_C$, ...], and $\mathbf{F}_S$, $\mathbf{F}'_S$, $\mathbf{F}''_S$, ...[$\mathbf{M}_S$, $\mathbf{M}'_S$, $\mathbf{M}''_S$, ...], respectively. A Chellas model $\mathcal{M}_C = \langle W, R, V\rangle$ is based on a Chellas frame $\mathcal{F}_C = \langle W', R'\rangle$ iff $W = W'$ and $R = R'$. A Segerberg model $\mathcal{M}_S = \langle W, R, P, V\rangle$ is based on a Segerberg frame $\mathcal{F}_S = \langle W', R', P'\rangle$ iff $W = W'$, $R = R'$, $P = P'$, and V is admissible in $\mathcal{F}_S$. A Segerberg frame $\mathcal{F}_S = \langle W, R, P\rangle$ [a Segerberg model $\mathcal{M}_S = \langle W, R, P, V\rangle$] is based on a Chellas frame $\mathcal{F}_S = \langle W', R'\rangle$ iff $W = W'$ and $R =\downarrow_P R'$. The expression $\downarrow_P R'$ refers to the restriction of R' to elements of P. A Chellas frame $\mathcal{F}_C = \langle W, R\rangle$ [a Segerberg frame $\mathcal{F}_S = \langle W, R, P\rangle$] is a frame for a system **L** iff all theorems of **L** are valid on $\mathcal{F}_C$ [$\mathcal{F}_S$].

A formula α is valid *in* a Chellas model $\mathcal{M}_C = \langle W, R, V\rangle$ (short: $\models_{\mathcal{M}_C} \alpha$) [*in* a Segerberg model $\mathcal{M}_S = \langle W, R, P, V\rangle$ (short: $\models_{\mathcal{M}_S} \alpha$)] iff for all $w \in W$: $\models_w^{\mathcal{M}_C} \alpha$ [$\models_w^{\mathcal{M}_S} \alpha$]. A formula α is valid *on* a Chellas frame $\mathcal{F}_C = \langle W, R\rangle$ (short: $\models_{\mathcal{F}_C} \alpha$) [*on* a Segerberg frame $\mathcal{F}_S = \langle W, R, P\rangle$ (short: $\models_{\mathcal{F}_S} \alpha$)] iff α is valid in all Chellas models based on $\mathcal{F}_C$ [in all Segerberg models based on $\mathcal{F}_S$]. A formula α is valid w.r.t. a class of Chellas models $\mathbf{M}_C$ (short: $\models_{\mathbf{M}_C} \alpha$) [w.r.t. a class of Segerberg models $\mathbf{M}_S$ (short: $\models_{\mathbf{M}_S} \alpha$)] iff $\models_{\mathcal{M}_C} \alpha$ for all $\mathcal{M}_C \in \mathbf{M}_C$ [$\models_{\mathcal{M}_S} \alpha$ for all $\mathcal{M}_S \in \mathbf{M}_S$]. A formula α is valid w.r.t. a class of Chellas frames $\mathbf{F}_C$ (short: $\models_{\mathbf{F}_C} \alpha$) [w.r.t. a class of Segerberg frames $\mathbf{F}_S$ (short: $\models_{\mathbf{F}_S} \alpha$)] iff $\models_{\mathcal{F}_C} \alpha$ for all $\mathcal{F}_C \in \mathbf{F}_C$ [$\models_{\mathcal{F}_S} \alpha$ for all $\mathcal{F}_S \in \mathbf{F}_S$].

A formula α follows from a formula set Γ in a world w in a Chellas model $\mathcal{M}_C$ (short: $\Gamma \models_w^{\mathcal{M}_C} \alpha$) [in a Segerberg model $\mathcal{M}_S$ (short: $\Gamma \models_w^{\mathcal{M}_S} \alpha$)] iff ($\forall\beta \in \Gamma$: $\models_w^{\mathcal{M}_C} \beta) \Rightarrow \models_w^{\mathcal{M}_C} \alpha$ [($\forall\beta \in \Gamma$: $\models_w^{\mathcal{M}_S} \beta) \Rightarrow \models_w^{\mathcal{M}_S} \alpha$].[6] A formula α follows

[6] Blackburn et al. (2001, p. 31f) define and discuss consequence relations in modal

from a formula set Γ in a Chellas model $\mathcal{M}_C$ (short: $\Gamma \models_{\mathcal{M}_C} \alpha$) [in a Segerberg model $\mathcal{M}_S$ (short: $\Gamma \models_{\mathcal{M}_S} \alpha$)] iff $\forall w$ in $\mathcal{M}_C$: $\Gamma \models_w^{\mathcal{M}_C} \alpha$ [$\forall w$ in $\mathcal{M}_S$: $\Gamma \models_w^{\mathcal{M}_S} \alpha$]. A formula α follows from a formula set Γ in a Chellas frame $\mathcal{F}_C$ (short: $\Gamma \models_{\mathcal{F}_C} \alpha$) [in a Segerberg frame $\mathcal{F}_S$ (short: $\Gamma \models_{\mathcal{F}_S} \alpha$)] iff $\forall \mathcal{M}_C$ based on $\mathcal{F}_C$: $\Gamma \models_{\mathcal{M}_C} \alpha$ [$\forall \mathcal{M}_S$ based on $\mathcal{F}_S$: $\Gamma \models_{\mathcal{M}_S} \alpha$]. A formula α follows from a formula set Γ in a class of Chellas models $\mathbf{M}_C$ (short: $\Gamma \models_{\mathbf{M}_C} \alpha$) [in a class of Segerberg models $\mathbf{M}_S$ (short: $\Gamma \models_{\mathbf{M}_S} \alpha$)] iff $\forall \mathcal{M}_C \in \mathbf{M}_C$: $\Gamma \models_{\mathcal{M}_C} \alpha$ [$\forall \mathcal{M}_S \in \mathbf{M}_S$: $\Gamma \models_{\mathcal{M}_S} \alpha$]. A formula α follows from a formula set Γ in a class of Chellas frames $\mathbf{F}_C$ (short: $\Gamma \models_{\mathbf{F}_C} \alpha$) [in a class of Segerberg frames $\mathbf{F}_S$ (short: $\Gamma \models_{\mathbf{F}_S} \alpha$)] iff $\forall \mathcal{F}_C \in \mathbf{F}_C$: $\Gamma \models_{\mathcal{F}_C} \alpha$ [$\forall \mathcal{F}_S \in \mathbf{F}_S$: $\Gamma \models_{\mathcal{F}_S} \alpha$]. Please note that the notions of logical consequence and logical validity (w.r.t. the same model, frame, class of models, and class of frames) in general do not have the same level of expressiveness.

Finally, let me define different notions of (simultaneous) satisfiability: A set of formulas Γ is (simultaneously) satisfiable in a Chellas model $\mathcal{M}_C$ [in a Segerberg model $\mathcal{M}_S$] iff all $\alpha \in \Gamma$ are true at at least one world w in $\mathcal{M}_C$ [in $\mathcal{M}_S$]. A set of formulas Γ is (simultaneously) satisfiable in a class of Chellas models $\mathbf{M}_C$ [in a class of Segerberg models $\mathbf{M}_S$] iff all $\alpha \in \Gamma$ are true at at least one world w in a model $\mathcal{M}_C \in \mathbf{M}_C$ [in a model $\mathcal{M}_S \in \mathbf{M}_S$]. A set of formulas Γ is (simultaneously) satisfiable in a Chellas frame $\mathcal{F}_C$ [in a Segerberg frame $\mathcal{F}_S$] iff all $\alpha \in \Gamma$ are satisfiable in at least one Chellas model $\mathcal{M}_C$ based on $\mathcal{F}_C$ [in at least one Segerberg model $\mathcal{M}_S$ based on $\mathcal{F}_S$]. A set of formulas Γ is (simultaneously) satisfiable in a class of Chellas frames $\mathbf{F}_C$ [in a class of Segerberg frames $\mathbf{F}_S$] iff all $\alpha \in \Gamma$ are satisfiable in at least one model $\mathcal{M}_C$ based on $\mathcal{F}_C$ such that $\mathcal{F}_C \in \mathbf{F}_C$ [in at least one model $\mathcal{M}_S$ based

logic. (This is not too often done in the modal logic literature.) The above notions correspond to their notion of local semantic consequence relation. This notion does not concur with their so-called global semantic consequence relation. The latter notion can, for example, be defined for validity in a Chellas frame [Segerberg frame] the following way: Γ implies α w.r.t. a Chellas frame $\mathcal{F}_C$ [w.r.t. a Segerberg frame $\mathcal{F}_S$] globally iff for all Chellas models $\mathcal{M}_C$ based on $\mathcal{F}_C$ it is the case that if $\forall \beta \in \Gamma(\models_{\mathcal{M}_C} \beta)$, then $\models_{\mathcal{M}_C} \alpha$ [for all Segerberg models $\mathcal{M}_S$ based on $\mathcal{F}_S$ it is the case that if $\forall \beta \in \Gamma(\models_{\mathcal{M}_S} \beta)$, then $\models_{\mathcal{M}_S} \alpha$]. One can also apply this notion to classes of Chellas and Segerberg frames and models.

on $\mathcal{F}_S$ such that $\mathcal{F}_S \in \mathbf{F}_S$].

4.3.5 Notions of Frame Correspondence

In this section, I describe and discuss notions of frame correspondence. These notions allow one to characterize Chellas [Segerberg] frames in terms of axiom schemata and frame conditions. I argue here for a characterization in terms of Chellas frames rather than Segerberg frames. The notion of Chellas (frame) correspondence, then, serves as basis for the correspondence proofs in Chapter 5.

Let us now focus on Chellas frame correspondence (C-correspondence) and Segerberg frame correspondence (S-correspondence). One can define both notions the following way:

Definition 4.20. *(C-Correspondence) A formula schema α C-corresponds to frame condition C_α iff for all Chellas frames $\mathcal{F}_C$ holds: $\models_{\mathcal{F}_C} \alpha \Leftrightarrow \models_{\mathcal{F}_C} C_\alpha$.*

Definition 4.21. *(S-Correspondence) A formula schema α S-corresponds to frame condition C_α iff for all Segerberg frames $\mathcal{F}_S$ holds: $\models_{\mathcal{F}_S} \alpha \Leftrightarrow \models_{\mathcal{F}_S} C_\alpha$.*

Here, $\models_{\mathcal{F}_C} C_\alpha$ and $\models_{\mathcal{F}_S} C_\alpha$ abbreviate the fact that frame condition C_α holds for the Chellas frame $\mathcal{F}_C = \langle W, R\rangle$ and the Segerberg frame $\mathcal{F}_S = \langle W, R, P\rangle$, respectively. Frame conditions for S-frames and C-frames can both draw on the parameters W and R, whereas for frame conditions for S-frames the additional parameter P is available. As I argued in the previous section, I do not have an original interest in the parameter P. For example, the frame condition C_{T} of the principle $(\alpha \mathrel{\Box\!\!\rightarrow} \beta) \rightarrow \beta$ ("T", see Table 5.2) for Chellas frames $\mathcal{F}_C = \langle W, R\rangle$ is $\forall X \subseteq W \forall w \in W(wR_X w)$ (see Table 5.4).[7] Rather for technical reasons (esp. for the completeness proof, see Chapter 6), one has to restrict, for example, the frame condition C_{T} to the frame condition C'_{T}

[7] I do not argue that T is a plausible principle for a conditional logic. (I think that Principle T is rather counter-intuitive for any standard account of conditionals.) I chose Principle T, since it corresponds to one of the most simple frame conditions among the principles discussed in Chapters 5–7 (see Tables 5.1 and 5.2).

$(\forall X \subseteq P \forall w \in W(wR_X w))$, which draws in addition on the parameter P that is specific to Segerberg frames $\mathcal{F}_S = \langle W, R, P \rangle$.

I prefer to give an account of frame conditions purely in terms of accessibility relations R and the set of worlds W, such as C_T. This makes it possible to describe classes of frames both by means of formula schemata and restrictions on the accessibility relation. I can, hence, discuss different conditional logic systems in terms of basic CS semantics plus frame conditions. Since this is my approach in Chapter 7, I therefore rather opt for C-correspondence. Furthermore, correspondence in terms of Chellas frames also gives us a handle on validity of formulas in terms of Segerberg frames, as the following lemma shows:

Lemma 4.22. *Let $\mathcal{F}_C = \langle W, R \rangle$ be a Chellas frame. Then, a formula α is valid on $\mathcal{F}_C$ iff it is valid on all Segerberg frames $\mathcal{F}_S$ based on $\mathcal{F}_C$.*

Proof. "⇒": Suppose that α is valid on $\mathcal{F}_C = \langle W, R \rangle$. So, we have to show that α is valid on any Segerberg frame based on $\mathcal{F}_C$. Let $\mathcal{F}_S = \langle W, R', P \rangle$ be an arbitrary Segerberg frame based on $\mathcal{F}_C$. Then, by definition $R' = \downarrow_P R$. Furthermore, let $\mathcal{M}_S = \langle W, R', P, V \rangle$ be a Segerberg model based on $\mathcal{F}_S$. So, $\mathcal{M}_S$ is based on $\mathcal{F}_C$. Since, $R' = \downarrow_P R$, due to Lemma 4.23 there is a Chellas model $\mathcal{M}_C = \langle W, R, V' \rangle$ such that $V' = V$. Hence for all formulas α in $\mathcal{L}_{\mathrm{CL}}$ and $w \in W$ holds: $V'(\alpha, w) = V(\alpha, w)$. $\mathcal{M}_C$ is based on $\mathcal{F}_C$ and, hence, by assumption the formula α is valid in $\mathcal{M}_C$. So, the formula α is valid in the Segerberg model $\mathcal{M}_S = \langle W, R', P, V \rangle$. As this holds for arbitrary Segerberg models based on $\mathcal{F}_C$, the formula α is valid in this class of Segerberg frames.

"⇐": Suppose that α is valid on all Segerberg frames $\mathcal{F}_S = \langle W, R, P \rangle$ based on the Chellas frame $\mathcal{F}_C = \langle W', R' \rangle$. Then, α is valid on the Segerberg frame $\mathcal{F}_S = \langle W', R', \mathrm{Pow}(W') \rangle$. This particular Segerberg frame is structurally equivalent to the Chellas frame $\mathcal{F}_C$, as (a) R is defined for all subsets of W and (b) valuations of Segerberg models based on $\mathcal{F}_S$ can take the same range of values as valuation functions of Chellas models based on $\mathcal{F}_C$. Hence, if α is valid in $\mathcal{F}_S$, then it is valid in $\mathcal{F}_C$. □

Lemma 4.23. *For any Segerberg model $\mathcal{M}_S = \langle W, R, P, V \rangle$ based on a Chel-*

las frame $\mathcal{F}_C = \langle W, R' \rangle$ there exists a Chellas model $\mathcal{M}_C = \langle W, R'', V' \rangle$ such that $\mathcal{M}_S$ is based on $\mathcal{F}_C$ and $V' = V$.

Proof. Let $\mathcal{M}_S = \langle W, R, P, V \rangle$ be a Segerberg model based on a Chellas frame $\mathcal{F}_C = \langle W, R' \rangle$. Then, $\mathcal{M}_C = \langle W, R', V \rangle$ is the desired Chellas model: By definition of $\mathcal{F}_C$ the accessibility relation R'_X is the expansion of R_X for arbitrary subsets X of W and, furthermore, by definition of $\mathcal{M}_C$, $\mathcal{M}_S$ is based on $\langle W, R' \rangle$. □

4.3.6 Standard and Non-Standard Chellas Models and Segerberg Frames

C-Correspondence for an axiom schema α and its frame condition C_α does not imply that any Chellas model which satisfies α also satisfies the frame condition C_α. Analogously to Kripke semantics case (Schurz, 2002/2005a, p. 451), there are so-called non-standard Chellas models $\mathcal{M}_C = \langle W, R, V \rangle$, for which $(\alpha \mathbin{\Box\!\!\rightarrow} \beta) \rightarrow \beta$, for example, is valid but the C-corresponding frame condition C_T – namely $\forall X \subseteq W \forall w \in W(wR_X w)$ – does not hold. Let $\mathcal{M}_C = \langle W, R, V \rangle$ be such that $W = \{w_1, w_2\}$, $R = \{\langle w_1, w_2, \varnothing \rangle, \langle w_1, w_2, \{w_1\} \rangle, \langle w_1, w_2, \{w_2\} \rangle, \langle w_1, w_2, W \rangle, \langle w_2, w_1, \varnothing \rangle, \langle w_2, w_1, \{w_1\} \rangle, \langle w_2, w_1, \{w_2\} \rangle, \langle w_1, w_2, W \rangle\}$ and let $V(p, w_1) = V(p, w_2)$ be the case for all $p \in \mathcal{AV}$. It is easy to see that $(\alpha \mathbin{\Box\!\!\rightarrow} \beta) \rightarrow \beta$ is valid in $\mathcal{M}_C$: By $V_{\Box\!\rightarrow}$ all formulas $(\alpha_1 \mathbin{\Box\!\!\rightarrow} \beta)$, $(\alpha_2 \mathbin{\Box\!\!\rightarrow} \beta)$, ... are true at a world w if β is true at all worlds accessible by R_X, where $X \subseteq W$. From our assumptions it follows that for all formulas β holds that β is true at world w_1 iff β is true at w_2. Hence $(\alpha \mathbin{\Box\!\!\rightarrow} \beta) \rightarrow \beta$ is the case, even if $\forall X \subseteq W \forall w \in W(wR_X w)$ does not hold.

In addition there are also non-standard Segerberg frames on which, for example, $(\alpha \mathbin{\Box\!\!\rightarrow} \beta) \rightarrow \beta$ is valid, but which do not satisfy the C-corresponding frame condition C_T: Let $\mathcal{F}_S = \langle W, R, P \rangle$ be an S-frame for which $\forall w_1, w_2 \in W(\forall X \in P(w_1 \in X \Leftrightarrow w_2 \in X) \wedge \forall Y \in P(w_1 \neq w_2 \rightarrow w_1 R_Y w_2))$ holds. It is easy to see (by the same considerations as described before) that $(\alpha \mathbin{\Box\!\!\rightarrow} \beta) \rightarrow \beta$ is valid on $\mathcal{F}_S$ although C_T does not hold for subsets of P.

Let us now define the notion of standard Segerberg frames:

Definition 4.24. *Let $\mathcal{F}_S = \langle W, R, P\rangle$ be a Segerberg frame and let C_α be a frame condition which C-corresponds to axiom schema α (see Definition 4.20). Then, $\mathcal{F}_S$ is a standard Segerberg frame for* **CK**+α *iff C_α holds for all $X \in P$.*

I exclude by Definition 4.24 cases such as the non-standard Segerberg frame described before. I furthermore take C-correspondence rather than S-correspondence here as basis, since I do not have a genuine interest in the parameter P of Segerberg frames (see previous section). Additionally, I sometime omit reference to **CK** and say that a frame $\mathcal{F}_S$ is standard w.r.t. α rather than w.r.t. **CK**+ α.

Let us now turn to our terminology regarding standard Segerberg frames: I abbreviate standard Segerberg frames and classes of standard Segerberg frames by $\mathcal{F}_S^{\mathrm{st}}, \mathcal{F}_S^{\mathrm{st}\prime}, \mathcal{F}_S^{\mathrm{st}\prime\prime}, \ldots$ and $\mathbf{F}_S^{\mathrm{st}}, \mathbf{F}_S^{\mathrm{st}\prime}, \mathbf{F}_S^{\mathrm{st}\prime\prime}, \ldots$, respectively.[8] Moreover, models based on standard Segerberg frames (short: standard Segerberg models) are referred to by $\mathcal{M}_C^{\mathrm{st}}, \mathcal{M}_C^{\mathrm{st}\prime}, \mathcal{M}_C^{\mathrm{st}\prime\prime}, \ldots$ Finally, the concepts of validity, logical consequence and (simultaneous) satisfiability regarding standard Segerberg frames, standard Segerberg models, and classes of standard Segerberg frames are to be defined analogously to the respective concepts for (simple) Segerberg frames, (simple) Segerberg models, and classes of (simple) Segerberg frames in Section 4.3.4, respectively.

4.3.7 Notions of Soundness and Completeness

In the previous sections I described a range of validity and satisfiability concepts w.r.t. Chellas and Segerberg frames [models] and standard Segerberg frames. Based on the concepts of derivability, validity, and logical consequence defined earlier, one can distinguish the notions of soundness and completeness described in Table 4.2. So, which among these notions are, then, the appropriate notions of soundness and completeness?

Before I address the above question, observe that I do not distinguish be-

[8] I do not define symbols for classes of standard Segerberg models, because these notions do not play any role here.

Table 4.2
Notions of Soundness and Completeness in CS Semantics

L is sound w.r.t. $\mathbf{M}_S$	iff	$\forall\alpha(\vdash_{\mathbf{L}} \alpha \Rightarrow \models_{\mathbf{M}_S} \alpha)$
L is sound w.r.t. $\mathbf{M}_C$	iff	$\forall\alpha(\vdash_{\mathbf{L}} \alpha \Rightarrow \models_{\mathbf{M}_C} \alpha)$
L is sound w.r.t. $\mathbf{F}_S$	iff	$\forall\alpha(\vdash_{\mathbf{L}} \alpha \Rightarrow \models_{\mathbf{F}_S} \alpha)$
L is sound w.r.t. $\mathbf{F}_C$	iff	$\forall\alpha(\vdash_{\mathbf{L}} \alpha \Rightarrow \models_{\mathbf{F}_C} \alpha)$
L is weakly (w.) complete w.r.t. $\mathbf{M}_S$ †	iff	$\forall\alpha(\models_{\mathbf{M}_S} \alpha \Rightarrow \vdash_{\mathbf{L}} \alpha)$
L is w. complete w.r.t. $\mathbf{M}_C$ †	iff	$\forall\alpha(\models_{\mathbf{M}_C} \alpha \Rightarrow \vdash_{\mathbf{L}} \alpha)$
L is w. complete w.r.t. $\mathbf{F}_S$ †	iff	$\forall\alpha(\models_{\mathbf{F}_S} \alpha \Rightarrow \vdash_{\mathbf{L}} \alpha)$
L is w. complete w.r.t. $\mathbf{F}_S^{st}$	iff	$\forall\alpha(\models_{\mathbf{F}_S^{st}} \alpha \Rightarrow \vdash_{\mathbf{L}} \alpha)$
L is w. complete w.r.t. $\mathbf{F}_C$	iff	$\forall\alpha(\models_{\mathbf{F}_C} \alpha \Rightarrow \vdash_{\mathbf{L}} \alpha)$
L is strongly (s.) complete w.r.t. $\mathbf{M}_S$ ‡	iff	$\forall\Gamma,\alpha(\Gamma \models_{\mathbf{M}_S} \alpha \Rightarrow \Gamma \vdash_{\mathbf{L}} \alpha)$
L is s. complete w.r.t. $\mathbf{M}_C$ ‡	iff	$\forall\Gamma,\alpha(\Gamma \models_{\mathbf{M}_C} \alpha \Rightarrow \Gamma \vdash_{\mathbf{L}} \alpha)$
L is s. complete w.r.t. $\mathbf{F}_S$ ‡	iff	$\forall\Gamma,\alpha(\Gamma \models_{\mathbf{F}_S} \alpha \Rightarrow \Gamma \vdash_{\mathbf{L}} \alpha)$
L is s. complete w.r.t. $\mathbf{F}_S^{st}$	iff	$\forall\Gamma,\alpha(\Gamma \models_{\mathbf{F}_S^{st}} \alpha \Rightarrow \Gamma \vdash_{\mathbf{L}} \alpha)$
L is s. complete w.r.t. $\mathbf{F}_C$	iff	$\forall\Gamma,\alpha(\Gamma \models_{\mathbf{F}_C} \alpha \Rightarrow \Gamma \vdash_{\mathbf{L}} \alpha)$

Note. The logic **L** is presupposed to be an extension of **CK**. I marked equivalent notions of completeness (see text) by † and ‡, respectively.

tween weak and strong versions of soundness, since for System **CK** and all its extensions investigated in Chapters 5–7 the Deduction Theorem holds (cf. Section 4.2.6): $\Gamma \vdash \alpha$ is the case iff $\vdash \bigwedge \Gamma_f \rightarrow \alpha$, where Γ_f represents a finite subset of a (possibly infinite) set of formulas Γ and $\bigwedge \Gamma$ stands for the conjunction of all elements of Γ (cf. Schurz, 2002/2005a, p. 453, see also Section 4.2.1). So, derivability of formulas from formula sets (strong version) is equivalent to derivability of formulas (weak version). Due to that fact, strong and weak versions of soundness are logically equivalent.

W.r.t. completeness concepts one can distinguish between strong and weak versions. All strong versions (for example w.r.t. $\mathbf{F}_C$) imply the respective weak versions but not vice versa (cf. Schurz, 2002/2005a, p. 454f). I shall, hence, focus on strong completeness versions in this section.

Strong completeness w.r.t. classes of Segerberg models $\mathbf{M}_S$ is equivalent to strong completeness w.r.t. classes of Chellas models: For every Segerberg model $\mathcal{M}_S = \langle W, R, P, V \rangle$ the model $\mathcal{M}_C = \langle W, R', V \rangle$ is, on the basis of Lemma 4.23, a Chellas model, where R'_X is the expansion of R_X to arbitrary subsets X of W. Moreover, for every Chellas model $\mathcal{M}_C = \langle W, R, V \rangle$ a Segerberg model $\mathcal{M}_S = \langle W, R, P_V, V \rangle$ exists (with the same parameters W, R, and V), where $P_V = \{||\alpha||^{\mathcal{M}_C} \mid \mathcal{M}_C = \langle W, R, V \rangle$ and α is a formula of $\mathcal{L}_{\mathrm{CL}}\}$. So, since the valuation function V remains the same in all these cases, it is the case that, if a logic $\mathbf{L}$ is strongly complete w.r.t. some class of Segerberg models $\mathbf{M}_S$, then it is strongly complete w.r.t. some class of Chellas models $\mathbf{M}_C$, as described above, and vice versa.

One can easily show that strong completeness w.r.t. classes of Chellas models $\mathbf{M}_C$ is trivial in the sense that every logic $\mathbf{L}$ which is an extension of **CK** is complete w.r.t. classes of Chellas models. This is due to the fact that for every logic $\mathbf{L}$ which is an extension of **CK** (except the inconsistent one), one can define a canonical model $\mathcal{M}^c$ (see Section 6.1.2; cf. Schurz, 2002/2005a, p. 455f) such that all formulas in $\mathbf{L}$ are valid in $\mathcal{M}^c$. So, all logics $\mathbf{L}$ are strongly complete w.r.t. a class of Chellas models, namely the class containing the respective canonical model (or in case of the inconsistent logic the empty class of Chellas models) (cf. Hughes & Cresswell, 1996/2003, p. 159).

At first glance one might think that strong completeness w.r.t. Segerberg frames is a stronger notion than strong completeness w.r.t. Chellas models. This is – as I shall prove below – not the case. A closer look reveals that this is not too surprising: Segerberg frames can in fact be viewed as generalizations of general frames in Kripke semantics (see Schurz, 2002/2005a, p. 461) and strong completeness w.r.t. general frames in Kripke semantics is equivalent to strong completeness w.r.t. Kripke models (see Schurz, 2002/2005a, p. 461).

I shall now prove a stronger result w.r.t. CS semantics, namely that a logic $\mathbf{L}$ is strongly characterized by a class of Segerberg frames $\mathbf{F}_S$ iff it is strongly characterized by a class of Chellas models $\mathbf{M}_C$, where strong characterization is defined as follows: A logic **L** as strongly characterized by a class of Chellas models $\mathbf{M}_C$ [by a class of Segerberg frames $\mathbf{F}_S$] iff **L** is sound and strongly complete w.r.t. $\mathbf{M}_C$ [w.r.t. $\mathbf{F}_S$]. Let us prove now the following theorem:

Theorem 4.25. *A logic* **L** *is strongly characterized by some class of Chellas models* $\mathbf{M}_C$ *(short: strong* $\mathbf{M}_C$*-characterization) iff* **L** *is strongly characterized by some class of Segerberg frames* $\mathbf{F}_S$ *(short: strong* $\mathbf{F}_S$*-characterization).*

Proof. We obtain Theorem 4.25 by Lemmata 4.26 and 4.27. □

Lemma 4.26. *A logic* **L** *is strongly characterized by some class of Chellas models* $\mathbf{M}_C$ *if* **L** *is strongly characterized by some class of Segerberg frames* $\mathbf{F}_S$.

Proof. To prove strong characterization of a logic **L** by a class of Chellas models ($\mathbf{M}_C$-characterization), we show that (a) arbitrary **L**-consistent formula sets are satisfiable in a Chellas model $\mathcal{M}_C$, and that (b) all formulas in **L** are valid in that given model $\mathcal{M}_C$. By (a), (b), and the Consistency Lemma (Schurz, 2002/2005a, p. 455; see also Lemma 6.4) this implies that (i) **L** is strongly complete w.r.t. a class of Chellas models $\mathbf{M}_C$ and that (ii) **L** is sound w.r.t. the same class of Chellas models $\mathbf{M}_C$. Hence, we shall conclude by the definition of strong $\mathbf{M}_C$-characterization that **L** is strongly characterized by a class of Chellas models $\mathbf{M}_C$.

Suppose that a logic **L** is strongly characterized by some class of Segerberg frames $\mathbf{F}_S$. By the notion of strong $\mathbf{F}_S$-characterization, logic **L** is sound w.r.t. $\mathbf{F}_S$ and strongly complete w.r.t. $\mathbf{F}_S$. This implies on the basis of the Consistency Lemma (Schurz, 2002/2005a, p. 455; see also Lemma 6.4) that (c) if Γ is an **L**-consistent formula set, then Γ is satisfiable in some Segerberg model $\mathcal{M}_S = \langle W, R, P, V\rangle$ based on a Segerberg frame $\mathcal{F}_S = \langle W, R, P\rangle$ so that (d) all formulas in **L** are valid on $\mathcal{F}_S$.

Let Γ be an **L**-consistent set of formulas. Then it follows by (c) that Γ is satisfiable in some Segerberg model $\mathcal{M}_S = \langle W, R, P, V\rangle$ based on a Segerberg frame $\mathcal{F}_S = \langle W, R', P\rangle$. This implies that $\langle W, R', V\rangle$ is a Chellas model, where R'_X is the expansion of R_X to arbitrary subsets X of W (see proof of Lemma 4.23). Hence, (a′) Γ is satisfiable in $\mathcal{M}_C = \langle W, R', V\rangle$. Moreover, by (d) all formulas in the formula set **L** are valid on $\mathcal{F}_S$ and, hence, in $\mathcal{M}_S$. Since the valuation function is the same for $\mathcal{M}_S$ and $\mathcal{M}_C$, it follows that (b′) all formulas in **L** are valid in $\mathcal{M}_C$. Points (a′) and (b′) imply by the above considerations that **L** is strongly $\mathbf{M}_C$-characterized. □

Lemma 4.27. *A logic* **L** *is strongly characterized by some class of Chellas models* $\mathbf{M}_C$ *only if* **L** *is strongly characterized by some class of Segerberg frames* $\mathbf{F}_S$.

Proof. To prove strong characterization of a logic **L** w.r.t. a class of Segerberg frames ($\mathbf{F}_S$-characterization), we show that (a) arbitrary **L**-consistent formula sets are satisfiable in a Segerberg model $\mathcal{M}_S = \langle W, R, P, V\rangle$ and that (b) all formulas in **L** are valid on the Segerberg frame $\mathcal{F}_S = \langle W, R', P\rangle$, on which $\mathcal{M}_S$ is based. By the Consistency Lemma (Schurz, 2002/2005a, p. 455; see also Lemma 6.4) points (a) and (b) imply that (i) **L** is strongly complete w.r.t. a class of Segerberg frames $\mathbf{F}_S$ and that (ii) **L** is sound w.r.t. the same class of Segerberg frames $\mathbf{F}_S$. Hence, we can conclude by definition of strong $\mathbf{F}_S$-characterization that **L** is strongly characterized by a class of Segerberg frames $\mathbf{F}_S$.

Let logic **L** be strongly characterized by some class of Chellas models $\mathbf{M}_C$. Then, on the basis of the notion of strong $\mathbf{M}_C$-characterization, it follows that logic **L** is sound w.r.t. $\mathbf{M}_S$ and strongly complete w.r.t. $\mathbf{M}_S$. This implies by the Consistency Lemma (Schurz, 2002/2005a, p. 455; see also Lemma 6.4) that (c) if Γ is a **L**-consistent formula set, then Γ is satisfiable in some Chellas model $\mathcal{M}_C$ so that (d) all formulas in **L** are valid in $\mathcal{M}_C$.

Suppose that Γ is a **L**-consistent set of formulas. Then, (c) implies that Γ is satisfiable in some Chellas model $\mathcal{M}_C = \langle W, R, V\rangle$. We now define P_V as $\{||\alpha||^{\mathcal{M}_C} \mid \mathcal{M}_C = \langle W, R, V\rangle$ and α is a formula of $\mathcal{L}_{\mathrm{CL}}\}$. It is, then, easy to show that $\mathcal{M}_S = \langle W, R', P_V, V'\rangle$ is a Segerberg model, where R'_X is the restriction of R_X to elements of P. Since $\mathcal{M}_S$ and $\mathcal{M}_C$ agree in the parameters W, R_X

$[R'_X]$ (for syntactically representable sets X), and V, it follows that (a′) Γ is satisfiable in $\mathcal{M}_S$. Hence, it remains to be shown that all formulas in $\mathbf{L}$ are valid on $\mathcal{F}_S$ where $\mathcal{F}_S = \langle W, R', P_V \rangle$.

Let $\mathcal{M}_S = \langle W, R', P_V, V' \rangle$ be an arbitrary Segerberg model based on $\mathcal{M}_S = \langle W, R, V \rangle$. Moreover, let α be an arbitrary formula such that $\alpha \in \mathbf{L}$. Then, α is by point (d) valid in $\mathcal{M}_C$. Since the definition of a logic implies that $\mathbf{L}$ is closed under substitution (see Section 4.2.2), it also follows that $s(\alpha)$ is valid in $\mathcal{M}_C = \langle W, R, V \rangle$ for any substitution instance $s(\alpha)$ of α. This means that for all substitution instances $s_1(\alpha)$, $s_2(\alpha)$, ... and worlds $w \in W$ holds that $V(\alpha, w) = V(s_1(\alpha), w) = V(s_2(\alpha), w) = \ldots$ Now, let V and V' be determined as above, and suppose that $(*)$ $V(s(\alpha), w) = V'(\alpha, w)$ holds for every world w in W, every formula α of $\mathcal{L}_{\mathrm{CL}}$ and some substitution instance $s(\alpha)$ of α. Then, it follows that $V(\alpha, w) = V'(\alpha, w)$ for all $w \in W$ and $\alpha \in \mathbf{L}$. So, all formulas in $\mathbf{L}$ are also valid in $\mathcal{M}_S = \langle W, R', P_V, V' \rangle$. Hence, $(*)$ implies that (b′) all formulas $\alpha \in \mathbf{L}$ are also valid on $\mathcal{F}_S$.

We shall now prove $(*)$ by induction on the construction of formulas. Rather than proving $(*)$ directly we show that $\|s(\alpha)\|^{\mathcal{M}_C} = \|\alpha\|^{\mathcal{M}_S}$ for some substitution instance $s(\alpha)$ of α, which – as one can easily establish – implies $(*)$ by the definitions of $\mathcal{M}_C$, $\mathcal{M}_S$, and $\mathrm{Def}_{\|\cdot\|}$.

Induction Basis: Let p_i be an arbitrary propositional formula ($i \in \mathbb{N}$). Then, $\|p_i\|^{\mathcal{M}_S} \in P_V$ by Definitions 4.17 and 4.18. P_V is by definition the subset of W such that for all elements X of P_V holds that $X = \|\alpha_i\|^{\mathcal{M}_C}$ for some formula α_i in language $\mathcal{L}_{\mathrm{CL}}$ ($i \in \mathbb{N}$). Hence, it is the case that $\|p_i\|^{\mathcal{M}_S} = \|\alpha_i\|^{\mathcal{M}_C}$ for some formula α_i of $\mathcal{L}_{\mathrm{CL}}$. Since every formula α_i of $\mathcal{L}_{\mathrm{CL}}$ is a substitution instance of an atomic formula p_i, it holds that $\|p_i\|^{\mathcal{M}_S} = \|s(p_i)\|^{\mathcal{M}_C}$ for some substitution instance of p_i.

Induction Steps: Due to the definitions of s, $V_\neg$, and $\mathrm{Def}_{\|\cdot\|}$ it is the case that $\|s(\neg\alpha)\|^{\mathcal{M}_C} = \|\neg s(\alpha)\|^{\mathcal{M}_C} = W - \|s(\alpha)\|^{\mathcal{M}_C}$. By the induction hypothesis it holds that $\|s(\alpha)\|^{\mathcal{M}_C} = \|\alpha\|^{\mathcal{M}_S}$. Thus, it follows that $\|s(\neg\alpha)\|^{\mathcal{M}_C} = W - \|\alpha\|^{\mathcal{M}_S}$. Since $V_\neg$ and $\mathrm{Def}_{\|\cdot\|}$ give us that $W - \|\alpha\|^{\mathcal{M}_S} = \|\neg\alpha\|^{\mathcal{M}_S}$, we get $\|s(\neg\alpha)\|^{\mathcal{M}_C} = \|\neg\alpha\|^{\mathcal{M}_S}$.

By definition of s, $V_\vee$, and $\mathrm{Def}_{\|\cdot\|}$ we have that $\|s(\alpha \vee \beta)\|^{\mathcal{M}_C} = \|s(\alpha) \vee$

$s(\beta)||^{\mathcal{M}_C} = ||s(\alpha)||^{\mathcal{M}_C} \cup ||s(\beta)||^{\mathcal{M}_C}$. By the induction hypothesis, we get $||s(\alpha)||^{\mathcal{M}_C} = ||\alpha||^{\mathcal{M}_S}$ and $||s(\beta)||^{\mathcal{M}_C} = ||\beta||^{\mathcal{M}_S}$. This implies that $||s(\alpha \vee \beta)||^{\mathcal{M}_C} = ||\alpha||^{\mathcal{M}_S} \cup ||\beta||^{\mathcal{M}_S}$. Since by $V_\vee$ and $\text{Def}_{||\cdot||}$ it is the case that $||\alpha||^{\mathcal{M}_S} \cup ||\beta||^{\mathcal{M}_S} = ||\alpha \vee \beta||^{\mathcal{M}_S}$, we get that $||s(\alpha \vee \beta)||^{\mathcal{M}_C} = ||\alpha \vee \beta||^{\mathcal{M}_S}$.

On the basis of the definition of s, $V_{\Box\!\!\rightarrow}$, and $\text{Def}_{||\cdot||}$ we have $||s(\alpha \Box\!\!\rightarrow \beta)||^{\mathcal{M}_C} = ||s(\alpha) \Box\!\!\rightarrow s(\beta)||^{\mathcal{M}_C} = \{w \,|\, \forall w'(wR_{||s(\alpha)||^{\mathcal{M}_C}} w' \Rightarrow w' \in ||s(\beta)||^{\mathcal{M}_C})\}$. The induction hypothesis gives us that $||s(\alpha)||^{\mathcal{M}_C} = ||\alpha||^{\mathcal{M}_S}$ and that $||s(\beta)||^{\mathcal{M}_C} = ||\beta||^{\mathcal{M}_S}$. Hence, we get $||s(\alpha \Box\!\!\rightarrow \beta)||^{\mathcal{M}_C} = \{w \,|\, \forall w'(wR_{||\alpha||^{\mathcal{M}_S}} w' \Rightarrow w' \in ||\beta||^{\mathcal{M}_S})\}$. As $V_{\Box\!\!\rightarrow}$ and $\text{Def}_{||\cdot||}$ imply that $\{w \,|\, \forall w'(wR_{||\alpha||^{\mathcal{M}_S}} w' \Rightarrow w' \in ||\beta||^{\mathcal{M}_S}) = ||\alpha \Box\!\!\rightarrow \beta||^{\mathcal{M}_S}$, we obtain $||s(\alpha \Box\!\!\rightarrow \beta)||^{\mathcal{M}_C} = ||\alpha \Box\!\!\rightarrow \beta||^{\mathcal{M}_S}$.

□

Let us now sum up the above discussion on appropriate notions of completeness. We saw earlier that (i) the strong versions are preferable to weak versions, that (ii) completeness w.r.t. classes of Segerberg models $\mathbf{M}_S$, classes of Chellas models $\mathbf{M}_C$ and classes of Segerberg frames $\mathbf{F}_S$ are equivalent, and (iii) that they are trivial in the sense that every logic **L** which is an extension of **CK** is then strongly frame complete. This leaves us with two further options: (a) strong completeness w.r.t. standard Segerberg frames $\mathbf{F}_C^{\text{st}}$ and (b) strong completeness w.r.t. classes of Chellas frames $\mathbf{F}_C$.

Completeness w.r.t. classes of Chellas frames restricts the characterization of logical systems to purely structural conditions on W and R. However, completeness w.r.t. classes of Segerberg frames takes in addition the parameter P into account, which allows us to limit the set of valuation functions of Segerberg models (hence Theorem 4.25). Since it is philosophically preferable for a logical system to admit all possible valuations of all non-logical symbols (Schurz, 2002/2005a, p. 447), notions of soundness and completeness w.r.t. classes Chellas frames are more appropriate. However, completeness w.r.t. classes of Chellas frames meets certain obstacles not present in Kripke semantics. In particular there are strictly more accessibility relations than possible worlds in any Chellas frame (point (a) of Section 4.3.2). Moreover, accessibility relations relativized to syntactically non-representable sets are defined (point (c2) of Section 4.3.2) although only accessibility relations relativized to syntactically representable sets impact the truth of formulas at possible worlds in a model (point (c1) of Section 4.3.2). It is exactly due to these obstacles that it seems doubtful a full characterization of the lattice of systems discussed in Chapter 5 is viable in terms of classes of Chellas frames.

Hence, our final option is completeness w.r.t. standard Segerberg frames. This notion of strong standard Segerberg completeness does not suffer from the problems associated with strong Chellas frame completeness (as I shall show in Chapter 6). Moreover, it is – as I will argue in the remainder of this section – non-trivial in the sense that not every logic **L** which is an extension of **CK** is complete w.r.t. standard Segerberg frames. I shall not prove this point here, but rather illustrate why Theorem 4.25 does not go through for completeness w.r.t. standard Segerberg frames (rather than w.r.t. simple Segerberg frames).

To prove Theorem 4.25, we relied on Lemma 4.27. In the proof for this lemma we constructed a Segerberg frame $\mathcal{F}_S = \langle W, R, P_V, V \rangle$ for each Chellas model $\mathcal{M}_C = \langle W, R, V \rangle$, where $P_V = \{||\alpha||^{\mathcal{M}_C} \mid \mathcal{M}_C = \langle W, R, V \rangle$ and α is a formula of $\mathcal{L}_{\mathrm{CL}}\}$. Such a move is not generally possible for standard Segerberg models. Take, for example, logic **CK**+T where T is $(\alpha \mathrel{\Box\!\!\rightarrow} \beta) \rightarrow \beta$. Given peculiar truth assignments, such as in the non-standard Chellas model $\mathcal{M}_C = \langle W, R, V \rangle$ described in Section 4.3.6, one cannot construct a standard Segerberg frame by using this definition of P_V. This is due to the fact that in $\mathcal{M}_C$ the relation R – whether restricted to P_V or not – does not satisfy C_{T}, namely it does not hold that for all worlds w and subsets X of the set of possible worlds W it is the case that wR_Xw (see Section 4.3.6).

Chapter 5

Frame Correspondence for a Lattice of Conditional Logics

In the present chapter I will discuss and provide formal results regarding frame correspondence w.r.t. CS semantics. I shall, for that purpose, focus exclusively on C-correspondence (i.e., correspondence w.r.t. Chellas frames), as defined in Section 4.3.5. I will employ the notion of C-correspondence rather than S-correspondence (i.e., correspondence w.r.t. Segerberg frames, see Section 4.3.5), since I intend to align systems based on the Chellas-Segerberg (CS) semantics both in terms axiom schemata and restrictions on the accessibility relation only. C-correspondence seems to be the more appropriate notion for this purpose. In addition, the formal results regarding C-correspondence proofs serve as basis for the philosophical discussion in Chapter 7.

Let me now give an example for C-correspondence. For instance, the axiom schema $(\alpha \mathrel{\Box\!\!\rightarrow} \beta) \rightarrow \beta$ (Principle T) "C-corresponds" to the frame condition C_{T}, namely $\forall X \subseteq W \forall w \in W(wR_X w)$, where X is a subset of the set of possible worlds W. This means (by Definition 4.20) that Principle T is valid on a Chellas frame $\mathcal{F}_C = \langle W, R \rangle$ iff the frame restriction C_{T} holds for $\mathcal{F}_C$.

Note that all (interesting) frame conditions in CS semantics must be second-order, in the sense that they involve also quantification over sets of possible worlds $X_1, X_2, \ldots$ in addition to quantification over possible worlds. This is due to fact that all accessibility relations R_X in CS semantics are relativized to subsets of the sets of possible worlds W. In contrast, in Kripke semantics this is not the case. In Kripke semantics important classes of axioms correspond to first-order frame conditions which involve only quantification over possible worlds (cf. van Benthem, 2001, Section 2.2; cf. also Schurz, 2002/2005a, p. 451).

Furthermore, in Kripke semantics some correspondence results are trivial in the sense that there is a translation procedure, which allows one to construct corresponding frame conditions from modal formula schemata (cf. van Benthem, 2001, Section 1). In CS semantics we can obtain an analogous result by specifying a translation function (see Definition 5.1) from conditional logic schemata to C-corresponding trivial frame conditions. These trivial frame conditions, however, are more complicated and from a philosophical standpoint less perspicuous than necessary. For example, the trivial frame condition which corresponds to principle $(\alpha \mathrel{\Box\!\!\rightarrow} \beta) \rightarrow \beta$ is *not* the frame condition C_T described above, but rather $\forall X, Y \forall w, w'((wR_Xw' \Rightarrow w' \in Y) \Rightarrow w \in Y)$. Here the reference to Y in the trivial frame condition is in a sense superfluous, since one can prove – as we shall do – that $(\alpha \mathrel{\Box\!\!\rightarrow} \beta) \rightarrow \beta$ C-corresponds to C_T which does not make reference to a second variable w.r.t. sets of possible worlds. The frame condition C_T is, hence, non-trivial insofar as it does not contain irrelevant occurrences of variables.

Note that I use the term 'trivial frame condition' in two non-equivalent senses. First, I refer to all frame conditions resulting from the translation function to be specified as 'trivial'. These "trivial" frame conditions need not be trivial in the sense discussed in the preceding parapraph, however. It should be clear from the context in which sense I use the term 'trivial frame condition'.

In this chapter I will first specify non-trivial frame conditions for the 29 conditional logic principles (see Section 5.1). I will then define a general translation procedure from axiom schemata to trivial frame conditions, show that this definition gives us the desired C-correspondence result and provide a definition of non-trivial frame conditions (see Section 5.2). Finally I will prove C-correspondence results for the 29 non-trivial frame conditions, where not already done by the trivial correspondence result (see Section 5.3). Note that I am at the present time neither aware of any definition or discussion of trivial and non-trivial frame conditions in the literature for CS semantics nor of any C-correspondence or S-correspondence proofs for any of the 29 conditional logic principles discussed here.

5.1 Non-Trivial Frame Conditions for a Lattice of Conditional Logics

Tables 5.1 and 5.2 list the 29 axiom schemata investigated in this and the following chapter. These conditional logic principles are mainly taken from indicative conditional logics systems, such as System **P** by Kraus et al. (1990) and Lehmann and Magidor (1992), and counterfactual conditional logic systems, such as D. Lewis (1973b). These 29 axiom schemata are not all logically independent from each other, but some of them are inter-derivable given the rules of System **CK** described in Section 4.2.6 (cf. Chapter 7). My terminology is in accordance with the literature in non-monotonic reasoning (i.e., Kraus et al., 1990; Lehmann & Magidor, 1992) rather than the conditional logic literature (e.g., Nute, 1980; Nute & Cross, 2001).

First, in Tables 5.1 and 5.2 axiom schemata are listed which are typically associated with System **P** and its extensions such as System $\mathbf{P}^{+}$ (see Section 3.6.1). The expressions 'Refl', 'CM', 'CC', and 'RM' stand for 'Reflexivity', 'Cautious Monotonicity', 'Cautious Cut', and 'Rational Monotonicity', respectively (see Schurz, 1998, p. 82; Kraus et al., 1990, p. 177f and p. 197; Lehmann & Magidor, 1992, p. 5f and p. 18). Thereby, the formula schema $\alpha \diamond\!\!\rightarrow \beta$ in the Principle RM abbreviates $\neg(\alpha \,\Box\!\!\rightarrow \neg\beta)$ ($\mathrm{Def}_{\diamond\rightarrow}$, Section 4.2.1). The labels 'Loop', 'Or', and 'S' are again taken from Kraus et al. (1990), but the authors do not explain what the names stand for. The respective axiom schemata suggests that these expressions abbreviate 'Loop of Antecedent and Consequent Formulas', 'Introduction of an "Or" in the Antecedent' and 'Shift of Conjuncts from the Antecedent to the Consequent',

Table 5.1
Axiom Schemata for Conditional Logics

System P	
$\alpha \mathrel{\Box\!\!\rightarrow} \alpha$	(Refl)
$(\alpha \mathrel{\Box\!\!\rightarrow} \gamma) \wedge (\alpha \mathrel{\Box\!\!\rightarrow} \beta) \rightarrow (\alpha \wedge \beta \mathrel{\Box\!\!\rightarrow} \gamma)$	(CM)
$(\alpha \wedge \beta \mathrel{\Box\!\!\rightarrow} \gamma) \wedge (\alpha \mathrel{\Box\!\!\rightarrow} \beta) \rightarrow (\alpha \mathrel{\Box\!\!\rightarrow} \gamma)$	(CC)
$(\alpha_0 \mathrel{\Box\!\!\rightarrow} \alpha_1) \wedge \ldots \wedge (\alpha_{k-1} \mathrel{\Box\!\!\rightarrow} \alpha_k) \wedge (\alpha_k \mathrel{\Box\!\!\rightarrow} \alpha_0) \rightarrow (\alpha_0 \mathrel{\Box\!\!\rightarrow} \alpha_k)\ (k \geq 2)$	(Loop)
$(\alpha \mathrel{\Box\!\!\rightarrow} \gamma) \wedge (\beta \mathrel{\Box\!\!\rightarrow} \gamma) \rightarrow (\alpha \vee \beta \mathrel{\Box\!\!\rightarrow} \gamma)$	(Or)
$(\alpha \wedge \beta \mathrel{\Box\!\!\rightarrow} \gamma) \rightarrow (\alpha \mathrel{\Box\!\!\rightarrow} (\beta \rightarrow \gamma))$	(S)
$(\neg\alpha \mathrel{\Box\!\!\rightarrow} \alpha) \rightarrow (\beta \mathrel{\Box\!\!\rightarrow} \alpha)$	(MOD)
Extensions of System P	
$(\alpha \mathrel{\Box\!\!\rightarrow} \gamma) \wedge (\alpha \mathrel{\Diamond\!\!\rightarrow} \beta) \rightarrow (\alpha \wedge \beta \mathrel{\Box\!\!\rightarrow} \gamma)$	(RM)
$(\alpha \mathrel{\Box\!\!\rightarrow} \beta) \vee (\alpha \mathrel{\Box\!\!\rightarrow} \neg\beta)$	(CEM)
Weak Probability Logic (Threshold Logic)	
$\neg(\top \mathrel{\Box\!\!\rightarrow} \bot)$	(P-Cons)
$(\alpha \wedge \beta \mathrel{\Box\!\!\rightarrow} \gamma) \wedge (\alpha \wedge \neg\beta \mathrel{\Box\!\!\rightarrow} \gamma) \rightarrow (\alpha \mathrel{\Box\!\!\rightarrow} \gamma)$	(WOR)
Monotonic Systems	
$(\alpha \wedge \beta \mathrel{\Box\!\!\rightarrow} \delta) \wedge (\gamma \mathrel{\Box\!\!\rightarrow} \beta) \rightarrow (\alpha \wedge \gamma \mathrel{\Box\!\!\rightarrow} \delta)$	(Cut)
$(\alpha \mathrel{\Box\!\!\rightarrow} \gamma) \rightarrow (\alpha \wedge \beta \mathrel{\Box\!\!\rightarrow} \gamma)$	(Mon)
$(\alpha \mathrel{\Box\!\!\rightarrow} \beta) \wedge (\beta \mathrel{\Box\!\!\rightarrow} \gamma) \rightarrow (\alpha \mathrel{\Box\!\!\rightarrow} \gamma)$	(Trans)
$(\alpha \mathrel{\Box\!\!\rightarrow} \beta) \rightarrow (\neg\beta \mathrel{\Box\!\!\rightarrow} \neg\alpha)$	(CP)
Bridge Principles	
$(\alpha \mathrel{\Box\!\!\rightarrow} \beta) \rightarrow (\alpha \rightarrow \beta)$	(MP)
$\alpha \wedge \beta \rightarrow (\alpha \mathrel{\Box\!\!\rightarrow} \beta)$	(CS)
$(\neg\alpha \mathrel{\Box\!\!\rightarrow} \alpha) \rightarrow \alpha$	(TR)
$(\top \mathrel{\Box\!\!\rightarrow} \alpha) \rightarrow \alpha$	(Det)
$\alpha \rightarrow (\top \mathrel{\Box\!\!\rightarrow} \alpha)$	(Cond)

For abbreviations see text.

Table 5.2
Axiom Schemata for Conditional Logics (Continued)

Collapse Conditions Material Implication	
$\beta \rightarrow (\alpha \Box\!\!\rightarrow \beta)$	(VEQ)
$\neg\alpha \rightarrow (\alpha \Box\!\!\rightarrow \beta)$	(EFQ)
Traditional Extensions	
$(\alpha \Box\!\!\rightarrow \beta) \rightarrow (\alpha \Diamond\!\!\rightarrow \beta)$	(D)
$(\alpha \Box\!\!\rightarrow \beta) \rightarrow \beta$	(T)
$\alpha \rightarrow (\alpha \Box\!\!\rightarrow (\alpha \Diamond\!\!\rightarrow \beta))$	(B)
$(\alpha \Box\!\!\rightarrow \beta) \rightarrow (\alpha \Box\!\!\rightarrow (\alpha \Box\!\!\rightarrow \beta))$	(4)
$(\alpha \Diamond\!\!\rightarrow \beta) \rightarrow (\alpha \Box\!\!\rightarrow (\alpha \Diamond\!\!\rightarrow \beta))$	(5)
Iterated Principles	
$(\alpha \wedge \beta \Box\!\!\rightarrow \gamma) \rightarrow (\alpha \Box\!\!\rightarrow (\beta \Box\!\!\rightarrow \gamma))$	(Ex)
$(\alpha \Box\!\!\rightarrow (\beta \Box\!\!\rightarrow \gamma)) \rightarrow (\alpha \wedge \beta \Box\!\!\rightarrow \gamma)$	(Im)

For abbreviations see text.

respectively.

Let us now focus on Principles MOD and CEM. Nute and Cross (2001, p. 10) and Nute (1980, p. 52f) do not motivate their names for the Principles MOD and CEM. The names 'CEM' and 'MOD', however, presumably abbreviate 'Modality' and 'Conditional Excluded Middle', respectively. The Principle MOD is also used in D. Lewis' (1973b) axiomatization of his counterfactual System **VC** (p. 132) and Principle CEM is specific to Stalnaker's (1968; Stalnaker & Thomason, 1970) conditional logic (D. Lewis, 1973b, p. 79).

The Principles P-Cons and WOR are discussed in the literature on weak probabilistic logics such as the threshold logic **O** (Hawthorne & Makinson, 2007, p. 252 and Footnote 2; cf. also Section 3.6.2) and the Principle P-Cons is also employed by Schurz (1998, p. 88 and p. 90). The name 'WOR' stands for 'Weak OR' (Hawthorne & Makinson, 2007, p. 252), whereas 'P-Cons' abbreviates 'Probabilistic Consistency'.

The names of the monotonic Principles 'Mon', 'Trans', and 'CP' stand for 'Monotonicity', 'Transitivity', and 'Contraposition', respectively. The Principle Cut does not correspond to "Cut" in Kraus et al. (1990, p. 177) and Lehmann and Magidor (1992, p. 6), the latter of which translates into our notion of Cautious Cut, denoted by 'CC'. My notion of Cut corresponds to Principle (9) in Lehmann and Magidor (1992, p. 6). The names of the bridge Principles MP, TR, Det, Cond, VEQ, and EFQ abbreviate 'Modus Ponens', 'Total Reflexivity', 'Detachment', 'Conditionalization', 'Verum Ex Quodlibet', and 'Ex Falso Quodlibet', respectively. The names 'MP' and 'CS' on the one hand and 'VEQ' and 'EFQ' on the other are taken from Nute (1980, p. 52 and p. 40; see also Nute & Cross, 2001, p. 10 and p. 14) and Weingartner and Schurz (1986, p. 10), respectively. Unfortunately, neither Nute (1980) nor Nute and Cross (2001) explain what 'CS' stands for, but 'CS' presumably abbreviates 'Conjunctive Sufficiency'. The use of the label 'TR' follows D. Lewis (1973b, p. 121). Furthermore, Schurz's (1998, p. 83) Principle Poss ("Possibility") is p.c. equivalent to TR. (We will not provide a proof for this fact here.)

Axiom Schemata D, T, B, 4, and 5 represent generalizations of the respective modal logic principles in Kripke semantics (cf. Schurz, 2002/2005a, p. 451). I generalized the corresponding frame conditions – namely the seriality, reflexivity, symmetry, transitivity, and euclidicity of the accessibility relation (cf. Schurz, 2002/2005a, p. 451) – to CS semantics (see Table 5.4) and identified the Conditional Logic Principles D, T, B, 4, and 5 (see Table 5.2). Finally, the names 'Ex' and 'Im' abbreviate 'Exportation' and 'Importation', respectively. The Principles Ex and Im are, for example, discussed by Arló-Costa (2001, 2007) and McGee (1989; see also McGee, 1985, p. 465).

The non-trivial frame conditions for the principles in Tables 5.1 and 5.2 can be found in Tables 5.3 and 5.4. A range of non-trivial frame conditions for conditional logic axioms were not specified in the conditional logic literature (i.e., Segerberg, 1989; Chellas, 1975). Chellas (1975, p. 142f) and Segerberg (1989, p. 163) identified non-trivial C-corresponding frame conditions for Axiom Schemata Refl, MP, and Or on the one hand and CM,

Table 5.3

Axioms and Corresponding Frame Conditions in CS Semantics

System P	
Refl	$\forall w, w'(wR_Xw' \Rightarrow w' \in X)$
CM	$\forall w(\forall w'(wR_Xw' \Rightarrow w' \in Y) \Rightarrow \forall w'(wR_{X\cap Y}w' \Rightarrow wR_Xw'))$
CC	$\forall w(\forall w'(wR_Xw' \Rightarrow w' \in Y) \Rightarrow \forall w'(wR_Xw' \Rightarrow wR_{X\cap Y}w'))$
Loop	$\forall w(\forall w'(wR_{X_0}w' \Rightarrow w' \in X_1) \wedge \ldots \wedge \forall w'(wR_{X_{k-1}}w' \Rightarrow w' \in X_k) \wedge$ $\forall w'(wR_{X_k}w' \Rightarrow w' \in X_0) \Rightarrow \forall w'(wR_{X_0}w' \Rightarrow w' \in X_k))\ (k \geq 2)$
Or	$\forall w, w'\,(wR_{X\cup Y}w' \Rightarrow wR_Xw' \vee wR_Yw')$
S	$\forall w, w'(wR_Xw' \wedge w' \in Y \Rightarrow wR_{X\cap Y}w')$
MOD	$\forall w(\forall w'(wR_{-X}w' \Rightarrow w' \in X) \Rightarrow \forall w'(wR_Yw' \Rightarrow w' \in X))$
Extensions of System P	
RM	$\forall w(\exists w'(wR_Xw' \wedge w' \in Y) \Rightarrow \forall w'(wR_{X\cap Y}w' \Rightarrow wR_Xw'))$
CEM	$\forall w, w', w''(wR_Xw' \wedge wR_Xw'' \Rightarrow w'' = w')$
Weak Probability Logic (Threshold Logic)	
P-Cons	$\forall w \exists w'(wR_Ww')$
WOR	$\forall w, w'(wR_Xw' \Rightarrow wR_{X\cap Y}w' \vee wR_{X\cap -Y}w')$
Monotonic Systems	
Cut	$\forall w(\forall w'(wR_Zw' \Rightarrow w' \in Y) \Rightarrow \forall w'(wR_{X\cap Z}w' \Rightarrow wR_{X\cap Y}w'))$
Mon	$\forall w, w'\,(wR_{X\cap Y}w' \Rightarrow wR_Xw')$
Trans	$\forall w(\forall w'(wR_Xw' \Rightarrow w' \in Y) \Rightarrow \forall w'(wR_Xw' \Rightarrow wR_Yw'))$
CP	$\forall w(\forall w'(wR_Xw' \Rightarrow w' \in Y) \Rightarrow \forall w'(wR_{-Y}w' \Rightarrow w' \in -X))$
Bridge Principles	
MP	$\forall w(w \in X \Rightarrow wR_Xw)$
CS	$\forall w(w \in X \Rightarrow \forall w'(wR_Xw' \Rightarrow w' = w))$
TR	$\forall w(\forall w'(wR_{-X}w' \Rightarrow w' \in X) \Rightarrow w \in X)$
Det	$\forall w(wR_Ww)$
Cond	$\forall w, w'(wR_Ww' \Rightarrow w' = w)$

For better readability outer universal quantifiers regarding sets of possible worlds $X, Y, \ldots, X_1, X_2, \ldots$ are omitted. For abbreviations see text.

Table 5.4
Axiom Schemata and Corresponding Frame Conditions in CS Semantics (Continued)

Collapse Conditions Material Implication	
VEQ	$\forall w, w'(wR_Xw' \Rightarrow w' = w)$
EFQ	$\forall w(w \in -X \Rightarrow \neg\exists w'(wR_Xw'))$
Traditional Extensions	
D	$\forall w \exists w'(wR_Xw')$
T	$\forall w(wR_Xw)$
B	$\forall w, w'(wR_Xw' \Rightarrow w'R_Xw)$
4	$\forall w, w', w''(wR_Xw' \wedge w'R_Xw'' \Rightarrow wR_Xw'')$
5	$\forall w, w', w''(wR_Xw' \wedge wR_Xw'' \Rightarrow w'R_Xw'')$
Iteration Principles	
Ex	$\forall w, w', w''(wR_Xw' \wedge w'R_Yw'' \Rightarrow wR_{X\cap Y}w'')$
Im	$\forall w, w'(wR_{X\cap Y}w' \Rightarrow \exists w''(wR_Xw'' \wedge w''R_Yw'))$

For better readability universal quantifiers regarding sets of possible worlds $X, Y, \ldots, X_1, X_2, \ldots$ are omitted. For abbreviations see text.

RM, S, Det, and Cond on the other hand. Segerberg (1989) discussed non-trivial frame conditions for two further frame conditions, namely his Axiom Schemata #2 and #6. Segerberg's Axiom Schema #2 is a variant to MOD from Table 5.1. Segerberg's version can be formulated as $(\neg\alpha \mathrel{\Box\!\!\rightarrow} \bot) \rightarrow (\beta \mathrel{\Box\!\!\rightarrow} \alpha)$. The Principle MOD and #2 are only equivalent given Refl. Thus, their corresponding frame conditions differ. Moreover, Segerberg's Axiom Schema #6 amounts to $(\alpha \mathrel{\Diamond\!\!\rightarrow} \beta) \wedge (\alpha \mathrel{\Box\!\!\rightarrow} (\beta \rightarrow \gamma)) \rightarrow (\alpha \wedge \beta \mathrel{\Box\!\!\rightarrow} \gamma)$. There is a related principle $(\alpha \mathrel{\Box\!\!\rightarrow} (\beta \rightarrow \gamma)) \rightarrow (\alpha \wedge \beta \mathrel{\Box\!\!\rightarrow} \gamma)$ (EHD, "Easy Half of the Deduction Theorem") sometimes discussed in the literature (e.g., Kraus et al., 1990, p. 180). I will, however, not cover these principles here.

I identified non-trivial frame conditions for a further 16 principles, namely for CC, Loop, MOD, CEM, P-Cons, WOR, Cut, Mon, Trans, CP, CS, TR, VEQ, EFQ, Ex, and Im. For the remaining 5 Principles D, T, B, 4, and 5 I took the reverse approach: I started with generalizations of frame conditions in Kripke semantics to CS semantics and identified the corresponding axiom schemata (see above).

The axiom schemata and frame conditions both specify a single lattice of conditional logic systems: The proof-theoretic side of this lattice is determined by System **CK** plus axiom schemata from Tables 5.1 and 5.2 and the model-theoretic side is described by Chellas frames (see Section 4.3.1) and (standard) Segerberg frames (see Section 4.3.3) plus frame conditions from Tables 5.3 and 5.4.

We identified each principle in Tables 5.1 and 5.2 with one of the non-trivial frame conditions in Tables 5.3 and 5.4. The upshot of our correspondence, soundness, and completeness proofs in Chapters 5 and 6 is that the model-theoretic side of the lattice and its proof-theoretic side concur in the following sense: Take any combination of principles from Tables 5.1 and 5.2 and their respective frame conditions from Tables 5.3 and 5.4 and add them to the basic proof theory and to the basic model theory, respectively. A proof theoretic system results which is (a) sound and strongly complete w.r.t. the class of standard Segerberg frames described by the model theoretic system (cf. Section 4.3.7). Moreover, (b) any of the principles chosen from Tables 5.1 and 5.2 and their semantic counterparts from Tables 5.3 and 5.4 C-correspond to each other (cf. Section 4.3.5). Note that we only have a completeness result for the whole lattice of system in terms of standard Segerberg frames rather than classes of Chellas frames. Hence, for the completeness property all variables pertaining to sets of possible worlds in the frame conditions in Tables 5.3 and 5.4 refer to elements of the additional parameter P rather than arbitrary sets of possible worlds.

5.2 The Notions of Trivial and Non-Trivial Frame Conditions

At the beginning of this chapter I illustrated the difference between trivial and non-trivial frame conditions by using the Principle T, namely (α $\Box\!\!\rightarrow$

$\beta) \rightarrow \beta$. Let me discuss here a further example, Principle CM, which is – unlike Principle T – a theorem of a range of conditional logics proposed in the literature (see Sections 7.2 and 7.3). CM is the following principle: $(\alpha \mathrel{\Box\!\!\rightarrow} \gamma) \wedge (\alpha \mathrel{\Box\!\!\rightarrow} \beta) \rightarrow (\alpha \wedge \beta \mathrel{\Box\!\!\rightarrow} \gamma)$ (see Table 5.1). Principle CM C-corresponds – as we shall prove in Section 5.3 – to the following non-trivial frame condition C_{CM} (see Table 5.3): $\forall X, Y \subseteq W \forall w \in W(\forall w'(wR_X w' \Rightarrow w' \in Y) \Rightarrow \forall w'(wR_{X\cap Y} w' \Rightarrow wR_X w'))$. C-correspondence gives us that whenever CM is valid in a Chellas frame $\mathcal{F}_C = \langle W, R\rangle$, then the frame condition C_{CM} holds for R, and vice versa.

The trivial counterpart for C_{CM} is the following: $\forall X, Y, Z \subseteq W \forall w \in W(\forall w' (wR_X w' \Rightarrow w' \in Z) \wedge \forall w'(wR_X w' \Rightarrow w' \in Y) \Rightarrow \forall w'(wR_{X\cap Y} w' \Rightarrow w' \in Z))$. The latter frame condition can be read off from axiom schemata by the following informal translation procedure (cf. van Benthem, 2001, Section 1): Use for any propositional connective the corresponding meta-language connective and translate any occurrence of a formula $\alpha \mathrel{\Box\!\!\rightarrow} \beta$ and α by $\forall w'(wR_X w' \Rightarrow w' \in Y)$ and $w \in X$, respectively, where for each axiom schema letter α, β, ... the same variable X, Y, ... is used and each new axiom schema letter is described by a new variable. Observe that the trivial frame conditions for Principles CM and T employ – compared to their non-trivial counterparts – instances of second-order variables in a spurious way. In the next section I shall specify this translation function in a formal way and prove that this translation function yields C-corresponding frame conditions. In Section 5.2.2 I will define and discuss the notion of non-trivial frame conditions.

5.2.1 A Translation Procedure from Axiom Schemata to Trivial Frame Conditions

Let me define the translations procedure which produces trivial frame conditions from formula schemata, where $\mathcal{L}_{\mathrm{FC}}$ refers to the first-order fragment for frame conditions informally described in Section 4.2.1. For that purpose I use the following linguistic conventions: First, I distinguish strictly between axiom schemata (w.r.t. the language $\mathcal{L}_{\mathrm{CL}}$, cf. Section 4.2.1) and axiom schema letters. While the former refers, for example, to all axiom schemata as described in Tables 5.1 and 5.2, the latter refers only to the letters A_1, A_2, ... Axiom schema letters are placeholders for arbitrary formulas $\alpha, \alpha_0, \alpha_1$, ... of $\mathcal{L}_{\text{CL}}$. Furthermore, $\alpha[A_1, \ldots, A_n]$ for $n \in \mathbb{N}$ is a name form variable (cf. Kleene, 1952, pp. 142–144; see also Schurz, 1997a, p. 45f) such that $A_1, \ldots, A_n$ are all and the only distinct axiom schema letters occurring in α. $\alpha[\alpha_1/A_1,\ldots,\alpha_n/A_n]$ denotes an instance of the axiom schema $\alpha[A_1,\ldots,A_n]$ in the language $\mathcal{L}_{\text{CL}}$ which results from replacing schematic letters A_i by formulas α_i of $\mathcal{L}_{\text{CL}}$ (for all $1 \leq i \leq n$).

Suppose that $\alpha = \alpha[A_1,\ldots,A_n]$ and that t_w is a translation function t_w from formulas of language $\mathcal{L}_{\text{CL}}$ to formulas of language $\mathcal{L}_{\text{FC}}$. Then, the universal closure $\mathcal{UC}(t_w(\alpha))$ is defined as the formula $\forall w \in W \forall X_1,\ldots,X_n \subseteq W(t_w(\alpha))$, where $w, X_1,\ldots,X_n$ occur free in $t_w(\alpha)$. Furthermore, in order to make the translations from axiom schemata to frame conditions more perspicuous I use Quinean corners (i.e., '$\ulcorner$' and '$\urcorner$') for the translations where $\ulcorner WZ \urcorner$ is defined as $\ulcorner W \urcorner \ulcorner Z \urcorner$ – that is the concatenation of the expressions W and Z – for expressions W and Z of language $\mathcal{L}_{\text{FC}}$.

Let me now specify the translation function, which gives us the corresponding frame conditions for formula schemata for language $\mathcal{L}_{\text{CL}}$:

Definition 5.1. *Let t_w for all w be a translation function from formulas of $\mathcal{L}_{\text{CL}}$ to formulas of $\mathcal{L}_{\text{FC}}$ (formally, for every w it holds that t_w: $\mathcal{L}_{\text{CL}} \rightarrow \mathcal{L}_{\text{FC}}$). Then, t determines trivial Chellas frame conditions for formulas of $\mathcal{L}_{\text{CL}}$ iff $t(\alpha) = \mathcal{UC}(t_w(\alpha))$, where t_w is determined the following way:*
(a) $t_w(A_j) = (w\ulcorner \in \urcorner X_j)$,
(b) $t_w(\neg\alpha) = \ulcorner \neg \urcorner t_w(\alpha)$,
(c) $t_w(\alpha \wedge \beta) = (t_w(\alpha)\ulcorner \vee \urcorner t_w(\beta))$, and
(d) $t_w(\alpha \mathrel{\Box\!\!\rightarrow} \beta) = \ulcorner \forall \urcorner w'(w \ulcorner R \urcorner_{\{w \in W \mid t_w(\alpha)\}} w' \ulcorner \Rightarrow \urcorner t_{w'}(\beta))$, where w' is a new variable.

Theorem 5.2. *Let $\mathcal{F}_C$ denote an arbitrary Chellas frame and let α denote an axiom schema for language $\mathcal{L}_{\text{CL}}$. Then, $\forall\alpha\forall\mathcal{F}_C(\models_{\mathcal{F}_C} \alpha$ iff $\models_{\mathcal{F}_C} t(\alpha))$.*

Proof. Let $\mathcal{F}_C = \langle W,R \rangle$ be an arbitrary Chellas frame. Furthermore, in the

following $\mathcal{M}_C^{\mathcal{F}_C} = \langle W, R, V\rangle$ refers to a Chellas model, which is based on $\mathcal{F}_C$. Then,
$\models_{\mathcal{F}_C} \alpha$ iff
$\forall \mathcal{M}_C^{\mathcal{F}_C} \forall \alpha_1, \ldots, \alpha_n \forall w(V(\alpha[\alpha_1/A_1, \ldots, \alpha_n/A_n], w) = 1)$ iff (by Lemma 5.4)[1]
$\forall \mathcal{M}_C^{\mathcal{F}_C} \forall \alpha_1, \ldots, \alpha_n \forall X_1, \ldots, X_n \forall w(X_1 = \{w \mid V(\alpha_1, w) = 1\} \wedge \ldots \wedge X_n = \{w \mid V(\alpha_n, w) = 1\} \Rightarrow V(\alpha[\alpha_1/A_1, \ldots, \alpha_n/A_n], w) = 1)$ iff (by Lemma 5.3)
$\forall \mathcal{M}_C^{\mathcal{F}_C} \forall \alpha_1, \ldots, \alpha_n \forall X_1, \ldots, X_n \forall w(X_1 = \{w \mid V(\alpha_1, w) = 1\} \wedge \ldots \wedge X_n = \{w \mid V(\alpha_n, w) = 1\} \Rightarrow t_w(\alpha[X_1, \ldots, X_n]))$ iff (by Lemma 5.5)[2]
$\forall X_1, \ldots, X_n \forall w(t_w(\alpha[X_1, \ldots, X_n]))$ iff
$\mathcal{UC}t_w(\alpha)$ iff (by Definition 5.1) $t(\alpha)$.

[1]The left-to-right direction holds trivially. For the right-to-left direction an extension of the f.o.l. valid principle $\forall X, Y(\alpha[X, Y] \Rightarrow \beta[Y]) \wedge \forall X \exists Y \alpha[X, Y] \Rightarrow \forall X \beta[X]$ to multiple Xs and Ys is used, where $\forall X \exists Y \alpha[X, Y]$ for the Ys being $X_1, \ldots, X_n$ is given by Lemma 5.4.

[2]The right-to-left direction holds trivially. For the left-to-right direction an extension of the f.o.l. valid principle $\forall X, Y(\alpha[X, Y] \Rightarrow \beta[Y]) \wedge \forall X \exists Y \alpha[X, Y] \Rightarrow \forall X \beta[X]$ to multiple Xs and Ys is used, where $\forall X \exists Y \alpha[X, Y]$ for Ys being $\mathcal{M}_C^{\mathcal{F}_C}$, $\alpha_1, \ldots, \alpha_n$ is given by Lemma 5.5. Here, V and $\alpha_1, \ldots, \alpha_n$ do not occur in the consequent, namely in $t_w(\alpha[X_1, \ldots, X_n])$. □

Lemma 5.3. *Let $\mathcal{M}_C$ be an arbitrary Chellas model such that $\mathcal{M}_C = \langle W, R, V\rangle$, and let $A_1, \ldots, A_n$ and $\alpha_1, \ldots, \alpha_n$ be arbitrary axiom schema letters and formulas of $\mathcal{L}_{\mathrm{CL}}$, respectively. Then, $\forall X_1, \ldots, X_n \subseteq W \forall \alpha_1, \ldots, \alpha_n \forall w \in W(X_1 = \{w \in W \mid V(\alpha_1, w) = 1\} \wedge \cdots \wedge X_n = \{w \in W \mid V(\alpha_n, w) = 1\} \Rightarrow (V(\alpha[\alpha_1/A_1, \ldots, \alpha_n/A_n], w) = 1 \Leftrightarrow t_w(\alpha[X_1, \ldots, X_n]))$.*

Proof. Let $X_1, \ldots, X_n$ be subsets of W such that $X_1 = \{w \in W \mid V(\alpha_1, w) = 1\}$, $\ldots$, $X_n = \{w \in W \mid V(\alpha_n, w) = 1\}$, and let $w, w', \ldots$ be worlds in W. Finally, let $j, k, l \in \mathbb{N}$. The proof is by induction on the construction of formulas.
(a) $V(\alpha_j, w) = 1$ iff $w \in \{w \in W \mid V(\alpha_j, w) = 1\}$ iff $(w \ulcorner\in\urcorner X_j)$ iff (by Def. 5.1.a) $t_w(\alpha_j)$.
(b) $V(\neg\beta_k, w) = 1$ iff (by $V_\neg$) $V(\beta_k, w) = 0$ iff (by the induction hypothesis)

$\ulcorner\neg\urcorner t_w(\beta_k)$ iff (by Def. 5.1.b) $t_w(\neg\beta_k)$.

(c) $V((\beta_k \vee \beta_l), w) = 1$ iff (by $V_\vee$) $V(\beta_k, w) = 1$ or $V(\beta_l, w) = 1$ iff (by the induction hypothesis) $(t_w(\beta_k)\ulcorner\vee\urcorner t_w(\beta_l))$ iff (by Def. 5.1.c) $t_w((\beta_k \vee \beta_l))$.

(d) $V((\beta_k \Box\!\!\rightarrow \beta_l), w) = 1$ iff (by $V_{\Box\!\rightarrow}$) $\forall w'(wR_{\{w\in W\,|\,V(\beta_k,w)=1\}}w' \Rightarrow V(\beta_l, w') = 1)$ iff (by the induction hypothesis) $\ulcorner\forall\urcorner w'(w\ulcorner R\urcorner_{\{w\in W\,|\,t_w(\beta_k)\}}w'\ulcorner\Rightarrow\urcorner t_{w'}(\beta_l))$ iff (by Def. 5.1.d) $t_w(\beta_k \Box\!\!\rightarrow \beta_l)$ □

Lemma 5.4. *Let $\mathcal{F}_C = \langle W, R\rangle$ be an arbitrary Chellas frame and let $\mathcal{M}_C^{\mathcal{F}_C} = \langle W, R, V\rangle$ be an arbitrary Chellas model based on $\mathcal{F}_C$. Let $X_1, \ldots, X_n$ for $n \in \mathbb{N}$ be subsets of W and let $\alpha_1, \ldots \alpha_n$ stand for arbitrary formulas of $\mathcal{L}_{\mathrm{CL}}$. Then, $\forall\mathcal{M}_C^{\mathcal{F}_C}\forall\alpha_1, \ldots, \alpha_n\exists X_1, \ldots, X_n(X_1 = \{w \in W\,|\,V(\alpha_1, w) = 1\} \wedge \ldots \wedge X_n = \{w \in W\,|\,V(\alpha_n, w) = 1\})$.*

Proof. Since the expressions $\alpha_1, \ldots, \alpha_n$ are placeholders for arbitrary formulas of $\mathcal{L}_{\mathrm{CL}}$ and all formulas are assigned on the basis of Definition 4.12 subsets of W, at which they are true, it follows that $\forall\mathcal{M}_C^{\mathcal{F}_C}\forall\alpha_1, \ldots, \alpha_n\exists X_1, \ldots, X_n(X_1 = \{w \in W\,|\,V(\alpha_1, w) = 1\} \wedge \ldots \wedge X_n = \{w \in W\,|\,V(\alpha_n, w) = 1\})$. □

Lemma 5.5. *Let $\mathcal{F}_C = \langle W, R\rangle$ be an arbitrary Chellas frame and let $\mathcal{M}_C^{\mathcal{F}_C} = \langle W, R, V\rangle$ be an arbitrary Chellas model based on $\mathcal{F}_C$. Moreover, let $X_1, \ldots, X_n$ for $n \in \mathbb{N}$ be subsets of W and let $\alpha_1, \ldots, \alpha_n$ be arbitrary formulas of $\mathcal{L}_{\mathrm{CL}}$. Then, $\forall X_1, \ldots, X_n\exists\mathcal{M}_C^{\mathcal{F}_C}\exists\alpha_1, \ldots, \alpha_n(X_1 = \{w \in W\,|\,V(\alpha_1, w) = 1\} \wedge \ldots \wedge X_n = \{w \in W\,|\,V(\alpha_n, w) = 1\})$.*

Proof. The expressions $\alpha_1, \ldots, \alpha_n$ are placeholders for formulas of $\mathcal{L}_{\mathrm{CL}}$. A fortiori, we can specify $\alpha_1, \ldots, \alpha_n$ in such a way that they are placeholders for the atomic propositions $p_1, \ldots, p_n$, respectively. Since the definition of Chellas models $\mathcal{M}_C^{\mathcal{F}_C} = \langle W, R, V\rangle$ (Definition 4.12) allows us to choose the worlds in W for the Vs, at which $p_1, \ldots, p_n$ are true in an arbitrary way, we can define V in such a way that $X_1 = \{w\,|\,V(p_1, w) = 1\}, \ldots, X_n = \{w\,|\,V(p_n, w) = 1\}$. Hence, it follows that $\forall X_1, \ldots, X_n\exists\mathcal{M}_C^{\mathcal{F}_C}\exists\alpha_1, \ldots, \alpha_n(X_1 = \{w \in W\,|\,V(A_1, w) = 1\} \wedge \ldots \wedge X_n = \{w \in W\,|\,V(A_n, w) = 1\})$ holds. □

5.2.2 A Non-Triviality Criterion

In the beginning we discussed of this section trivial and non-trivial frame conditions and argued that trivial frame conditions – compared to their non-trivial counterparts – use the second-order language $\mathcal{L}_{\text{FC}}$ in a spurious way. In order to describe what we mean by "spurious way", let us repeat trivial and non-trivial frame conditions for the Principle CM, namely $(\alpha \mathrel{\Box\!\!\rightarrow} \gamma) \wedge (\alpha \mathrel{\Box\!\!\rightarrow} \beta) \rightarrow (\alpha \wedge \beta \mathrel{\Box\!\!\rightarrow} \gamma)$ (see Table 5.1):

C_{CM} $\forall X,Y \subseteq W \forall w \in W(\forall w'(wR_X w' \Rightarrow w' \in Y) \Rightarrow \forall w'(wR_{X\cap Y}w' \Rightarrow wR_X w'))$.

$C^{\text{tr}}_{\text{CM}}$ $\forall X,Y,Z \subseteq W \forall w \in W(\forall w'(wR_X w' \Rightarrow w' \in Z) \wedge \forall w'(wR_X w' \Rightarrow w' \in Y) \Rightarrow \forall w'(wR_{X\cap Y}w' \Rightarrow w' \in Z))$

The expression 'C^{tr}' indicates the trivial version of a frame condition C. The difference between C_{CM} on the one hand and $C^{\text{tr}}_{\text{CM}}$ on the other lies in the fact that $C^{\text{tr}}_{\text{CM}}$ employs the addition variable Z. Variable Z is not a functional part of an accessibility relation symbol R. In other words no accessibility relation R described in $C^{\text{tr}}_{\text{CM}}$ is relativized to the subset Z. Let us also repeat trivial and non-trivial frame conditions for the axiom $(\alpha \mathrel{\Box\!\!\rightarrow} \beta) \rightarrow \beta$ (Principle T, see the beginning section of this chapter):

C_{T} $\forall X \subseteq W \forall w \in W(wR_X w)$

C^{tr}_{T} $\forall X,Y \subseteq W \forall w,w'((wR_X w' \Rightarrow w' \in Y) \Rightarrow w \in Y)$

Again the trivial frame condition C^{tr}_{T} but not the frame condition C_{T} uses a variable, namely variable Y, which is not a functional part of the accessibility relation symbol R. Let us, hence, define the notion of non-trivial frame conditions the following way:

Definition 5.6. *A frame condition C_α for a Chellas frame $\mathcal{F}_C = \langle W,R \rangle$ is non-trivial iff in C_α no reference to a variable Y over subsets of W is made unless variable Y is a functional part of the index of some accessibility relation symbols in C_α.*

One might argue that Definition 5.6 is not sufficiently strict, since then, for example, the following frame condition for CM would also be non-trivial:

$\forall w(\forall w'(wR_X w' \Rightarrow w' \in Y \wedge (w' \in X \vee w' \in -X)) \Rightarrow \forall w'(wR_{X\cap Y}w' \Rightarrow wR_X w'))$. We could exclude this type of frame condition by requiring for a non-trivial frame condition C_α in addition that (a) there is no equivalent frame condition C'_α such that C'_α has less references to a variable Y over subsets of W than C_α. We will not strengthen the definition of non-trivial frame conditions in that way, since it is in general extremely hard to prove that a frame condition satisfies criterion (a).

5.3 Chellas Frame Correspondence Proofs

In this section we shall give C-correspondence proofs (see Definition 4.20) for the principles in Tables 5.1 and 5.2 w.r.t. the respective frame conditions in Tables 5.3 and 5.4, but only do so for axiom schemata and frame conditions whose C-correspondence is not already established by Theorem 5.2. So, we do not prove C-correspondence for Principles Refl, Loop, MOD, CP, and TR. We also prove here only C-correspondence. Hence the frame restrictions refer to arbitrary subsets of W of any Chellas frame $\mathcal{F}_C = \langle W, R\rangle$.

In order to prove the right-to-left direction ("$\Leftarrow$") of C-correspondence we show for all worlds w that α is true at w in any Chellas model $\mathcal{M}_C = \langle W, R, V\rangle$ based on any Chellas frame $\mathcal{F}_C = \langle W, R\rangle$ for which C_α holds. Conversely, to prove the left-to-right direction ("$\Rightarrow$"), we show by contraposition that, whenever C_α does not hold in a Chellas frame $\mathcal{F}_C = \langle W, R\rangle$, α is not valid on the frame $\mathcal{F}_C$. To establish the latter result, we construct a Chellas model $\mathcal{M}_C$ based on frame $\mathcal{F}_C$ in which some instance of the axiom schema α is not true at some world w in $\mathcal{M}_C$.

5.3.1 System P

Axiom Schema CM. ("$\Leftarrow$") Let $\mathcal{F}_C = \langle W, R\rangle$ be a frame such that C_{CM} holds. Thus, for any world $w \in W$ and sets $X, Y \subseteq W$, $\forall w'(wR_X w' \Rightarrow w' \in Y) \Rightarrow \forall w'(wR_{X\cap Y}w' \Rightarrow wR_X w')$. Let $\mathcal{M}_C = \langle W, R, V\rangle$ be any model based on $\mathcal{F}_C$ such that $\models_w^{\mathcal{M}_C} (\alpha \,\Box\!\!\rightarrow \gamma) \wedge (\alpha \,\Box\!\!\rightarrow \beta)$. Due to $V_{\Box\!\rightarrow}$ it follows that $\forall w'(wR_{||\alpha||}w' \Rightarrow w' \in ||\gamma||)$ and $\forall w'(wR_{||\alpha||}w' \Rightarrow w' \in ||\beta||)$. Moreover, as $\forall w'(wR_X w' \Rightarrow w' \in$

$Y) \Rightarrow \forall w'(wR_{X\cap Y}w' \Rightarrow wR_Xw')$ holds for any $X, Y \subseteq W$, we can infer $\forall w'$ $(wR_{||\alpha||}w' \Rightarrow w' \in ||\beta||) \Rightarrow \forall w'(wR_{||\alpha||\cap||\beta||}w' \Rightarrow wR_{||\alpha||}w')$ for any $||\alpha||, ||\beta|| \subseteq W$. The latter result and $\forall w'(wR_{||\alpha||}w' \Rightarrow w' \in ||\beta||)$ imply $\forall w'(wR_{||\alpha||\cap||\beta||}w' \Rightarrow wR_{||\alpha||}w')$. Moreover, due to $\text{Def}_{||.||}$ we get $\forall w'(wR_{||\alpha\wedge\beta||}w' \Rightarrow wR_{||\alpha||}w')$. From this and $\forall w'(wR_{||\alpha||}w' \Rightarrow w' \in ||\gamma||)$ we can conclude that $\forall w'(wR_{||\alpha\wedge\beta||}w' \Rightarrow w' \in ||\gamma||)$. Thus, by $V_{\Box\!\!\rightarrow}$ we get $\models_w^{\mathcal{M}_C} \alpha \wedge \beta \Box\!\!\rightarrow \gamma$.

("⇒") Let $\mathcal{F}_C = \langle W, R\rangle$ be any frame such that C_{CM} does not hold. For this to be the case, there have to be worlds $w, w' \in W$ and sets $X, Y \subseteq W$ such that $\forall w''(wR_Xw'' \Rightarrow w'' \in Y)$, $wR_{X\cap Y}w'$, and not wR_Xw'. Let $\mathcal{M}_C = \langle W, R, V\rangle$ be a model based on $\mathcal{F}_C$ such that $X = ||\alpha||$, $Y = ||\beta||$, $\forall w''(wR_{||\alpha||}w'' \Rightarrow w'' \in ||\gamma||)$ and $w' \notin ||\gamma||$. It follows that $\forall w''(wR_{||\alpha||}w'' \Rightarrow w'' \in ||\beta||)$, $wR_{||\alpha||\cap||\beta||}w'$, and not $wR_{||\alpha||}w'$. This assignment is possible for any frame $\mathcal{F}_C$, as by assumption w' is not among the w''s, such that $wR_{||\alpha||}w''$. Since $\forall w''(wR_{||\alpha||}w'' \Rightarrow w'' \in ||\gamma||)$ and $\forall w''(wR_{||\alpha||}w'' \Rightarrow w'' \in ||\beta||)$, we obtain by $V_{\Box\!\!\rightarrow}$ that $\models_w^{\mathcal{M}_C} \alpha \Box\!\!\rightarrow \gamma$ and $\models_w^{\mathcal{M}_C} \alpha \Box\!\!\rightarrow \beta$ and, consequently, $\models_w^{\mathcal{M}_C} (\alpha \Box\!\!\rightarrow \gamma) \wedge (\alpha \Box\!\!\rightarrow \beta)$. Moreover, since $wR_{||\alpha||\cap||\beta||}w'$ it follows by $\text{Def}_{||.||}$ that $wR_{||\alpha\wedge\beta||}w'$ and $w' \notin ||\gamma||$. Thus, by $V_{\Box\!\!\rightarrow}$ we get $\not\models_w^{\mathcal{M}_C} \alpha \wedge \beta \Box\!\!\rightarrow \gamma$. □

Axiom Schema CC. ("⇐") Let $\mathcal{F}_C = \langle W, R\rangle$ be a frame such that C_{CC} holds. Then, for any world $w \in W$ and sets $X, Y \subseteq W$, $\forall w'(wR_Xw' \Rightarrow w' \in Y) \Rightarrow \forall w'(wR_Xw' \Rightarrow wR_{X\cap Y}w')$. Let $\mathcal{M}_C = \langle W, R, V\rangle$ be a model based on $\mathcal{F}_C$ such that $\models_w^{\mathcal{M}_C} (\alpha \wedge \beta \Box\!\!\rightarrow \gamma) \wedge (\alpha \Box\!\!\rightarrow \beta)$. By $V_{\Box\!\!\rightarrow}$ it follows that $\forall w'(wR_{||\alpha\wedge\beta||}$ $w' \Rightarrow w' \in ||\gamma||)$ and $\forall w'(wR_{||\alpha||}w' \Rightarrow w' \in ||\beta||)$. By $\text{Def}_{||.||}$ we obtain that $\forall w'(wR_{||\alpha||\cap||\beta||}w' \Rightarrow w' \in ||\gamma||)$. Since $\forall w'(wR_Xw' \Rightarrow w' \in Y) \Rightarrow \forall w'(wR_Xw' \Rightarrow wR_{X\cap Y}w')$ for any $X, Y \subseteq W$, we get $\forall w'(wR_{||\alpha||}w' \Rightarrow w' \in ||\beta||) \Rightarrow \forall w'(wR_{||\alpha||}w'$ $\Rightarrow wR_{||\alpha||\cap||\beta||}w')$ for every $||\alpha||, ||\beta|| \subseteq W$. As we have $\forall w'(wR_{||\alpha||}w' \Rightarrow w' \in ||\beta||)$, we obtain $\forall w'(wR_{||\alpha||}w' \Rightarrow wR_{||\alpha||\cap||\beta||}w')$. From that and $\forall w'(wR_{||\alpha||\cap||\beta||}w'$ $\Rightarrow w' \in ||\gamma||)$, it follows that $\forall w'(wR_{||\alpha||}w' \Rightarrow w' \in ||\gamma||)$. This implies by $V_{\Box\!\!\rightarrow}$ that $\models_w^{\mathcal{M}_C} \alpha \Box\!\!\rightarrow \gamma$.

("⇒") Let $\mathcal{F}_C = \langle W, R\rangle$ be a frame such that C_{CC} does not hold. Then, there exist two worlds $w, w' \in W$ and sets $X, Y \subseteq W$ such that $\forall w''(wR_Xw'' \Rightarrow w'' \in Y)$, wR_Xw', and not $wR_{X\cap Y}w'$. Let $\mathcal{M}_C = \langle W, R, V\rangle$ be a model based on

$\mathcal{F}_C$ such that $X = ||\alpha||$, $Y = ||\beta||$, $\forall w''(wR_{||\alpha|| \cap ||\beta||}w'' \Rightarrow w'' \in ||\gamma||)$, and $w' \notin ||\gamma||$. Then, $\forall w''(wR_{||\alpha||}w'' \Rightarrow w'' \in ||\beta||)$, $wR_{||\alpha||}w'$, and not $wR_{||\alpha|| \cap ||\beta||}w'$. This assignment is possible for any frame $\mathcal{F}_C$, since by assumption w' is not among the w''s such that $wR_{||\alpha|| \cap ||\beta||}w''$. Due to $\mathrm{Def}_{||.||}$ it follows that $\forall w''(wR_{||\alpha \wedge \beta||}w'' \Rightarrow w'' \in ||\gamma||)$. From that and $\forall w''(wR_{||\alpha||}w'' \Rightarrow w'' \in ||\beta||)$, we get by $V_{\Box\!\!\rightarrow}$ that $\models_w^{\mathcal{M}_C} \alpha \wedge \beta \Box\!\!\rightarrow \gamma$ and $\models_w^{\mathcal{M}_C} \alpha \Box\!\!\rightarrow \beta$. Since $wR_{||\alpha||}w'$ and $w' \notin ||\gamma||$, by $V_{\Box\!\!\rightarrow}$ we can infer $\not\models_w^{\mathcal{M}_C} \alpha \Box\!\!\rightarrow \gamma$. □

Axiom Schema Or. ($\Leftarrow$) Let $\mathcal{F}_C = \langle W, R\rangle$ be a frame such that C_{Or} holds. Then, for any world $w \in W$ and sets $X, Y \subseteq W$, $\forall w'(wR_{X \cup Y}w' \Rightarrow wR_Xw' \vee wR_Yw')$. Let $\mathcal{M}_C = \langle W, R, V\rangle$ be any model based on $\mathcal{F}_C$ such that $\models_w^{\mathcal{M}_C} (\alpha \Box\!\!\rightarrow \gamma) \wedge (\beta \Box\!\!\rightarrow \gamma)$. According to $V_{\Box\!\!\rightarrow}$ we have $\forall w'(wR_{||\alpha||}w' \Rightarrow w' \in ||\gamma||)$ and $\forall w'(wR_{||\beta||}w' \Rightarrow w' \in ||\gamma||)$. From that we can infer $\forall w'(wR_{||\alpha||}w' \vee wR_{||\beta||}w' \Rightarrow w' \in ||\gamma||)$. As $\forall w'(wR_{X \cup Y}w' \Rightarrow wR_Xw' \vee wR_Yw')$ holds for any $X, Y \subseteq W$, this implies that $\forall w'(wR_{||\alpha|| \cup ||\beta||}w' \Rightarrow wR_{||\alpha||}w' \vee wR_{||\beta||}w')$ for every $||\alpha||, ||\beta|| \subseteq W$. Since $\forall w'(wR_{||\alpha||}w' \vee wR_{||\beta||}w' \Rightarrow w' \in ||\gamma||)$ holds, we get $\forall w'(wR_{||\alpha|| \cup ||\beta||}w' \Rightarrow w' \in ||\gamma||)$. By $\mathrm{Def}_{||.||}$ it follows that $\forall w'(wR_{||\alpha \vee \beta||}w' \Rightarrow w' \in ||\gamma||)$. Due to $V_{\Box\!\!\rightarrow}$ this implies that $\models_w^{\mathcal{M}_C} \alpha \vee \beta \Box\!\!\rightarrow \gamma$.

("$\Rightarrow$") Let $\mathcal{F}_C = \langle W, R\rangle$ be a frame such that C_{Or} does not hold. Then, there exist two worlds $w, w' \in W$, and sets $X, Y \subseteq W$ such that $wR_{X \cup Y}w'$, and neither wR_Xw' nor wR_Yw'. Thus, for C_{Or} to fail, w' is not allowed to be among the w''s such that wR_Xw'' or wR_Yw''. Let $\mathcal{M}_C = \langle W, R, V\rangle$ be a model based on $\mathcal{F}_C$ such that $X = ||\alpha||$, $Y = ||\beta||$, $\forall w''(wR_{||\alpha||}w'' \Rightarrow w'' \in ||\gamma||)$, $\forall w''(wR_{||\beta||}w'' \Rightarrow w'' \in ||\gamma||)$, and $w' \notin ||\gamma||$. It follows that $wR_{||\alpha|| \cup ||\beta||}w'$. This assignment is possible for any frame $\mathcal{F}_C$, as by assumption w' is not among the w''s such that $wR_{||\alpha||}w''$ or $wR_{||\beta||}w''$. Since $\forall w''(wR_{||\alpha||}w'' \Rightarrow w'' \in ||\gamma||)$ and $\forall w''(wR_{||\beta||}w'' \Rightarrow w'' \in ||\gamma||)$ are the case, due to $V_{\Box\!\!\rightarrow}$ we get $\models_w^{\mathcal{M}_C} \alpha \Box\!\!\rightarrow \gamma$ and $\models_w^{\mathcal{M}_C} \beta \Box\!\!\rightarrow \gamma$ and thus, $\models_w^{\mathcal{M}_C} (\alpha \Box\!\!\rightarrow \gamma) \wedge (\beta \Box\!\!\rightarrow \gamma)$. By $\mathrm{Def}_{||.||}$ we can infer $wR_{||\alpha \vee \beta||}w'$. As $w' \notin ||\gamma||$, we can conclude by $V_{\Box\!\!\rightarrow}$ that $\not\models_w^{\mathcal{M}_C} \alpha \vee \beta \Box\!\!\rightarrow \gamma$. □

Axiom Schema S. ("$\Leftarrow$"). Let $\mathcal{F}_C = \langle W, R\rangle$ be a frame such that C_S holds. Then, for any $w \in W$ and $X, Y \subseteq W$, $\forall w'(wR_Xw' \wedge w' \in Y \Rightarrow wR_{X \cap Y}w')$. Let $\mathcal{M}_C = \langle W, R, V\rangle$ be any model based on $\mathcal{F}_C$ such that $\models_w^{\mathcal{M}_C} \alpha \wedge \beta \Box\!\!\rightarrow \gamma$ holds.

Then, by $V_{\Box\!\!\rightarrow}$ we have $\forall w'(wR_{||\alpha\wedge\beta||}w' \Rightarrow w' \in ||\gamma||)$. Due to $\mathrm{Def}_{||.||}$ we get $\forall w'(wR_{||\alpha||\cap||\beta||}w' \Rightarrow w' \in ||\gamma||)$. Moreover, as $\forall w'(wR_Xw' \wedge w' \in Y \Rightarrow wR_{X\cap Y}w')$ is the case for any $X,Y \subseteq W$, it follows that $\forall w'(wR_{||\alpha||}w' \wedge w' \in ||\beta|| \Rightarrow wR_{||\alpha||\cap||\beta||}w')$ for every $||\alpha||,||\beta|| \subseteq W$. With $\forall w'(wR_{||\alpha||\cap||\beta||}w' \Rightarrow w' \in ||\gamma||)$ this implies $\forall w'(wR_{||\alpha||}w' \wedge w' \in ||\beta|| \Rightarrow w' \in ||\gamma||)$. Hence, $\forall w'(wR_{||\alpha||}w' \Rightarrow w' \notin ||\beta|| \vee w' \in ||\gamma||)$ and thus, due to $\mathrm{Def}_{||.||}$ we get $\forall w'(wR_{||\alpha||}w' \Rightarrow w' \in ||\neg\beta|| \vee w' \in ||\gamma||)$ and thus, $\forall w'(wR_{||\alpha||}w' \Rightarrow w' \in ||\beta \rightarrow \gamma||)$. Hence, we have by $V_{\Box\!\!\rightarrow}$ that $\models_w^{\mathcal{M}_C} \alpha \Box\!\!\rightarrow (\beta \rightarrow \gamma)$.

("$\Rightarrow$") Let $\mathcal{F}_C = \langle W,R\rangle$ be a frame such that C_S does not hold. Then, there are some $w,w' \in W$ and $X,Y \subseteq W$ such that wR_Xw', $w' \in Y$, and $\neg wR_{X\cap Y}w'$. Let $\mathcal{M}_C = \langle W,R,V\rangle$ be a model based on $\mathcal{F}_C$ such that $X = ||\alpha||$, $Y = ||\beta||$, $\forall w''(wR_{||\alpha||\cap||\beta||}w'' \Rightarrow w'' \in ||\gamma||)$, and $w' \notin ||\gamma||$. Then, $wR_{||\alpha||}w'$, $w' \in ||\beta||$, and $\neg wR_{||\alpha||\cap||\beta||}w'$. This assignment is possible for any frame $\mathcal{F}_C$, since by assumption w' is not among the w''s such that $wR_{||\alpha||\cap||\beta||}w''$. By $\mathrm{Def}_{||.||}$ it follows from $\forall w''(wR_{||\alpha||\cap||\beta||}w'' \Rightarrow w'' \in ||\gamma||)$ that $\forall w''(wR_{||\alpha\wedge\beta||}w'' \Rightarrow w'' \in ||\gamma||)$. This implies by $V_{\Box\!\!\rightarrow}$ that $\models_w^{\mathcal{M}_C} \alpha \wedge \beta \Box\!\!\rightarrow \gamma$. As $w' \in ||\beta||$ and $w' \notin ||\gamma||$, we have by $\mathrm{Def}_{||.||}$ that $w' \notin ||\beta \rightarrow \gamma||$. Since $wR_{||\alpha||}w'$, by we get due to $V_{\Box\!\!\rightarrow}$ that $\not\models_w^{\mathcal{M}_C} \alpha \Box\!\!\rightarrow (\beta \rightarrow \gamma)$. □

5.3.2 Extensions of System P

Axiom Schema RM. ("$\Leftarrow$") Let $\mathcal{F}_C = \langle W,R\rangle$ be any frame such that C_{RM} is the case. Hence, for any world $w \in W$, and sets $X,Y \subseteq W$, it holds that $\exists w'(wR_Xw' \wedge w' \in Y) \Rightarrow \forall w'(wR_{X\cap Y}w' \Rightarrow wR_Xw')$. Let $\mathcal{M}_C = \langle W,R,V\rangle$ be any model based on $\mathcal{F}_C$ such that $\models_w^{\mathcal{M}_C} (\alpha \Box\!\!\rightarrow \gamma) \wedge (\alpha \Diamond\!\!\rightarrow \beta)$. By $V_{\Box\!\!\rightarrow}$, $\mathrm{Def}_{||.||}$, and $\mathrm{Def}_{\Diamond\!\!\rightarrow}$ it follows that $\forall w'(wR_{||\alpha||}w' \Rightarrow w' \in ||\gamma||)$ and $\exists w'(wR_{||\alpha||}w' \wedge w' \in ||\beta||)$. As $\exists w'(wR_Xw' \wedge w' \in Y) \Rightarrow \forall w'(wR_{X\cap Y}w' \Rightarrow wR_Xw')$ holds for any $X,Y \subseteq W$, we obtain that $\exists w'(wR_{||\alpha||}w' \wedge w' \in ||\beta||) \Rightarrow \forall w'(wR_{||\alpha||\cap||\beta||}w' \Rightarrow wR_{||\alpha||}w')$ for every $||\alpha||,||\beta|| \subseteq W$. Since $\exists w'(wR_{||\alpha||}w' \wedge w' \in ||\beta||)$, we have $\forall w'(wR_{||\alpha||\cap||\beta||}w' \Rightarrow wR_{||\alpha||}w')$. Moreover, as $\forall w'(wR_{||\alpha||}w' \Rightarrow w' \in ||\gamma||)$, it follows that $\forall w'(wR_{||\alpha||\cap||\beta||}w' \Rightarrow w' \in ||\gamma||)$. This gives us by $V_{||.||}$ that $\forall w'(wR_{||\alpha\wedge\beta||} w' \Rightarrow w' \in ||\gamma||)$, and due to $V_{\Box\!\!\rightarrow}$ we get that $\models_w^{\mathcal{M}_C} \alpha \wedge \beta \Box\!\!\rightarrow \gamma$.

("$\Rightarrow$") Let $\mathcal{F}_C = \langle W,R\rangle$ be a frame such that C_{RM} does not hold. Then, there are worlds $w, w' \in W$ and sets $X, Y \subseteq W$ such that $\exists w''(wR_X w'' \wedge w'' \in Y)$, $wR_{X \cap Y} w'$, and not $wR_X w'$. Let $\mathcal{M}_C = \langle W, R, V\rangle$ be a model based on $\mathcal{F}_C$ such that $X = ||\alpha||$, $Y = ||\beta||$, $\forall w''(wR_{||\alpha||} w'' \Rightarrow w'' \in ||\gamma||)$, and $w' \notin ||\gamma||$. Then, $\exists w''(wR_{||\alpha||} w'' \wedge w'' \in ||\beta||)$, $wR_{||\alpha|| \cap ||\beta||} w'$, and not $wR_{||\alpha||} w'$. This assignment is possible for any frame $\mathcal{F}_C$, as by assumption w' is not among the w''s such that $wR_{||\alpha||} w''$. Since $\forall w''(wR_{||\alpha||} w'' \Rightarrow w'' \in |\gamma||)$, we have by $V_{\Box\!\!\rightarrow}$ that $\models_w^{\mathcal{M}_C} \alpha \Box\!\!\rightarrow \gamma$. As $\exists w''(wR_{||\alpha||} w'' \wedge w'' \in ||\beta||)$, we get by $V_{\Box\!\!\rightarrow}$, $\text{Def}_{||.||}$, and $\text{Def}_{\Diamond\!\!\rightarrow}$ that $\models_w^{\mathcal{M}_C} \alpha \Diamond\!\!\rightarrow \beta$. Hence, $\models_w^{\mathcal{M}_C} (\alpha \Box\!\!\rightarrow \gamma) \wedge (\alpha \Diamond\!\!\rightarrow \beta)$. Moreover, as $wR_{||\alpha|| \cap ||\beta||} w'$, by $V_{||.||}$ this implies $wR_{||\alpha \wedge \beta||} w'$. As $w' \notin ||\gamma||$, due to $V_{\Box\!\!\rightarrow}$ it follows that $\not\models_w^{\mathcal{M}_C} \alpha \wedge \beta \Box\!\!\rightarrow \gamma$. □

Axiom Schema CEM. ("⇐") In order to show that CEM is valid in all Chellas frames for which C_{CEM} holds, we proceed as follows: We suppose that $\models_w^{\mathcal{M}_C} \neg(\alpha \Box\!\!\rightarrow \neg\beta)$ holds for a world w in model $\mathcal{M}_C$, based on such a frame, and demonstrate that $\models_w^{\mathcal{M}_C} \alpha \Box\!\!\rightarrow \beta$. Let $\mathcal{F}_C = \langle W, R\rangle$ be a frame such that C_{CEM} holds. Then, for any $w \in W$, $X \subseteq W$, $\forall w', w''(wR_X w' \wedge wR_X w'' \Rightarrow w'' = w')$. Let $\mathcal{M}_C = \langle W, R, V\rangle$ be any model based on $\mathcal{F}_C$ and suppose that $\models_w^{\mathcal{M}_C} \neg(\alpha \Box\!\!\rightarrow \neg\beta)$. Then, by $V_{\Box\!\!\rightarrow}$ and $\text{Def}_{||.||}$ it follows that there is a world w' such that $wR_{||\alpha||} w'$ and $\models_{w'}^{\mathcal{M}_C} \beta$. As $\forall w', w''(wR_X w' \wedge wR_X w'' \Rightarrow w'' = w')$ for any $X \subseteq W$, it follows that $\forall w', w''(wR_{||\alpha||} w' \wedge wR_{||\alpha||} w'' \Rightarrow w'' = w')$ for every $||\alpha|| \subseteq W$. Since there is a world w' such that $wR_{||\alpha||} w'$ and $\models_{w'}^{\mathcal{M}_C} \beta$, this implies that $\forall w''(wR_{||\alpha||} w'' \Rightarrow w'' \in ||\beta||)$. Thus, we yield by $V_{\Box\!\!\rightarrow}$ that $\models_w^{\mathcal{M}_C} \alpha \Box\!\!\rightarrow \beta$.

("⇒") Let $\mathcal{F}_C = \langle W, R\rangle$ be a model such that C_{CEM} does not hold. Then, there are $w, w', w'' \in W$ and $X \subseteq W$ such that $wR_X w', wR_X w''$, and $w'' \neq w'$. Let $\mathcal{M}_C = \langle W, R, V\rangle$ be a model based on $\mathcal{F}_C$ such that $X = ||\alpha||$, $w' \notin ||\beta||$, and $w'' \in ||\beta||$. Then, $wR_{||\alpha||} w'$ and $wR_{||\alpha||} w''$. This assignment is possible for any frame $\mathcal{F}_C$, since by assumption $w'' \neq w'$. As there is a world w' such that $wR_{||\alpha||} w'$ and $w' \notin ||\beta||$, we have by $V_{\Box\!\!\rightarrow}$ that $\not\models_w^{\mathcal{M}_C} \alpha \Box\!\!\rightarrow \beta$. Since $w'' \in ||\beta||$, it follows by $\text{Def}_{||.||}$ that $w'' \notin ||\neg\beta||$. By $V_{\Box\!\!\rightarrow}$ we get $\not\models_w^{\mathcal{M}_C} \alpha \Box\!\!\rightarrow \neg\beta$. As $\not\models_w^{\mathcal{M}_C} \alpha \Box\!\!\rightarrow \beta$ and $\not\models_w^{\mathcal{M}_C} \alpha \Box\!\!\rightarrow \neg\beta$, it follows that $\not\models_w^{\mathcal{M}_C} (\alpha \Box\!\!\rightarrow \beta) \vee (\alpha \Box\!\!\rightarrow \neg\beta)$. □

5.3.3 Weak Probability Logic (Threshold Logic)

Axiom Schema P-Cons. ("$\Leftarrow$"). Let $\mathcal{F}_C = \langle W, R\rangle$ be a frame such that the frame condition $C_{\text{P-Cons}}$ holds. Then, for any $w \in W$ there is a $w' \in W$ such that $wR_W w'$. Let $\mathcal{M}_C = \langle W, R, V\rangle$ be any model based on $\mathcal{F}_C$. Then, since $W = ||\top||$ for any model, we have $wR_{||\top||}w'$. As $\emptyset = ||\bot||$, it follows that $w' \notin ||\bot||$ and, hence, by $V_{\Box\!\!\rightarrow}$ this implies that $\not\models_w^{\mathcal{M}_C} \top \Box\!\!\rightarrow \bot$ and, hence, $\models_w^{\mathcal{M}_C} \neg(\top \Box\!\!\rightarrow \bot)$.

("$\Rightarrow$") Let $\mathcal{F}_C = \langle W, R\rangle$ be a frame such that $C_{\text{P-Cons}}$ does not hold. Then, for some $w \in W$ there is no w' such that $wR_W w'$. Let $\mathcal{M}_C = \langle W, R, V\rangle$ be a model based on $\mathcal{F}_C$. Since $W = ||\top||$ for any model, this implies that there is no w', such that $wR_{||\top||}w'$. Hence, by $V_{\Box\!\!\rightarrow}$ it follows trivially that $\models_w^{\mathcal{M}_C} \top \Box\!\!\rightarrow \bot$. Hence, we obtain $\not\models_w^{\mathcal{M}_C} \neg(\top \Box\!\!\rightarrow \bot)$. □

Axiom Schema WOR. ("$\Leftarrow$"). Let $\mathcal{F}_C = \langle W, R\rangle$ be a frame such that C_{WOR} holds. Then, for any $w \in W$ and $X, Y \subseteq W$, $\forall w'(wR_X w' \Rightarrow wR_{X\cap Y}w' \vee wR_{X\cap -Y}w')$. Let $\mathcal{M}_C = \langle W, R, V\rangle$ be any model based on $\mathcal{F}_C$ such that $\models_w^{\mathcal{M}_C} (\alpha \wedge \beta \Box\!\!\rightarrow \gamma) \wedge (\alpha \wedge \neg\beta \Box\!\!\rightarrow \gamma)$ holds. Then, by $V_{\Box\!\!\rightarrow}$ we have $\forall w'(wR_{||\alpha\wedge\beta||}w' \Rightarrow w' \in ||\gamma||)$ and $\forall w'(wR_{||\alpha\wedge\neg\beta||}w' \Rightarrow w' \in ||\gamma||)$. This implies that $\forall w'(wR_{||\alpha\wedge\beta||}w' \vee wR_{||\alpha\wedge\neg\beta||}w' \Rightarrow w' \in ||\gamma||)$. By $\text{Def}_{||.||}$ it follows that $\forall w'(wR_{||\alpha||\cap||\beta||}w' \vee wR_{||\alpha||\cap-||\beta||}w' \Rightarrow w' \in ||\gamma||)$. Moreover, since $\forall w'(wR_X w' \Rightarrow wR_{X\cap Y}w' \vee wR_{X\cap -Y}w')$ holds for any $X, Y \subseteq W$, we have $\forall w'(wR_{||\alpha||}w' \Rightarrow wR_{||\alpha||\cap||\beta||}w' \vee wR_{||\alpha||\cap-||\beta||}w')$ for every $||\alpha||, ||\beta|| \subseteq W$. As $\forall w'(wR_{||\alpha||\cap||\beta||}w' \vee wR_{||\alpha||\cap-||\beta||}w' \Rightarrow w' \in ||\gamma||)$, this implies $\forall w'(wR_{||\alpha||}w' \Rightarrow w' \in ||\gamma||)$. By $V_{\Box\!\!\rightarrow}$ it follows that $\models_w^{\mathcal{M}_C} \alpha \Box\!\!\rightarrow \gamma$.

("$\Rightarrow$") Let $\mathcal{F}_C = \langle W, R\rangle$ be a frame such that C_{WOR} does not hold. Then, for some $w, w' \in W$ and $X, Y \subseteq W$ we have $wR_X w'$, and neither $wR_{X\cap Y}w'$ nor $wR_{X\cap -Y}w'$. Let $\mathcal{M}_C = \langle W, R, V\rangle$ be a model based on $\mathcal{F}_C$ such that $X = ||\alpha||$, $Y = ||\beta||$, $\forall w''(wR_{||\alpha||\cap||\beta||}w'' \Rightarrow w'' \in ||\gamma||)$, and $\forall w''(wR_{||\alpha||\cap-||\beta||}w'' \Rightarrow w'' \in ||\gamma||)$, and $w' \notin ||\gamma||$. Then, $wR_{||\alpha||}w'$, and neither $wR_{||\alpha||\cap||\beta||}w'$ nor $wR_{||\alpha||\cap||\beta||}w'$. This assignment is possible for any frame $\mathcal{F}_C$, since by assumption w' is not among the w''s such that $wR_{||\alpha||\cap||\beta||}w''$ or $wR_{||\alpha||\cap-||\beta||}w''$. As $\forall w''(wR_{||\alpha||\cap||\beta||}w'' \Rightarrow w'' \in ||\gamma||)$ and $\forall w''(wR_{||\alpha||\cap-||\beta||}w'' \Rightarrow w'' \in ||\gamma||)$ it follows by $\text{Def}_{||.||}$ that

$\forall w''(wR_{||\alpha\wedge\beta||}w'' \Rightarrow w'' \in ||\gamma||)$ and $\forall w''(wR_{||\alpha\wedge\neg\beta||}w'' \Rightarrow w'' \in ||\gamma||)$. By $V_{\Box\!\rightarrow}$ this implies $\models_w^{\mathcal{M}_C} \alpha\wedge\beta \Box\!\!\rightarrow \gamma$ and $\models_w^{\mathcal{M}_C} \alpha\wedge\neg\beta \Box\!\!\rightarrow \gamma$, and, thus, $\models_w^{\mathcal{M}_C} (\alpha\wedge\beta \Box\!\!\rightarrow \gamma)\wedge(\alpha\wedge\neg\beta \Box\!\!\rightarrow \gamma)$. Since $wR_{||\alpha||}w'$ and $w' \notin ||\gamma||$, we obtain by $V_{\Box\!\rightarrow}$ that $\not\models_w^{\mathcal{M}_C} \alpha \Box\!\!\rightarrow \gamma$. □

5.3.4 Monotonic Principles

Axiom Schema Cut. Let $\mathcal{F}_C = \langle W, R\rangle$ be a frame such that C_{Cut} holds. Then, for any world $w \in W$ and sets $X, Y, Z \subseteq W$, $\forall w'(wR_Z w' \Rightarrow w' \in Y) \Rightarrow \forall w'(wR_{X\cap Z}w' \Rightarrow wR_{X\cap Y}w')$. Let $\mathcal{M}_C = \langle W, R, V\rangle$ be any model based on $\mathcal{F}_C$ such that $\models_w^{\mathcal{M}_C} (\alpha\wedge\beta \Box\!\!\rightarrow \delta)\wedge(\gamma \Box\!\!\rightarrow \beta)$. Due to $V_{\Box\!\rightarrow}$ it follows that $\forall w'(wR_{||\alpha\wedge\beta||}w' \Rightarrow w' \in ||\delta||)$ and $\forall w'(wR_{||\gamma||}w' \Rightarrow w' \in ||\beta||)$. As $\forall w'(wR_Z w' \Rightarrow w' \in Y) \Rightarrow \forall w'(wR_{X\cap Z}w' \Rightarrow wR_{X\cap Y}w')$ holds for any $X, Y, Z \subseteq W$, we get $\forall w'(wR_{||\gamma||}w' \Rightarrow w' \in ||\beta||) \Rightarrow \forall w'(wR_{||\alpha||\cap||\gamma||}w' \Rightarrow wR_{||\alpha||\cap||\beta||}w')$ for every $||\alpha||, ||\beta||, ||\gamma|| \subseteq W$. Since $\forall w'(wR_{||\gamma||}w' \Rightarrow w' \in ||\beta||)$ holds by assumption, this implies that $\forall w'(wR_{||\alpha||\cap||\gamma||}w' \Rightarrow wR_{||\alpha||\cap||\beta||}w')$. By $\text{Def}_{||.||}$ we have that $\forall w'(wR_{||\alpha\wedge\gamma||}w' \Rightarrow wR_{||\alpha\wedge\beta||}w')$. Moreover, since $\forall w'(wR_{||\alpha\wedge\beta||}w' \Rightarrow w' \in ||\delta||)$, we obtain $\forall w'(wR_{||\alpha\wedge\gamma||}w' \Rightarrow w' \in ||\delta||)$, and, hence, due to $V_{\Box\!\rightarrow}$ this implies that $\models_w^{\mathcal{M}_C} \alpha\wedge\gamma \Box\!\!\rightarrow \delta$.

("⇒") Let $\mathcal{F}_C = \langle W, R\rangle$ be a frame such that C_{Cut} does not hold. Then, there are two worlds $w, w' \in W$ and sets $X, Y, Z \subseteq W$ such that $\forall w''(wR_Z w'' \Rightarrow w'' \in Y)$, $wR_{X\cap Z}w'$, and not $wR_{X\cap Y}w'$. Let $\mathcal{M}_C = \langle W, R, V\rangle$ be a model based on $\mathcal{F}_C$ such that $X = ||\alpha||$, $Y = ||\beta||$, $Z = ||\gamma||$, $\forall w''(wR_{||\alpha||\cap||\beta||}w'' \Rightarrow w'' \in ||\delta||)$, and $w' \notin ||\delta||$. Then, it follows that $\forall w''(wR_{||\gamma||}w'' \Rightarrow w'' \in ||\beta||)$, $wR_{||\alpha||\cap||\gamma||}w'$, and not $wR_{||\alpha||\cap||\beta||}w'$. This assignment is possible for any frame $\mathcal{F}_C$, since by assumption w' is not among the w''s such that $wR_{||\alpha||\cap||\beta||}w''$. Since $\forall w''(wR_{||\alpha||\cap||\beta||}w'' \Rightarrow w'' \in ||\delta||)$, by $\text{Def}_{||.||}$ we obtain that $\forall w''(wR_{||\alpha\wedge\beta||}w'' \Rightarrow w'' \in ||\delta||)$. Moreover, as $\forall w''(wR_{||\gamma||}w'' \Rightarrow w'' \in ||\beta||)$, it follows by $V_{\Box\!\rightarrow}$ that $\models_w^{\mathcal{M}_C} \alpha\wedge\beta \Box\!\!\rightarrow \delta$ and $\models_w^{\mathcal{M}_C} \gamma \Box\!\!\rightarrow \beta$, respectively, and, consequently, $\models_w^{\mathcal{M}_C} (\alpha\wedge\beta \Box\!\!\rightarrow \delta)\wedge(\gamma \Box\!\!\rightarrow \beta)$ holds. Moreover, by $\text{Def}_{||.||}$ $wR_{||\alpha\wedge\gamma||}w'$. Since $w' \notin ||\delta||$, this implies by $V_{\Box\!\rightarrow}$ that $\not\models_w^{\mathcal{M}_C} \alpha\wedge\gamma \Box\!\!\rightarrow \delta$.

□

Axiom Schema Mon. Let $\mathcal{F}_C = \langle W,R \rangle$ be a frame such that C_{Mon} holds. Then, for any world $w \in W$ and sets $X,Y \subseteq W$, $\forall w'(wR_{X\cap Y}w' \Rightarrow wR_X w')$. Let $\mathcal{M}_C = \langle W,R,V \rangle$ be any model based on $\mathcal{F}_C$ such that $\models_w^{\mathcal{M}_C} \alpha \mathrel{\Box\!\!\rightarrow} \gamma$. By $V_{\Box\rightarrow}$ we obtain that $\forall w'(wR_{||\alpha||}w' \Rightarrow w' \in ||\gamma||)$. As $\forall w'(wR_{X\cap Y}w' \Rightarrow wR_X w')$ holds for any $X,Y \subseteq W$, we can infer that $\forall w'(wR_{||\alpha||\cap||\beta||}w' \Rightarrow wR_{||\alpha||}w')$ for every $||\alpha||, ||\beta|| \subseteq W$. Since $\forall w'(wR_{||\alpha||}w' \Rightarrow w' \in ||\gamma||)$, it follows that $\forall w'(wR_{||\alpha||\cap||\beta||}w' \Rightarrow w' \in ||\gamma||)$. By $\mathrm{Def}_{||.||}$ we can conclude that $\forall w'(wR_{||\alpha\wedge\beta||}w' \Rightarrow w' \in ||\gamma||)$, and due to $V_{\Box\rightarrow}$ this implies $\models_w^{\mathcal{M}_C} \alpha \wedge \beta \mathrel{\Box\!\!\rightarrow} \gamma$.

("$\Rightarrow$") Let $\mathcal{F}_C = \langle W,R \rangle$ be a frame such that C_{Mon} does not hold. Then, there are two worlds $w, w' \in W$ and sets $X,Y \subseteq W$ such that $wR_{X\cap Y}w'$, and not $wR_X w'$. Let $\mathcal{M}_C = \langle W,R,V \rangle$ be a model based on $\mathcal{F}_C$ such that $X = ||\alpha||$, $Y = ||\beta||$, $\forall w''(wR_{||\alpha||}w'' \Rightarrow w'' \in ||\gamma||)$, and $w' \notin ||\gamma||$. Then, $wR_{||\alpha||\cap||\beta||}w'$, and not $wR_{||\alpha||}w'$. This assignment is possible for any frame $\mathcal{F}_C$, since by assumption w' is not among the w''s such that $wR_{||\alpha||}w''$. As $\forall w''(wR_{||\alpha||}w'' \Rightarrow w'' \in ||\gamma||)$ is the case, it follows by $V_{\Box\rightarrow}$ that $\models_w^{\mathcal{M}_C} \alpha \mathrel{\Box\!\!\rightarrow} \gamma$. Since $wR_{||\alpha||\cap||\beta||}w'$, $\mathrm{Def}_{||.||}$ gives us that $wR_{||\alpha\wedge\beta||}w'$. Since $w' \notin ||\gamma||$ this implies by $V_{\Box\rightarrow}$ that $\not\models_w^{\mathcal{M}_C} \alpha \wedge \beta \mathrel{\Box\!\!\rightarrow} \gamma$. □

Axiom Schema Trans. Let $\mathcal{F}_C = \langle W,R \rangle$ be a frame such that C_{Trans} holds. Then, for any world $w \in W$ and sets $X,Y \subseteq W$, $\forall w'(wR_X w' \Rightarrow w' \in Y) \Rightarrow \forall w'(wR_X w' \Rightarrow wR_Y w')$. Let $\mathcal{M}_C = \langle W,R,V \rangle$ be any model based on $\mathcal{F}_C$ such that $\models_w^{\mathcal{M}_C} (\alpha \mathrel{\Box\!\!\rightarrow} \beta) \wedge (\beta \mathrel{\Box\!\!\rightarrow} \gamma)$. It follows by $V_{\Box\rightarrow}$ that $\forall w'(wR_{||\alpha||}w' \Rightarrow w' \in ||\beta||)$ and $\forall w'(wR_{||\beta||}w' \Rightarrow w' \in ||\gamma||)$. As $\forall w'(wR_X w' \Rightarrow w' \in Y) \Rightarrow \forall w'(wR_X w' \Rightarrow wR_Y w')$ holds for any $X,Y \subseteq W$, we have $\forall w'(wR_{||\alpha||}w' \Rightarrow w' \in ||\beta||) \Rightarrow \forall w'(wR_{||\alpha||}w' \Rightarrow wR_{||\beta||}w')$ for every $||\alpha||, ||\beta|| \subseteq W$. Since $\forall w'(wR_{||\alpha||}w' \Rightarrow w' \in ||\beta||)$, this implies $\forall w'(wR_{||\alpha||}w' \Rightarrow wR_{||\beta||}w')$. Moreover, since $\forall w'(wR_{||\beta||}w' \Rightarrow w' \in ||\gamma||)$, it follows that $\forall w'(wR_{||\alpha||}w' \Rightarrow w' \in ||\gamma||)$. By $V_{\Box\rightarrow}$ we get $\models_w^{\mathcal{M}_C} \alpha \mathrel{\Box\!\!\rightarrow} \gamma$.

("$\Rightarrow$") Let $\mathcal{F}_C = \langle W,R \rangle$ be a frame such that C_{Trans} does not hold. Then there are worlds $w, w' \in W$ and sets $X,Y \subseteq W$ such that $\forall w''(wR_X w'' \Rightarrow w'' \in Y)$, $wR_X w'$, and not $wR_Y w'$. Let $\mathcal{M}_C = \langle W,R,V \rangle$ be a model based on $\mathcal{F}_C$ such that $X = ||\alpha||$, $Y = ||\beta||$, $\forall w''(wR_{||\beta||}w'' \Rightarrow w'' \in ||\gamma||)$, and $w' \notin ||\gamma||$. Then, we have $\forall w''(wR_{||\alpha||}w'' \Rightarrow w'' \in ||\beta||)$, $wR_{||\alpha||}w'$, and not $wR_{||\beta||}w'$. This assignment is possible for any frame $\mathcal{F}_C$, since by assumption w' is not among the w''s such that $wR_{||\beta||}w''$. As $\forall w''(wR_{||\alpha||}w'' \Rightarrow w'' \in ||\beta||)$ and $\forall w''(wR_{||\beta||}w'' \Rightarrow w'' \in ||\gamma||)$, it follows by $V_{\Box\!\!\rightarrow}$ that $\models_w^{\mathcal{M}_C} \alpha \Box\!\!\rightarrow \beta$ and $\models_w^{\mathcal{M}_C} \beta \Box\!\!\rightarrow \gamma$ and, hence, $\models_w^{\mathcal{M}_C} (\alpha \Box\!\!\rightarrow \beta) \wedge (\beta \Box\!\!\rightarrow \gamma)$. Since $wR_{||\alpha||}w'$ and $w' \notin ||\gamma||$, we get that $\not\models_w^{\mathcal{M}_C} \alpha \Box\!\!\rightarrow \gamma$ due to $V_{\Box\!\!\rightarrow}$.

□

5.3.5 Bridge Principles

Axiom Schema MP. ("⇐") Let $\mathcal{F}_C = \langle W, R\rangle$ be a frame such that C_{MP} holds. Then, for any $w \in W$ and $X \subseteq W$, $w \in X \Rightarrow wR_Xw$. Let $\mathcal{M}_C = \langle W, R, V\rangle$ be a model based on $\mathcal{F}_C$ such that $\models_w^{\mathcal{M}_C} \alpha \Box\!\!\rightarrow \beta$. Moreover, suppose that $\models_w^{\mathcal{M}_C} \alpha$. Then, by $V_{\Box\!\!\rightarrow}$ we get $\forall w'(wR_{||\alpha||}w' \Rightarrow w' \in ||\beta||)$ and $w \in ||\alpha||$. Since $w \in X \Rightarrow wR_Xw$ for any $X \subseteq W$, it follows that $w \in ||\alpha|| \Rightarrow wR_{||\alpha||}w$ for every $||\alpha|| \subseteq W$, and, hence, $wR_{||\alpha||}w$. As $\forall w'(wR_{||\alpha||}w' \Rightarrow w' \in ||\beta||)$, this implies that $w \in ||\beta||$ and we get $\models_w^{\mathcal{M}_C} \beta$.

("⇒") Let $\mathcal{F}_C = \langle W, R\rangle$ be a frame such that C_{MP} does not hold. Then, for some $w \in W$ and $X \subseteq W$, $w \in X$ and $\neg wR_Xw$. Let $\mathcal{M}_C = \langle W, R, V\rangle$ be a model based on $\mathcal{F}_C$ such that $X = ||\alpha||$, $\forall w'(wR_{||\alpha||}w' \Rightarrow w' \in ||\beta||)$, and $w \notin ||\beta||$. Then, $w \in ||\alpha||$ and $\neg wR_{||\alpha||}w$. This assignment is possible for any frame $\mathcal{F}_C$, since by assumption w is not among the w's such that $wR_{||\alpha||}w'$. Since $\forall w'(wR_{||\alpha||}w' \Rightarrow w' \in ||\beta||)$, we obtain that $\models_w^{\mathcal{M}_C} \alpha \Box\!\!\rightarrow \beta$ due to $V_{\Box\!\!\rightarrow}$. Since by assumption $w \in ||\alpha||$ and $w \notin ||\beta||$, it follows that $\not\models_w^{\mathcal{M}_C} \alpha \rightarrow \beta$ by $\mathrm{Def}_{||.||}$. □

Axiom Schema CS. ("⇐"). Let $\mathcal{F}_C = \langle W, R\rangle$ be a frame such that C_{CS} holds. Then, for any $w \in W$ and $X \subseteq W$, $w \in X \Rightarrow \forall w'(wR_Xw' \Rightarrow w' = w)$. Let $\mathcal{M}_C = \langle W, R, V\rangle$ be any model based on $\mathcal{F}_C$ such that $\models_w^{\mathcal{M}_C} \alpha \wedge \beta$. This implies that $w \in ||\alpha||$ and $w \in ||\beta||$. Since, $w \in X \Rightarrow \forall w'(wR_Xw' \Rightarrow w' = w)$ holds for any $X \subseteq W$, it follows that $w \in ||\alpha|| \Rightarrow \forall w'(wR_{||\alpha||}w' \Rightarrow w' = w)$ for every $||\alpha|| \subseteq W$. As we have $w \in ||\alpha||$, we can infer $\forall w'(wR_{||\alpha||}w' \Rightarrow w' = w)$. Two cases are possible for $w \in W$: (a) World w cannot see any world $w' \in W$ by $R_{||\alpha||}$. Then, by $V_{\Box\!\!\rightarrow}$ we trivially get $\models_w^{\mathcal{M}_C} \alpha \Box\!\!\rightarrow \beta$. (b) World w can see some world $w' \in W$.

In this case, as $w' = w$ for all w' such $wR_{||\alpha||}w'$ and $w \in ||\beta||$, it follows by $V_{\Box\!\!\rightarrow}$ that $\models_w^{\mathcal{M}_C} \alpha \Box\!\!\rightarrow \beta$. Hence, $\models_w^{\mathcal{M}_C} \alpha \Box\!\!\rightarrow \beta$ holds in both cases.

("$\Rightarrow$") Let $\mathcal{F}_C = \langle W, R\rangle$ be a frame such that C_{CS} does not hold. Then, there are $w, w' \in W$ and $X \subseteq W$ such that $w \in X$, wR_Xw', and $w' \neq w$. Let $\mathcal{M}_C = \langle W, R, V\rangle$ be a model based on $\mathcal{F}_C$ such that $X = ||\alpha||$, $w \in ||\beta||$, and $w' \notin ||\beta||$. Then, $w \in ||\alpha||$ and $wR_{||\alpha||}w'$. This assignment is possible for any frame $\mathcal{F}_C$, since $w' \neq w$. As $w \in ||\alpha||$ and $w \in ||\beta||$, it follows by $\text{Def}_{||.||}$ that $\models_w^{\mathcal{M}_C} \alpha \wedge \beta$. Since $wR_{||\alpha||}w'$ and $w' \notin ||\beta||$, we get by $V_{\Box\!\!\rightarrow}$ that $\not\models_w^{\mathcal{M}_C} \alpha \Box\!\!\rightarrow \beta$. □

Axiom Schema Det. ("$\Leftarrow$"). Let $\mathcal{F}_C = \langle W, R\rangle$ be a frame such that C_{Det} holds. Then, for any $w \in W$, wR_Ww. Let $\mathcal{M}_C = \langle W, R, V\rangle$ be any model based on $\mathcal{F}_C$ such that $\models_w^{\mathcal{M}_C} \top \Box\!\!\rightarrow \alpha$. Then, by $V_{\Box\!\!\rightarrow}$ we have $\forall w'(wR_{||\top||}w' \Rightarrow w' \in ||\alpha||)$. By $\text{Def}_{||.||}$ it holds $||\top|| = W$. Thus, from wR_Ww it follows that $wR_{||\top||}w$ and, hence, $w \in ||\alpha||$. Thus, $\models_w^{\mathcal{M}_C} \alpha$.

("$\Rightarrow$") Let $\mathcal{F}_C = \langle W, R\rangle$ be a frame such that C_{Det} does not hold. Then, there is some $w \in W$ such that $\neg wR_Ww$. Let $\mathcal{M}_C = \langle W, R, V\rangle$ be a model based on $\mathcal{F}_C$ such that $\forall w'(wR_{||\top||}w' \Rightarrow w' \in ||\alpha||)$ and $w \notin ||\alpha||$. Thus, since by $\text{Def}_{||.||}$ $W = ||\top||$, we have $\neg wR_{||\top||}w$. This assignment is possible for any frame $\mathcal{F}_C$, as by assumption w is not among the w's such that $wR_{||\top||}w'$. Hence, since $\forall w'(wR_{||\top||}w' \Rightarrow w' \in ||\alpha||)$, we have by $V_{\Box\!\!\rightarrow}$ that $\models_w^{\mathcal{M}_C} \top \Box\!\!\rightarrow \alpha$. Since $w \notin ||\alpha||$, this implies $\not\models_w^{\mathcal{M}_C} \alpha$. □

Axiom Schema Cond. ("$\Leftarrow$"). Let $\mathcal{F}_C = \langle W, R\rangle$ be a frame such that C_{Cond} holds. Then, for any $w \in W$, $\forall w'(wR_Ww' \Rightarrow w' = w)$. Let $\mathcal{M}_C = \langle W, R, V\rangle$ be any model based on $\mathcal{F}_C$ such that $\models_w^{\mathcal{M}_C} \alpha$. Thus, $w \in ||\alpha||$. Since $\forall w'(wR_Ww' \Rightarrow w' = w)$ and $W = ||\top||$, we have $\forall w'(wR_{||\top||}w' \Rightarrow w' = w)$. Hence, w can see by $R_{||\top||}$ at most w. So it can see by $R_{||\top||}$ either no other world or only w. In both cases, since $w \in ||\alpha||$, we have, thus, $\forall w'(wR_{||\top||}w' \Rightarrow w' \in ||\alpha||)$, and by $V_{\Box\!\!\rightarrow}$ it follows that $\models_w^{\mathcal{M}_C} \top \Box\!\!\rightarrow \alpha$.

("$\Rightarrow$") Let $\mathcal{F}_C = \langle W, R\rangle$ be a frame such that C_{Cond} does not hold. Then, there are $w, w' \in W$ such that wR_Ww' and $w' \neq w$. Let $\mathcal{M}_C = \langle W, R, V\rangle$ be a model based on $\mathcal{F}_C$ such that $w \in ||\alpha||$ and $w' \notin ||\alpha||$. Since by $\text{Def}_{||.||}$ it holds that $W = ||\top||$, we have $wR_{||\top||}w'$. This assignment is possible for any frame

$\mathcal{F}_C$, since by assumption $w' \neq w$. As $w \in ||\alpha||$, we have $\models_w^{\mathcal{M}_C} \alpha$. Since $w' \notin ||\alpha||$ and $wR_{||\top||}w'$, it follows by $V_{\Box\!\!\rightarrow}$ that $\not\models_w^{\mathcal{M}_C} \top \Box\!\!\rightarrow \alpha$. □

5.3.6 Collapse Conditions Material Implication

Axiom Schema VEQ. ("⇐"). Let $\mathcal{F}_C = \langle W, R\rangle$ be a frame such that C_{VEQ} holds. Then, for any world $w \in W$ and set $X \subseteq W$, $\forall w'(wR_Xw' \Rightarrow w' = w)$. Let $\mathcal{M}_C = \langle W, R, V\rangle$ be any model based on $\mathcal{F}_C$ such that $\models_w^{\mathcal{M}_C} \beta$. Hence, $w \in ||\beta||$. Since $\forall w'(wR_Xw' \Rightarrow w' = w)$ holds for any $X \subseteq W$, we have $\forall w'(wR_{||\alpha||}w' \Rightarrow w' = w)$ for every $||\alpha|| \subseteq W$. There are two possible cases for w: (a) There is no world $w' \in W$ such that $wR_{||\alpha||}w'$. Then, by $V_{\Box\!\!\rightarrow}$ we trivially obtain that $\models_w^{\mathcal{M}_C} \alpha \Box\!\!\rightarrow \beta$. (b) There is a world $w' \in W$ such that $wR_{||\alpha||}w'$. Since $\forall w'(wR_{||\alpha||}w' \Rightarrow w' = w)$, this implies that any world w can see by $R_{||\alpha||}$ is identical with w. Hence, $w \in ||\beta||$ and it follows by $V_{\Box\!\!\rightarrow}$ that $\models_w^{\mathcal{M}_C} \alpha \Box\!\!\rightarrow \beta$. Therefore, it holds in both cases that $\models_w^{\mathcal{M}_C} \alpha \Box\!\!\rightarrow \beta$.

("⇒") Let $\mathcal{F}_C = \langle W, R\rangle$ be a frame such that C_{VEQ} does not hold. Then, there are $w, w' \in W$ and $X \subseteq W$ such that wR_Xw' and $w' \neq w$. Let $\mathcal{M}_C = \langle W, R, V\rangle$ be a model based on $\mathcal{F}_C$ such that $X = ||\alpha||$ and $w' \notin ||\beta||$. Then, $wR_{||\alpha||}w'$. This assignment is trivially possible for any $\mathcal{F}_C$. As $w \in ||\alpha||$, it follows that $\models_w^{\mathcal{M}_C} \alpha$. Since $w' \notin ||\beta||$ and $wR_{||\alpha||}w'$, this implies by $V_{\Box\!\!\rightarrow}$ that $\not\models_w^{\mathcal{M}_C} \alpha \Box\!\!\rightarrow \beta$. □

Axiom Schema EFQ. ("⇐"). Let $\mathcal{F}_C = \langle W, R\rangle$ be a frame such that C_{EFQ} holds. Then, for any $w \in W$ and $X \subseteq W$, $w \in -X \Rightarrow \neg\exists w'(wR_Xw')$. Let $\mathcal{M}_C = \langle W, R, V\rangle$ be any model based on $\mathcal{F}_C$ such that $\models_w^{\mathcal{M}_C} \neg\alpha$. Hence, by $\text{Def}_{||.||}$ $w \in -||\alpha||$. Since $w \in -X \Rightarrow \neg\exists w'(wR_Xw')$ holds for any $X \subseteq W$, it follows that $w \in -||\alpha|| \Rightarrow \neg\exists w'(wR_{||\alpha||}w')$ for every $||\alpha|| \subseteq W$. Moreover, as $w \in -||\alpha||$ is the case, we get $\neg\exists w'(wR_{||\alpha||}w')$ and due to $V_{\Box\!\!\rightarrow}$ it trivially follows that $\models_w^{\mathcal{M}_C} \alpha \Box\!\!\rightarrow \beta$ for any β.

("⇒") Let $\mathcal{F}_C = \langle W, R\rangle$ be a frame such that C_{EFQ} does not hold. Then, there are $w, w' \in W$ and $X \subseteq W$ such that $w \in -X$ and wR_Xw'. Let $\mathcal{M}_C = \langle W, R, V\rangle$ be a model based on $\mathcal{F}_C$ such that $X = ||\alpha||$ and $w' \notin ||\beta||$. Then, $w \in -||\alpha||$ and $wR_{||\alpha||}w'$. This assignment is trivially possible for any frame $\mathcal{F}_C$. It follows by $\mathrm{Def}_{||.||}$ that $\models_w^{\mathcal{M}_C} \neg\alpha$. Since $wR_{||\alpha||}w'$ and $w' \notin ||\beta||$, this implies by $V_{\Box\!\!\rightarrow}$ that $\not\models_w^{\mathcal{M}_C} \alpha \Box\!\!\rightarrow \beta$.

□

5.3.7 Traditional Extensions

Axiom Schema D. ("⇐") Let $\mathcal{F}_C = \langle W,R \rangle$ be a frame such that C_{D} holds. Then, for any $w \in W$ and any $X \subseteq W$, there exists a $w' \in W$ such that wR_Xw'. Let $\mathcal{M}_C = \langle W,R,V \rangle$ be any model based on $\mathcal{F}_C$ such that $\models_w^{\mathcal{M}_C} \alpha \Box\!\!\rightarrow \beta$. Then, by $V_{\Box\!\!\rightarrow}$ we get $\forall w''(wR_{||\alpha||}w'' \Rightarrow w'' \in ||\beta||)$. As for any $X \subseteq W$ there is a w' such that wR_Xw', it follows that $wR_{||\alpha||}w'$ for every $||\alpha|| \subseteq W$. Moreover, as $\forall w''(wR_{||\alpha||}w'' \Rightarrow w'' \in ||\beta||)$ is the case, we get $w' \in ||\beta||$. Since $wR_{||\alpha||}w'$, it follows by $V_{\Box\!\!\rightarrow}$, $\mathrm{Def}_{||.||}$, and $\mathrm{Def}_{\Diamond\!\!\rightarrow}$ that $\models_w^{\mathcal{M}_C} \alpha \Diamond\!\!\rightarrow \beta$.

("⇒") Let $\mathcal{F}_C = \langle W,R \rangle$ be a frame such that C_{D} does not hold. Then, there exists a world $w \in W$ and a set $X \subseteq W$ such that there is no world w' with wR_Xw'. Let $\mathcal{M}_C = \langle W,R,V \rangle$ be a model based on $\mathcal{F}_C$ such that $X = ||\alpha||$. Then, there is no world w' such that $wR_{||\alpha||}w'$. It trivially follows by $V_{\Box\!\!\rightarrow}$ that $\models_w^{\mathcal{M}_C} \alpha \Box\!\!\rightarrow \beta$. Moreover, as by assumption there is no world w' such that $wR_{||\alpha||}w'$, on the basis of by $V_{\Box\!\!\rightarrow}$, $\mathrm{Def}_{||.||}$, and $\mathrm{Def}_{\Diamond\!\!\rightarrow}$ this implies that $\not\models_w^{\mathcal{M}_C} \alpha \Diamond\!\!\rightarrow \beta$. □

Axiom Schema T. ("⇐") Let $\mathcal{F}_C = \langle W,R \rangle$ be a frame such that C_{T} holds. Then, for any w and $X \subseteq W$, wR_Xw. Let $\mathcal{M}_C = \langle W,R,V \rangle$ be any model based on $\mathcal{F}_C$ such that $\models_w^{\mathcal{M}_C} \alpha \Box\!\!\rightarrow \beta$. By $V_{\Box\!\!\rightarrow}$ it follows that $\forall w'(wR_{||\alpha||}w' \Rightarrow w' \in ||\beta||)$. As wR_Xw for any $X \subseteq W$, this implies that $wR_{||\alpha||}w$ for every $||\alpha|| \subseteq W$. Hence, w is among the w's so that $w \in ||\beta||$. Thus, $\models_w^{\mathcal{M}_C} \beta$.

("⇒") Let $\langle W,R \rangle$ be a frame such that C_{T} does not hold. Then, there exists a world w and a set $X \subseteq W$ such that $\neg wR_Xw$. Let $\mathcal{M}_C = \langle W,R,V \rangle$ be a model based on $\mathcal{F}_C$ such that $X = ||\alpha||$, $\forall w'(wR_{||\alpha||}w' \Rightarrow w' \in ||\beta||)$, and $w \notin ||\beta||$. Then, $\neg wR_{||\alpha||}w$. This assignment is possible for any frame $\mathcal{F}_C$, since by assumption w is not among the w's such that $wR_{||\alpha||}w'$. Since $\forall w'(wR_{||\alpha||}w' \Rightarrow w' \in ||\beta||)$, by $V_{\Box\!\!\rightarrow}$ it follows that $\models_w^{\mathcal{M}_C} \alpha \Box\!\!\rightarrow \beta$. As $w \notin ||\beta||$, we have that $\not\models_w^{\mathcal{M}_C} \beta$ and, hence, that $\not\models_w^{\mathcal{M}_C} (\alpha \Box\!\!\rightarrow \beta) \rightarrow \beta$. □

Axiom Schema B. ("$\Leftarrow$") Let $\mathcal{F}_C = \langle W, R\rangle$ be a frame such that C_{B} holds. Then, for any $w \in W$ and $X \subseteq W$, $\forall w'(wR_X w' \Rightarrow w'R_X w)$. Let $\mathcal{M}_C = \langle W, R, V\rangle$ be any model based on $\mathcal{F}_C$ such that $\models_w^{\mathcal{M}_C} \alpha$. Then, $w \in ||\alpha||$. As $\forall w'(wR_X w' \Rightarrow w'R_X w)$ holds for any $X \subseteq W$, it follows that $\forall w'(wR_{||\alpha||}w' \Rightarrow w'R_{||\alpha||}w)$ for every $||\alpha|| \subseteq W$. Concerning w there are two possible cases: (a) There is no $w' \in W$ such that $wR_{||\alpha||}w'$. Then, by $V_{\Box\!\!\rightarrow}$ we trivially obtain that $\models_w^{\mathcal{M}_C} \alpha \Box\!\!\rightarrow (\alpha \Diamond\!\!\rightarrow \beta)$. (b) There is a $w' \in W$ such that $wR_{||\alpha||}w'$. Then, for all w such that $wR_{||\alpha||}w'$, it holds that $w'R_{||\alpha||}w$. Since $w \in ||\beta||$, we get by $V_{\Box\!\!\rightarrow}$, $\mathrm{Def}_{||.||}$, and $\mathrm{Def}_{\Diamond\!\!\rightarrow}$ that $\models_{w'}^{\mathcal{M}_C} \alpha \Diamond\!\!\rightarrow \beta$. As this holds for any w' such that $wR_{||\alpha||}w'$, it follows by $V_{\Box\!\!\rightarrow}$ that $\models_w^{\mathcal{M}_C} \alpha \Box\!\!\rightarrow (\alpha \Diamond\!\!\rightarrow \beta)$. Hence, in both cases, we get $\models_w^{\mathcal{M}_C} \alpha \Box\!\!\rightarrow (\alpha \Diamond\!\!\rightarrow \beta)$.

("$\Rightarrow$") Let $\mathcal{F}_C = \langle W, R\rangle$ be a frame such that C_{B} does not hold. Then, there exist worlds $w, w' \in W$ and a set $X \subseteq W$ such that $wR_X w'$ and $\neg w'R_X w$. Let $\mathcal{M}_C = \langle W, R, V\rangle$ be any model based on $\mathcal{F}_C$ such that $X = ||\alpha||$, $\neg\exists w''(w'R_{||\alpha||}w'' \wedge w'' \in ||\beta||)$ and $w \in ||\alpha||$. Then, $wR_{||\alpha||}w'$ and not $w'R_{||\alpha||}w$. This assignment is trivially possible for any frame $\mathcal{F}_C$. As $w \in ||\alpha||$, we get $\models_w^{\mathcal{M}_C} \alpha$, and, since $\neg\exists w''(w'R_{||\alpha||}w'' \wedge w'' \in ||\beta||)$, it follows by $V_{\Box\!\!\rightarrow}$, $\mathrm{Def}_{||.||}$, and $\mathrm{Def}_{\Diamond\!\!\rightarrow}$ that $\not\models_{w'}^{\mathcal{M}_C} \alpha \Diamond\!\!\rightarrow \beta$. Since $wR_{||\alpha||}w'$, by $V_{\Box\!\!\rightarrow}$ we get $\not\models_w^{\mathcal{M}_C} \alpha \Box\!\!\rightarrow (\alpha \Diamond\!\!\rightarrow \beta)$. □

Axiom Schema 4. ("$\Leftarrow$") Let $\mathcal{F}_C = \langle W, R\rangle$ be a frame such that C_4 holds. Then, for any world w and set $X \subseteq W$, $\forall w', w''(wR_X w' \wedge w'R_X w'' \Rightarrow wR_X w'')$. Let $\mathcal{M}_C = \langle W, R, V\rangle$ be any model based on $\mathcal{F}_C$ such that $\models_w^{\mathcal{M}_C} \alpha \Box\!\!\rightarrow \beta$. Then, by $V_{\Box\!\!\rightarrow}$ it follows that $\forall w'(wR_{||\alpha||}w' \Rightarrow w' \in ||\beta||)$. As $\forall w', w''(wR_X w' \wedge w'R_X w'' \Rightarrow wR_X w'')$ holds for any $X \subseteq W$, we obtain that $\forall w', w''(wR_{||\alpha||}w' \wedge w'R_{||\alpha||}w'' \Rightarrow wR_{||\alpha||}w'')$ for every $||\alpha|| \subseteq W$. Since $\forall w'(wR_{||\alpha||}w' \Rightarrow w' \in ||\beta||)$ is the case, it follows that $\forall w', w''(wR_{||\alpha||}w' \wedge w'R_{||\alpha||}w'' \Rightarrow w'' \in ||\beta||)$. There are two possible cases for w: (a) There is no w' such that $wR_{||\alpha||}w'$. Then, by $V_{\Box\!\!\rightarrow}$ it holds trivially that $\models_w^{\mathcal{M}_C} \alpha \Box\!\!\rightarrow (\alpha \Box\!\!\rightarrow \beta)$. (b) There are worlds w' such that $wR_{||\alpha||}w'$. As $\forall w', w''(wR_{||\alpha||}w' \wedge w'R_{||\alpha||}w'' \Rightarrow w'' \in ||\beta||)$, we have $\forall w''(w'R_{||\alpha||}w'' \Rightarrow w'' \in ||\beta||)$ for any such w'. Hence, by $V_{\Box\!\!\rightarrow}$ it follows that $\models_{w'}^{\mathcal{M}_C} \alpha \Box\!\!\rightarrow \beta$. Since this is the case for any w' such that $wR_{||\alpha||}w'$, we get by $V_{\Box\!\!\rightarrow}$ that $\models_w^{\mathcal{M}_C} \alpha \Box\!\!\rightarrow (\alpha \Box\!\!\rightarrow \beta)$. So, we obtain $\models_w^{\mathcal{M}_C} \alpha \Box\!\!\rightarrow (\alpha \Box\!\!\rightarrow \beta)$ in both cases.

("$\Rightarrow$") Let $\mathcal{F}_C = \langle W, R\rangle$ be a frame such that C_4 does not hold. Then, there exist worlds $w, w', w'' \in W$ and a set $X \subseteq W$ such that wR_Xw', $w'R_Xw''$, and not wR_Xw''. Let $\mathcal{M}_C = \langle W, R, V\rangle$ be a model based on $\mathcal{F}_C$, such that $X = ||\alpha||$, $\forall w'(wR_{||\alpha||}w' \Rightarrow w' \in ||\beta||)$, and $w'' \notin ||\beta||$. Then, $wR_{||\alpha||}w'$, $w'R_{||\alpha||}w''$, and not $wR_{||\alpha||}w''$. This assignment is possible for any frame $\mathcal{F}_C$, as by assumption w'' is not among the w's such that $wR_{||\alpha||}w'$. Since $\forall w'(wR_{||\alpha||}w' \Rightarrow w' \in ||\beta||)$, by $V_{\Box\!\!\rightarrow}$ we obtain that $\models_w^{\mathcal{M}_C} \alpha \Box\!\!\rightarrow \beta$. Since $w'R_{||\alpha||}w''$ and $w'' \notin ||\beta||$, by $V_{\Box\!\!\rightarrow}$ we get $\not\models_{w'}^{\mathcal{M}_C} \alpha \Box\!\!\rightarrow \beta$. As $wR_{||\alpha||}w'$, due to $V_{\Box\!\!\rightarrow}$ it follows that $\not\models_w^{\mathcal{M}_C} \alpha \Box\!\!\rightarrow (\alpha \Box\!\!\rightarrow \beta)$. □

Axiom Schema 5. ("$\Leftarrow$") Let $\mathcal{F}_C = \langle W, R\rangle$ be a frame such that C_5 holds. Then, for any world $w \in W$ and set $X \subseteq W$, $\forall w', w''(wR_Xw' \wedge wR_Xw'' \Rightarrow w'R_Xw'')$. Let $\mathcal{M}_C = \langle W, R, V\rangle$ be a model based on $\mathcal{F}_C$ such that $\models_w^{\mathcal{M}_C} \alpha \Diamond\!\!\rightarrow \beta$. Then, by $V_{\Box\!\!\rightarrow}$, $\mathrm{Def}_{||.||}$, and $\mathrm{Def}_{\Diamond\!\!\rightarrow}$ it follows that there is a world w''' such that $wR_{||\alpha||}w'''$ and $w''' \in ||\beta||$. Since $\forall w', w''(wR_Xw' \wedge wR_Xw'' \Rightarrow w'R_Xw'')$ for any $X \subseteq W$, this yields $\forall w', w''(wR_{||\alpha||}w' \wedge wR_{||\alpha||}w'' \Rightarrow w'R_{||\alpha||}w'')$ for every $||\alpha|| \subseteq W$. This implies that any worlds $w', w'' \in W$ such that $wR_{||\alpha||}w'$ and $wR_{||\alpha||}w''$ can see each other including themselves. Hence, as $wR_{||\alpha||}w'''$ and $w''' \in ||\beta||$, any world w' such that $wR_{||\alpha||}w'$ can see world w''' by $R_{||\alpha||}$. Thus, since $w''' \in ||\beta||$, this implies by $V_{\Box\!\!\rightarrow}$, $\mathrm{Def}_{||.||}$, and $\mathrm{Def}_{\Diamond\!\!\rightarrow}$ that $\models_{w'}^{\mathcal{M}_C} \alpha \Diamond\!\!\rightarrow \beta$. Since this holds for any w' such that $wR_{||\alpha||}w'$, we have by $V_{\Box\!\!\rightarrow}$ that $\models_w^{\mathcal{M}_C} \alpha \Box\!\!\rightarrow (\alpha \Diamond\!\!\rightarrow \beta)$.

("$\Rightarrow$") Let $\mathcal{F}_C = \langle W, R\rangle$ be a frame such that C_5 does not hold. Then, there exist worlds $w, w' \in W$ and a set $X \subseteq W$ such that wR_Xw', wR_Xw'', and not $w'R_Xw''$. Let $\mathcal{M}_C = \langle W, R, V\rangle$ be a model based on $\mathcal{F}_C$ such that $X = ||\alpha||$, $\neg\exists w'''(w'R_{||\alpha||}w''' \wedge w''' \in ||\beta||)$ and $w'' \in ||\beta||$. Then, $wR_{||\alpha||}w'$, $wR_{||\alpha||}w''$, and not $w'R_{||\alpha||}w''$. This assignment is possible for any frame $\mathcal{F}_C$, since by assumption $\neg w'R_{||\alpha||}w''$. As $wR_{||\alpha||}w''$ and $w'' \in ||\beta||$, it follows by $V_{\Box\!\!\rightarrow}$, $\mathrm{Def}_{||.||}$, and $\mathrm{Def}_{\Diamond\!\!\rightarrow}$ that $\models_w^{\mathcal{M}_C} \alpha \Diamond\!\!\rightarrow \beta$. Moreover, since $\neg\exists w'''(w'R_{||\alpha||}w''' \wedge w''' \in ||\beta||)$, by $V_{\Box\!\!\rightarrow}$, $\mathrm{Def}_{||.||}$, and $\mathrm{Def}_{\Diamond\!\!\rightarrow}$ we obtain that $\not\models_{w'}^{\mathcal{M}_C} \alpha \Diamond\!\!\rightarrow \beta$, and, since $wR_{||\alpha||}w'$, by $V_{\Box\!\!\rightarrow}$ we have $\not\models_w^{\mathcal{M}_C} \alpha \Box\!\!\rightarrow (\alpha \Diamond\!\!\rightarrow \beta)$. □

5.3.8 Iteration Principles

Axiom Schema Ex. ("⇐") Let $\mathcal{F}_C = \langle W, R\rangle$ be a frame such that C_{Ex} holds. Then, for any world $w \in W$ and sets $X, Y \subseteq W$, $\forall w', w''(wR_X w' \wedge w' R_Y w'' \Rightarrow wR_{X\cap Y}w'')$. Let $\mathcal{M}_C = \langle W, R, V\rangle$ be a model based on $\mathcal{F}_C$ such that $\models_w^{\mathcal{M}_C} \alpha \wedge \beta \,\Box\!\!\rightarrow \gamma$. By $V_{\Box\!\rightarrow}$ we have $\forall w'(wR_{||\alpha\wedge\beta||}w' \Rightarrow w' \in ||\gamma||)$. Due to $\text{Def}_{||.||}$, $\forall w'(wR_{||\alpha||\cap||\beta||}w' \Rightarrow w' \in ||\gamma||)$. As $\forall w', w''(wR_X w' \wedge w' R_Y w'' \Rightarrow wR_{X\cap Y}w'')$ for any $X, Y \subseteq W$, we have $\forall w', w''(wR_{||\alpha||}w' \wedge w' R_{||\beta||}w'' \Rightarrow wR_{||\alpha||\cap||\beta||}w'')$ for every $||\alpha||, ||\beta|| \subseteq W$. Since $\forall w'(wR_{||\alpha||\cap||\beta||}w' \Rightarrow w' \in ||\gamma||)$, this implies $\forall w', w''(wR_{||\alpha||}w' \wedge w' R_{||\beta||}w'' \Rightarrow w'' \in ||\gamma||)$ and hence $\forall w'(wR_{||\alpha||}w' \Rightarrow \forall w''(w' R_{||\beta||}w'' \Rightarrow w'' \in ||\gamma||))$. Thus, by $V_{\Box\!\rightarrow}$ this results in $\forall w'(wR_{||\alpha||}w' \Rightarrow \models_{w'}^{\mathcal{M}_C} \beta \,\Box\!\!\rightarrow \gamma)$ and $\models_w^{\mathcal{M}_C} \alpha \,\Box\!\!\rightarrow (\beta \,\Box\!\!\rightarrow \gamma)$.

("⇒") Let $\mathcal{F}_C = \langle W, R\rangle$ be a frame such that C_{Ex} does not hold. Then, there exist worlds $w, w', w'' \in W$ and sets $X, Y \subseteq W$ such that $wR_X w'$, $w' R_Y w''$, and not $wR_{X\cap Y}w''$. Let $\mathcal{M}_C = \langle W, R, V\rangle$ be a model based on $\mathcal{F}_C$ such that $X = ||\alpha||$, $Y = ||\beta||$, $\forall w'''(wR_{||\alpha||\cap||\beta||}w''' \Rightarrow w''' \in ||\gamma||)$, and $w'' \notin ||\gamma||$. Then, $wR_{||\alpha||}w'$, $w' R_{||\beta||}w''$, and not $wR_{||\alpha||\cap||\beta||}w''$. This assignment is possible for any frame $\mathcal{F}_C$, since by assumption w'' is not among the w'''s such that $wR_{||\alpha||\cap||\beta||}w'''$. By $\text{Def}_{||.||}$ it follows that $\forall w'''(wR_{||\alpha\wedge\beta||}w''' \Rightarrow w''' \in ||\gamma||)$ and hence, due to $V_{\Box\!\rightarrow}$ we have $\models_w^{\mathcal{M}_C} \alpha \wedge \beta \,\Box\!\!\rightarrow \gamma$. Moreover, since $w' R_{||\beta||}w''$ and $w'' \notin ||\gamma||$, by $V_{\Box\!\rightarrow}$ this implies $\not\models_{w'}^{\mathcal{M}_C} \beta \,\Box\!\!\rightarrow \gamma$. Since $wR_{||\alpha||}w'$, by $V_{\Box\!\rightarrow}$ it follows that $\not\models_w^{\mathcal{M}_C} \alpha \,\Box\!\!\rightarrow (\beta \,\Box\!\!\rightarrow \gamma)$. □

Axiom Schema Im. ("⇐") Let $\mathcal{F}_C = \langle W, R\rangle$ be a frame such that C_{Im} holds. Then, for any world $w \in W$ and sets $X, Y \subseteq W$, $\forall w'(wR_{X\cap Y}w' \Rightarrow \exists w''(wR_X w'' \wedge w'' R_Y w'))$. Let $\mathcal{M}_C = \langle W, R, V\rangle$ be a model based on $\mathcal{F}_C$ such that $\models_w^{\mathcal{M}_C} \alpha \,\Box\!\!\rightarrow (\beta \,\Box\!\!\rightarrow \gamma)$. By $V_{\Box\!\rightarrow}$ we have $\forall w'(wR_{||\alpha||}w' \Rightarrow \forall w''(w' R_{||\beta||}w'' \Rightarrow w'' \in ||\gamma||))$ and hence $\forall w', w''(wR_{||\alpha||}w' \wedge w' R_{||\beta||}w'' \Rightarrow w'' \in ||\gamma||)$. Since $\forall w'(wR_{X\cap Y}w' \Rightarrow \exists w''(wR_X w'' \wedge w'' R_Y w'))$ holds for any $X, Y \subseteq W$, we get $\forall w'(wR_{||\alpha||\cap||\beta||}w' \Rightarrow \exists w''(wR_{||\alpha||}w'' \wedge w'' R_{||\beta||}w'))$ for every $||\alpha||, ||\beta|| \subseteq W$. Let $wR_{||\alpha||\cap||\beta||}w'$ be the case. Then, there exists a world $w'' \in W$ such that $wR_{||\alpha||}w''$ and $w'' R_{||\beta||}w'$. Since $\forall w', w''(wR_{||\alpha||}w' \wedge w' R_{||\beta||}w'' \Rightarrow w'' \in ||\gamma||)$ and $wR_{||\alpha||}w'' \wedge w'' R_{||\beta||}w'$, we get $w' \in ||\gamma||$. This implies $\forall w'(wR_{||\alpha||\cap||\beta||}w' \Rightarrow w' \in ||\gamma||)$. Then by $V_{||\cdot||}$

it follows that $\forall w'(wR_{||\alpha\wedge\beta||}w' \Rightarrow w' \in ||\gamma||)$. Due to $V_{\Box\!\!\rightarrow}$ we obtain $\models^{\mathcal{M}_C}_{w} \alpha\wedge\beta \Box\!\!\rightarrow \gamma$.

("$\Rightarrow$") Let $\mathcal{F}_C = \langle W, R\rangle$ be a frame such that C_{Im} does not hold. Then, there exist worlds $w, w' \in W$ and sets $X, Y \subseteq W$ such that $wR_{X\cap Y}w'$ and $\neg\exists w''(wR_X w'' \wedge w''R_Y w')$. Let $\mathcal{M}_C = \langle W, R, V\rangle$ be a model based on $\mathcal{F}_C$ such that $X = ||\alpha||$, $Y = ||\beta||$, $\forall w''(wR_{||\alpha||}w'' \Rightarrow \forall w'''(w''R_{||\beta||}w''' \Rightarrow w''' \in ||\gamma||))$ and $w' \notin ||\gamma||$. Then, $wR_{||\alpha||\cap||\beta||}w'$ and $\neg\exists w''(wR_{||\alpha||}w'' \wedge w''R_{||\beta||}w')$. This assignment is possible for any frame $\mathcal{F}_C$, since if there is a $w'' \in W$ such that $wR_{||\alpha||}w''$, then not $w''R_{||\beta||}w'$. Hence, w' is not among the w'''s such that $w''' \in ||\gamma||$. Regarding w there are two possible cases: (a) There exists a world w'' such that $wR_{||\alpha||}w''$. Since $\forall w''(wR_{||\alpha||}w'' \Rightarrow \forall w'''(w''R_{||\beta||}w''' \Rightarrow w''' \in ||\gamma||)$, by $V_{\Box\!\!\rightarrow}$ we have $\forall w''(wR_{||\alpha||}w'' \Rightarrow \models^{\mathcal{M}_C}_{w''} \beta \Box\!\!\rightarrow \gamma)$. (b) There exists no w'' such that $wR_{||\alpha||}w''$. Then, trivially $\forall w''(wR_{||\alpha||}w'' \Rightarrow w'' \in ||\beta \Box\!\!\rightarrow \gamma||)$. Hence, in both cases we have $\forall w''(wR_{||\alpha||}w'' \Rightarrow w'' \models^{\mathcal{M}_C}_{w''} \beta \Box\!\!\rightarrow \gamma)$. From that it follows due to $V_{\Box\!\!\rightarrow}$ that $\models^{\mathcal{M}_C}_{w} \alpha \Box\!\!\rightarrow (\beta \Box\!\!\rightarrow \gamma)$. By $\mathrm{Def}_{||.||}$ and $wR_{||\alpha||\cap||\beta||}w'$, we get that $wR_{||\alpha\wedge\beta||}w'$. As $w' \notin ||\beta||$, by $\mathrm{Def}_{||.||}$ this implies $\not\models^{\mathcal{M}_C}_{w} \alpha\wedge\beta \Box\!\!\rightarrow \gamma$. □

Chapter 6

Soundness and Completeness for a Lattice of Conditional Logics

6.1 General Overview

In this chapter I will provide soundness and completeness proofs for the lattice of conditional logics described by System **CK** (see Section 4.2.6) plus axiom schemata from Tables 5.1 and 5.2. As model theoretic basis for the soundness proofs I shall use Chellas frames (see Section 4.3.1) and for the proofs of strong completeness I will employ classes of standard Segerberg frames (see Sections 4.3.6 and 4.3.7) plus the frame conditions in Tables 5.3 and 5.4. Furthermore, I will discuss singleton frames in Section 6.2 which provide easy means to show that the axiom schemata from Tables 5.1 and 5.2 are indeed satisfiable and, therefore, Chellas and Segerberg frames for logics containing these axiom schemata exist. Let me first discuss the notion of completeness established in this chapter.

6.1.1 Focus of the Completeness Proofs

My completeness result draws on Chellas (1975) and Segerberg (1989). Chellas (1975) gave a completeness proof in terms of Chellas frames for System **CK** (pp. 139–141), System **CK**+Refl, System **CK**+MP, and System **CK**+Refl+MP (pp. 141–143). Chellas (1975, p. 143) correctly pointed out that his canonical model approach does not work for a system corresponding to the Monotonic System **M** without the bridge principles, where **M**=**CK**+Refl+Mon+Or (cf. Section 7.2.6). Since System **M** is a point in the lattice of conditional logics described in Section 5.1, this implies that Kripke frame completeness for this lattice of conditional logic systems cannot be established by Chellas' (1975) approach.

Segerberg (1989, p. 162) proves a strong completeness result for System **CK** and conjectures that such a completeness result also obtains for some points in the lattice, namely for those systems which result from **CK** plus a selection of the Axiom Schemata Refl, CM, Or, S, RM, Det, and Cond. [1] Segerberg (1989) does so in terms of strong completeness w.r.t. (simple) Segerberg frames. For this type of completeness result it is essential that all subsets described in the frame condition refer to elements X of P. Hence, not only the accessibility relations R_X are exclusively defined w.r.t. elements X of P, but also any other set of possible worlds X used in this frame condition is required to be an element of P.

Strong (simple) Segerberg frame completeness, while applicable to the whole lattice of conditional logics described above, has a serious drawback. It is trivial in the sense that any logic which is an extension of System **CK** is by Theorem 4.25 strongly complete w.r.t. some class of Segerberg frames. I hence propose here a non-trivial alternative notion of completeness, namely *strong standard Segerberg frame completeness* (see Section 4.3.7).

Let me specify here what I mean by strong standard Segerberg frame completeness. This notion of completeness is based on strong (simple) Segerberg frame completeness, but requires the class of frames w.r.t. which the logic in question is complete to be standard Segerberg frames in the sense of Definition 4.24: [2] Let **L** be a logic such that $\mathbf{L} = \mathbf{CK}+\alpha_1+\ldots+\alpha_n$ for $n \in \mathbb{N}$ and α_1,

[1] Segerberg (1989, p. 163) also discussed a completeness result involving two further axiom schemata, his Axiom Schemata #2 and #6 (see Section 5.1 for further details). Segerberg, in addition, identified non-trivial C-corresponding frame conditions for all of these principles.

[2] Segerberg (1989) does not discuss any of these difficulties. His conjectures can,

$\ldots, \alpha_n$ being axiom schemata and let $C_{\alpha_1}, \ldots, C_{\alpha_n}$ be frame conditions which C-correspond to $\alpha_1, \ldots, \alpha_n$, respectively. Furthermore, let $\mathcal{F}_S^{\mathrm{st}} = \langle W, R, P \rangle$ be an element of the frame class $\mathbf{F}_S^{\mathrm{st}}$, w.r.t. which **L** is strongly frame complete. Then, $\mathcal{F}_S^{\mathrm{st}}$ has to satisfy the frame conditions $C_{\alpha_1}, \ldots, C_{\alpha_n}$ for all elements X of the parameter P (see Section 4.3.6).

In this chapter I shall prove strong completeness in terms of strong *standard* Segerberg frame completeness (see Section 4.3.7). I will also extend Segerberg's completeness result in as much as I provide a strong standard Segerberg frame completeness result for the full lattice of conditional logics defined by Tables 5.1 and 5.2. For that purpose I will draw on frame conditions for an additional 20 axiom schemata, which I identifed in Chapter 5 (see Tables 5.3 and 5.4).

6.1.2 Proofs for Segerberg Frame Completeness and Chellas Frames Completeness

In this section I shall discuss strong completeness proofs for classes of standard Segerberg frames and contrast them with strong completeness proofs for classes of Chellas frames. To prove strong completeness w.r.t. classes of standard Segerberg frames, one can employ the canonical model technique (cf. Hughes & Cresswell, 1996/2003, Chapter 6). For that purpose a canonical model $\mathcal{M}^c = \langle W^c, R^c, P^c, V^c \rangle$ for a given logic **L** is constructed, where logic **L** is System **CK** plus any number of axiom schemata $\alpha_1, \ldots, \alpha_n$ ($n \in \mathbb{N}$) from Tables 5.1 and 5.2. The canonical model is, then, used to demonstrate that any **L**-consistent formula set is satisfiable in a standard Segerberg frame which satisfies the frame conditions $C_{\alpha_1}, \ldots, C_{\alpha_n}$ for elements of P^c so that the latter of which C-correspond to $\alpha_1, \ldots, \alpha_n$, respectively. (The latter result is a reformulation of the completeness property based on the Consistency Lemma, see Lemma 6.4). We do so by showing that (i) the purported canonical model is in fact a Segerberg model and (ii) that it is based on a standard

however, in hindsight be interpreted as conjectures regarding strong *standard* Segerberg frame completeness (p. 163).

Segerberg frame is a frame which satisfies $C_{\alpha_1}, \ldots, C_{\alpha_n}$. Condition (ii) is also referred to as the canonicity property (see Definition 6.1 for a formal characterization of this property).

The set W^c in a canonical model $\mathcal{M}^c = \langle W^c, R^c, P^c, V^c \rangle$ for a given logic **L** is, then, identified with the set of all *maximally* **L**-consistent sets. The valuation function V^c gives us that for all atomic propositions $p \in \mathcal{AV}$ and all worlds $w \in W^c$ holds: $V^c(p,w) = 1$ iff $p \in w$. In $\mathcal{M}^c$ the parameter P^c is specified to be the set of semantically representable subsets of W^c (cf. Segerberg, 1989, p. 162; cf. Section 4.3.2), where the set of semantically representable subsets of W^c is the set of all sets $|\alpha|^{\mathcal{M}^c} = \{w \in W^c \mid \alpha \in w\}$ for arbitrary formulas α of language $\mathcal{L}_{\mathrm{CL}}$. The notion of semantically representable subsets is analogous to the one for syntactically representable subsets insofar as membership in a world is taken as basic rather than truth at a world (see Definition 6.8). The valuation function of the canonical model is, then, defined in such a way that the set of semantically representable subsets of W^c is identical with the set of syntactically representable subsets of W^c (by means of the Truth Lemma; see Lemma 6.9). According to the definition of $\mathcal{M}^c$ the acessibility relations R^c_X are also defined only w.r.t. the set of semantically representable subsets X of W^c and, hence, w.r.t. the set of syntactically representable subsets of W^c. As we saw in Section 4.3.2, only accessibility relations relativized to syntactically (or semantically) representable subsets are relevant for the truth of formulas at possible worlds in a model.

The canonical Chellas model (for proofs of Chellas frame completeness) $\mathcal{M}^c_C = \langle W^c_C, R'^c_C, V^c_C \rangle$ for a logic **L**, which is an extension of System **CK**, agrees with the canonical standard Segerberg model $\mathcal{M}^c = \langle W^c, R^c, P^c, V^c \rangle$ for the same logic **L** w.r.t. the parameters W^c and V^c (cf. Chellas, 1975, p. 140; Segerberg, 1989, p. 162). Both types of canonical models, however, differ in the accessibility relation. In the canonical Chellas model $\mathcal{M}^c_C$ the accessibility relation has to be relativized to all subsets of W^c. Hence, R'^c_{CX} has to be defined for all subsets X whether they are syntactically representable or not, whereas in the canonical standard Segerberg model $\mathcal{M}^c$ the accessibility relation R^c_X is by means of P^c restricted to the syntactically representable subsets of W. For syntactically representable subsets X of W^c_C, then R^c_X and R'^c_X agree (cf. Chellas, 1975, p. 140). Due to that fact the specification of a canonical Chellas model $\mathcal{M}^c_C$ is not unique and one can choose any canonical model in this class of canonical Chellas models. Chellas (1975, p. 140) calls any such canonical model 'proper'.

The problem for a completeness result in terms of Chellas frames are now exactly the accessibility relations w.r.t. syntactically non-representable subsets of W^c_C. We saw earlier that this type of accessibility relations is in a sense irrelevant as it does not bear on the truth of formulas in worlds in the canonical model (see Lemma 4.15). Despite this fact, these accessibility relations have to be specified in any Chellas model. For Chellas frame completeness of extensions of System **CK** the canonical model must satisfy the respective frame conditions from Tables 5.3 and 5.4, which correspond to the axiom schemata from Tables 5.1 and 5.2 added to System **CK**. This holds for all sets X in the canonical model whether syntactically representable or not. On a general basis the addition of an axiom schema of a system **L** does not affect the accessibility relations R^c_{CX} in the canonical model relativized to syntactically non-representable sets X. Moreover, there seems to be no general way available to define this type of accessibility relation in such a way that it complies with any frame condition required by the lattice of systems. In the case of standard Segerberg frame completeness no such problem arises, since one can restrict oneself to the set of accessibility relations which are only relativized to syntactically (and thus semantically) representable subsets of W^c_C.

Finally let me in this section discuss the notion of canonicity (see point (ii) earlier), on which the completeness proofs in the next sections draw:

Definition 6.1. *Let* **L** *be an extension of System* **CK** *such that* **L** = **CK**+α *and let* C_α *be a frame condition which C-corresponds to* α *and where* α *is an axiom schema. Moreover, let* $\mathcal{M}^c = \langle W^c, R^c, P^c, V^c\rangle$ *be the canonical model of* **L**. *Then* **L** *is canonical (w.r.t. classes of standard Segerberg frames) iff* $\langle W^c, R^c, P^c\rangle$ *satisfies* C_α *for all elements* X *of* P^c.

The canonicity property of a logic **CK** + α gives us that the canonical model of **L** is based on a frame, for which the corresponding frame condition C_α

from Tables 5.3 and 5.4 holds. The canonicity property generalizes the following way:

Theorem 6.2. *Let* **L** *be an extension of System* **CK** *such that* $\mathbf{L} = \mathbf{CK}+\alpha_1+\ldots+\alpha_n$ *(*$n \in \mathbb{N}$*) for* $\alpha_1, \ldots, \alpha_n$ *being axiom schemata and let* $\mathbf{CK}+\alpha_1, \ldots, \mathbf{CK}+\alpha_n$ *be canonical (w.r.t. classes of standard Segerberg frames). Then,* **L** *is canonical (w.r.t. classes of standard Segerberg frames).*

Proof. Let (a) $\mathbf{CK}+\alpha_1, \ldots, \mathbf{CK}+\alpha_n$ be canonical and let (b) $C_{\alpha_1}, \ldots, C_{\alpha_n}$ be frame conditions C-corresponding to $\alpha_1, \ldots, \alpha_n$, respectively. That such $C_{\alpha_1}, \ldots, C_{\alpha_n}$ always exist for axiom schemata $\alpha_1, \ldots, \alpha_n$ is given by Theorem 5.2. Since by (a) $\mathbf{CK}+\alpha_1, \ldots, \mathbf{CK}+\alpha_n$ are canonical (w.r.t. classes of standard Segerberg frames), the frame of the canonical model $\mathcal{F}^c = \langle W^c, R^c, P^c, R^c\rangle$ for these systems satisfies $C_{\alpha_1}, \ldots, C_{\alpha_n}$ (restricted to P^c), respectively. Moreover, by (b) $C_{\alpha_1}, \ldots, C_{\alpha_n}$ C-correspond to $\alpha_1, \ldots, \alpha_n$, respectively. Thus, (i) $\alpha_1 \wedge \ldots \wedge \alpha_n$ C-corresponds to $C_{\alpha_1} \wedge \ldots \wedge C_{\alpha_n}$. Furthermore, System **CK** does not contain non-monotonic rules by Definition 4.1, and $\alpha_1, \ldots, \alpha_n$ are axiom schemata and, hence, (monotonic) one-step rules (see Section 4.2.3). Thus, the set of possible worlds of the respective canonical model – which is determined by the set of **L**-consistent formula sets of the logic **L** in question – is never increased. Hence, every frame condition C_{α_i} (for $1 \leq i \leq n$), which holds in frame of the canonical model $\mathcal{M}^c$ for $\mathbf{CK}+\alpha_i$ (restricted to P^c), holds also for the frame $\mathcal{F}^c_{\mathbf{L}}$ of the canonical model $\mathcal{M}^c_{\mathbf{L}} = \langle W^c, R^c, P^c, V^c\rangle$ for **L** (restricted to P^c). Since $C_{\alpha_1}, \ldots, C_{\alpha_n}$ apply to $\mathcal{F}^c_{\mathbf{L}}$ (restricted to P^c), (ii) also $C_{\alpha_1} \wedge \ldots \wedge C_{\alpha_n}$ apply also to $\mathcal{F}_{\mathbf{L}}$ (restricted to P^c). By (i) and (ii) it follows that **L** is canonical (w.r.t. classes of standard Segerberg frames). □

We shall, hence, only prove the canonicity property for System **CK** plus individual principles from Tables 5.1 and 5.2 (cf. Blackburn et al., 2001, p. 202; see also below), since Theorem 6.2 gives us that the canonicity results extend to the whole lattice of systems described by Tables 5.1 and 5.2 (cf. Section 5.1). Since any systems we discuss here has this property, I shall sometimes say that a principle is canonical rather than the system which contains it.

In the remainder of this chapter I will proceed as follows: First, in Section 6.2 I will describe singleton frames (frames that contain a single world) for CS semantics. Then, in Section 6.3 I will discuss the soundness result for the lattice of systems. Finally, in Section 6.4 I shall describe a completeness result in terms of standard Segerberg frames for the lattice of systems defined by Tables 5.1 and 5.2.

6.2 Singleton Frames for CS Semantics

Before I focus on the soundness and completeness results, let me first discuss singleton Chellas frames. By singleton Chellas frames $\mathcal{F}_C = \langle W, R\rangle$ I mean frames in which the set of possible worlds consists only of a single world w (formally: $W = \{w\}$). Analogous to Kripke semantics, singleton frames in CS semantics allow one to specify collapse conditions. Moreover, by the use of singleton Chellas frames one can easily demonstrate that there exist Chellas models (and Segerberg models) which satisfy the axiom schemata listed in Tables 5.1 and 5.2. Due to the correspondence results, it suffices to check whether one of the singleton frames to be specified satisfies the respective frame condition in Tables 5.3 and 5.4.

Let me now describe the singleton Chellas frames. We can distinguish between four (incompatible) singleton frames $\mathcal{F}_C = \langle W, R\rangle$, based on the following specifications of the accessibility relation R:

$$\begin{aligned}
&\mathcal{F}_C^1 = \langle W, R_1\rangle, \quad \text{with } R_1 = \{\langle w, w, W\rangle, \langle w, w, \varnothing\rangle\}\\
&\mathcal{F}_C^2 = \langle W, R_2\rangle, \quad \text{with } R_2 = \{\langle w, w, W\rangle\}\\
&\mathcal{F}_C^3 = \langle W, R_3\rangle, \quad \text{with } R_3 = \{\langle w, w, \varnothing\rangle\}\\
&\mathcal{F}_C^4 = \langle W, R_4\rangle, \quad \text{with } R_4 = \varnothing
\end{aligned}$$

One can easily show – although we shall not prove this result here – that the following characteristic Axiom Schemata CA_1, CA_2, CA_3, and CA_4 are valid on the respective singleton frames $\mathcal{F}_C^1$, $\mathcal{F}_C^2$, $\mathcal{F}_C^3$, and $\mathcal{F}_C^4$:

CA$_1$: $(\alpha \mathrel{\Box\!\!\rightarrow} \beta) \leftrightarrow \beta$
CA$_2$: $(\alpha \mathrel{\Box\!\!\rightarrow} \beta) \leftrightarrow (\alpha \rightarrow \beta)$
CA$_3$: $(\alpha \mathrel{\Box\!\!\rightarrow} \beta) \leftrightarrow (\alpha \vee \beta)$
CA$_4$: $\alpha \mathrel{\Box\!\!\rightarrow} \beta$

$\mathcal{F}_C^1$ and $\mathcal{F}_C^4$ represent generalizations of the Systems Triv and Ver in Kripke semantics (e.g., Hughes & Cresswell, 1996/2003, pp. 64–68), respectively. We saw earlier (Section 4.3.2) that the conditional operator, when relativized to an antecedent formula α, can be interpreted as a modal operator $[\alpha]$. On the basis of that terminology the Principles CA$_1$ and CA$_4$ can, then, be formulated as $[\alpha]\beta \leftrightarrow \beta$ and $[\alpha]\beta$, respectively, which correspond again directly to the Principles Triv and Ver in Kripke semantics, respectively (see Hughes & Cresswell, 1996/2003, p. 65 and p. 67). Moreover, in the singleton Kripke frame for Triv the world w can see only itself by the accessibility relation R (see Hughes & Cresswell, 1996/2003, p. 64f), whereas in the singleton Kripke frame for Ver no world can see any world at all (see Hughes & Cresswell, 1996/2003, p. 66). The frames $\mathcal{F}_C^1$ and $\mathcal{F}_C^4$ correspond to the singleton frames for Triv and Ver insofar as the world w can see only itself by all accessibility relations (namely $R_\varnothing$ and R_W) in the case of $\mathcal{F}_C^1$ and no world by an accessibility relation in the case of $\mathcal{F}_C^4$. It is important to note that system **CK**+CA$_4$ is not the inconsistent system: Although $\bot$ is naturally not valid on $\mathcal{F}_C^4$, any conditional formula is, including $\top \mathrel{\Box\!\!\rightarrow} \bot$. This is possible, since no bridge principle, including $\neg(\top \mathrel{\Box\!\!\rightarrow} \bot)$ (P-Cons, see Table 5.1) is valid on $\mathcal{F}_C^4$ either.

The singleton Chellas frame, which is most interesting from a conditional logic perspective is $\mathcal{F}_C^2$. Except for the Principles D and T – which are valid on $\mathcal{F}_C^1$ – all principles from Tables 5.1 and 5.2 are valid on $\mathcal{F}_C^2$. Moreover, as we shall see in Section 7.3.4, all principles of the monotonic collapse of conditional logics with bridge principles – which contains all conditional logics investigated in Chapter 7 – are also valid in $\mathcal{F}_C^2$. It is, finally, interesting to see that in the case of the singleton frame $\mathcal{F}_C^3$ the two place modal operator $\alpha \mathrel{\Box\!\!\rightarrow} \beta$ is characterized by the disjunction rather than the material implication.

6.3 Soundness w.r.t. Classes of Chellas Frames

To provide soundness results for a logic **L** – where **L** is an extension of System **CK** by axiom schemata from Tables 5.1 and 5.2 – w.r.t. Chellas frames, it suffices to prove the following two points: (a) All axiom schemata and rules of System **CK** (see Definition 4.1) are valid in the class of all Chellas models, and (b) when a frame condition from Tables 5.3 and 5.4 holds for a Chellas frame, then the corresponding principle from Tables 5.1 and 5.2 is valid on that frame.

I will not provide a proof for (a), since (a) holds trivially for arbitrary Chellas models. Moreover, (b) is exactly the right-to-left direction of C-correspondence (see Definition 4.20) and C-correspondence proofs are given in Chapter 5 for all principles in Tables 5.1 and 5.2. Note that (a) also holds trivially for all Segerberg models and that the proofs for the right-to-left direction of C-correspondence in Chapter 5 go through also w.r.t. (standard) Segerberg frames $\mathcal{F}_S = \langle W, R, P \rangle$ when the frame conditions are restricted to elements of the parameter P.

In Chapter 5 we proved (b) only for individual principles from Tables 5.1 and 5.2 (i.e. by adding a single principle α to **CK**), but not for combinations of principles $\alpha_1, \ldots, \alpha_n$ ($n \in \mathbb{N}$) from those tables (i.e., by adding $\alpha_1, \ldots, \alpha_n$ to **CK**). Despite this fact, the soundness proofs generalize to the whole lattice of systems defined by Tables 5.1 and 5.2 by the following theorem:

Theorem 6.3. *Let* **L** *be an extension of System* **CK** *such that* **L** = **CK**+α_1+... +α_n *for* $\alpha_1, \ldots, \alpha_n$ *being axiom schemata* ($n \in \mathbb{N}$) *and let* **CK**+α_1, ..., **CK**+α_n *be sound (w.r.t. classes of Chellas frames). Then,* **L** *is sound (w.r.t. classes of Chellas frames).*

Proof. Let $C_{\alpha_1}, \ldots, C_{\alpha_n}$ be frame conditions which C-correspond to $\alpha_1, \ldots, \alpha_n$, respectively. That such $C_{\alpha_1}, \ldots, C_{\alpha_n}$ always exist for axiom schemata $\alpha_1, \ldots, \alpha_n$ is given by Theorem 5.2. Then, by the right-to-left direction of C-correspondence (Definition 4.20) it follows that all elements of **CK**+α_1,

..., **CK**+α_n are valid in the classes of Chellas frames which satisfy C_{α_1}, ..., C_{α_n}, respectively. Since C_{α_1}, ..., C_{α_n} C-correspond to axiom schemata rather than (non-monotonic) rules, C_{α_1}, ..., C_{α_n} do not depend on certain formulas being non-valid in a Chellas model or a Chellas frame. Thus, a class of Chellas frames $\mathcal{F}_C$ cannot increase when $\mathcal{F}_C$ is restricted by an additional frame condition C_{α_i} (for $1 \leq i \leq n$). Hence, when $C_{\alpha_1}, \ldots, C_{\alpha_n}$ hold conjointly for a class of Chellas frames $\mathbf{F}_C$, then $\mathbf{F}_C \subseteq \mathbf{F}_C^{C_{\alpha_i}}$ for each C_{α_i} where $\mathbf{F}_C^{C_{\alpha_i}}$ is the set of Chellas frames, for which the frame condition C_{α_i} holds. Hence, by the right-to-left direction of C-correspondence (Definition 4.20) we have that all elements of **CK**+α_i are valid in $\mathbf{F}_C^{C_{\alpha_i}}$ and, hence, in $\mathbf{F}_C$. Thus, $\mathbf{L} =$ **CK**+α_1+...+α_n is sound (w.r.t. classes of Chellas frames). □

6.4 Standard Segerberg Frame Completeness

In this section we will prove strong completeness w.r.t. classes of standard Segerberg frames. I shall first discuss general lemmata for the completeness result and then focus on canonical models for this type of semantics.

6.4.1 General Principles

I directly state the following lemmata and omit some of the proofs, since they can be obtained by a standard procedure:

Lemma 6.4. *(Consistency Lemma)*
$\mathbf{L}$ *is strongly complete w.r.t.* $\mathbf{F}_S^{\mathrm{st}}$ *iff every* $\mathbf{L}$*-consistent formula set* Γ *is satisfiable in* $\mathbf{F}_S^{\mathrm{st}}$.

Lemma 6.5. *For any maximally* $\mathbf{L}$*-consistent set* Δ *and formulas* α *and* β *the following properties hold:*
(a) if $\alpha \in \Delta$ *and* $\vdash_{\mathbf{L}} \alpha \rightarrow \beta$, *then* $\beta \in \Delta$, *(Deductive Closure)*
(b) either $\alpha \in \Delta$ *or* $\neg\alpha \in \Delta$, *(Maximality)*
(c) if $\alpha \in \Delta$, *then* $\neg\alpha \notin \Delta$, *and (Consistency)*
(d) $\alpha \vee \beta \in \Delta$ *iff* $\alpha \in \Delta$ *or* $\beta \in \Delta$. *(Primeness)*

Proof. For proofs of Lemmata 6.5.a, 6.5.b, and 6.5.d see proofs of Lemmata 2.1f, 2.1a, and 2.1b by Hughes and Cresswell (1984, p. 18f), respectively. Lemma 6.5c follows by the consistency of Δ. □

Lemma 6.6. *(Lindenbaum Lemma) Let Γ be a* **CK***-consistent formula set. Then there is a maximally* **CK***-consistent formula set Δ such that $\Gamma \subseteq \Delta$.*

Proof. See proof of Theorem 2.2 by Hughes and Cresswell (1984, p. 19f). □

The Consistency Lemma (Lemma 6.4) shows that in order to prove strong completeness of **CK** w.r.t. classes of standard Segerberg frames $\mathbf{F}_S^{st}$, it suffices to demonstrate the following: Every **CK**-consistent formula set Γ is satisfiable in $\mathbf{F}_S^{st}$. To prove this result we construct a canonical model $\mathcal{M}^c = \langle W^c, R^c, P^c, V^c \rangle$ along the lines of Segerberg (1989). This canonical model specifies W^c as the set of all maximally **CK**-consistent formula sets. The Lindenbaum Lemma (Lemma 6.6) gives us that every **CK**-consistent formula set is a subset of a maximally **CK**-consistent set. Since the possible worlds of the canonical model are by definition all (and only) maximally **CK**-consistent sets, any **CK**-consistent set is, hence, a subset of a possible world in $\mathcal{M}^c$. The next building block of the completeness proof is the Truth Lemma (Lemma 6.9). This lemma establishes that a formula α is true at a world w iff it is a member of w. Hence, any **CK**-consistent formula set Γ is a subset of a possible world w in $\mathcal{M}^c$. Since any element of a possible world w is true at w, it follows that all formulas in Γ are true at w. Hence, there is a model, namely the canonical model $\mathcal{M}^c$, in which the formula set Γ is true. Thus, Γ is $\mathbf{F}_S^{st}$-satisfiable. Finally, for basic System **CK** we would have to show that the frame of the canonical model is a frame on which all axiom schemata of **CK** are valid. This is trivial, since – as one can easily see – all Chellas and (standard) Segerberg frames are frames for logic **CK**.

6.4.2 Canonical Models

In this section we will specify the canonical model for extensions of System **CK** as defined by Tables 5.1 and 5.2. This definition and the following proofs draw on Segerberg (1989, p. 162f) and the standard technique for providing completeness results as described in Hughes and Cresswell (1996/2003, Chapter 6). We use the following definition: $|\alpha|^{\mathcal{M}^c} =_{def} \{w \in W^c \mid \alpha \in w\}$ ($\text{Def}_{|.|}$), where W^c refers to the set of possible worlds in a canonical model $\mathcal{M}^c = \langle W^c, R^c, P^c, V^c \rangle$ to be specified. We will also say that α is semantically representable (in a canonical model $\mathcal{M}^c$) iff there is a set of possible worlds $X \subseteq W^c$ such that $X = |\alpha|^{\mathcal{M}^c}$. In the following we will leave out reference to the canonical model $\mathcal{M}^c$ and simplify expressions, such as $|\alpha|^{\mathcal{M}^c}$ to $|\alpha|$, since those always refer to respective canonical model. Moreover, all formulas $\alpha, \beta, \ldots$ will be formulas of the language $\mathcal{L}_{\text{CL}}$.

Definition 6.7. *A Segerberg model $\mathcal{M}^c = \langle W^c, R^c, P^c, V^c \rangle$ is the canonical model for an extension* **L** *of* **CK** *iff*

(a) *W^c is the class of all maximally* **L**-*consistent formula sets,* (Def_{W^c})

(b) *$P^c = \{X \subseteq W^c \mid \exists\alpha$ is such that $X = |\alpha|\}$,* (Def_{P^c})

(c) *$\forall\alpha\forall w, w' \in W^c \forall X \in P^c$ such that $X = |\alpha|$: $wR^c_X w'$ iff $\forall\beta(\alpha \mathrel{\Box\!\!\rightarrow} \beta \in w \Rightarrow \beta \in w')$, and* ($Def_{R^c}$)

(d) *$\forall p \in \mathcal{AV}, w \in W^c$: $V^c(p,w) = 1$ iff $p \in w$.* (Def_{V^c})

Point (c) of Definition 6.7 gives us that R_X is defined for all elements $X \in P^c$ since by Definition 6.7.b every such element X is semantically representable. Observe that a restriction of the definition of R_X to semantically (or syntactically) representable subsets X of W^c – as in the case of standard Segerberg frames – is not possible for Chellas frame completeness. Chellas (1975, p. 139), thus, uses the notion of a *class* of canonical (standard) models rather than the notion of a *single* canonical model. Chellas characterizes this class of canonical (standard) models by specifying the accessibility relation R_X only for semantically representable subsets X of W^c (p. 139). Canonical (standard) models in Chellas' account differ, then, in the specification of R_X for semantically non-representable subsets of W^c.

Let us now officially define the notion of semantically representable subsets of W^c.

Definition 6.8. *Let $\mathcal{M}^c = \langle W^c, R^c, P^c, V^c \rangle$ be the canonical model of an extension* **L** *of System* **CK** *and let $X \subseteq W^c$ be the case. Then, X is semantically* *representable in model $\mathcal{M}^c$ (by formulas of $\mathcal{L}_{\mathrm{CL}}$) iff there exists a formula α (of $\mathcal{L}_{\mathrm{CL}}$) such that $X = |\alpha|^{\mathcal{M}^c}$.*

The expression P^c refers to the set of subsets of W^c as described in Definition 6.7. We will now focus on the proof for Lemma 6.9:

Lemma 6.9. *(Truth Lemma) For the canonical model $\mathcal{M}^c = \langle W^c, R^c, P^c, V^c\rangle$ of an extensions* **L** *of System* **CK** *and for all α and all $w \in W^c$ holds: $\alpha \in w$ iff $\models_w^{\mathcal{M}^c} \alpha$.*

Proof. The proof is by induction on the construction of formulas. The proof of the non-modal cases are standard and can be found in textbooks for modal logic (e.g., Hughes & Cresswell, 1984, p. 23f). The only case that differs from the standard procedure is the modal one. Hence, we restrict our proof to this case and have to prove for arbitrary formulas α, β and $w \in W^c$ that $\alpha \mathrel{\Box\!\!\rightarrow} \beta \in w$ iff $\models_w^{\mathcal{M}^c} \alpha \mathrel{\Box\!\!\rightarrow} \beta$, based on the following induction hypotheses: (a) $\forall w \in W^c(\alpha \in w$ iff $\models_w^{\mathcal{M}^c} \alpha)$ and (b) $\forall w \in W^c(\beta \in w$ iff $\models_w^{\mathcal{M}^c} \beta)$.

"$\Rightarrow$": Let $\alpha \mathrel{\Box\!\!\rightarrow} \beta \in w$ be the case. Then, by Lemma 6.10 it follows that for all $w' \in W^c$ holds: $wR^c_{|\alpha|}w' \Rightarrow \beta \in w'$. By the induction hypothesis (a), $\mathrm{Def}_{|.|}$, and $\mathrm{Def}_{||.||}$ we get $|\alpha| = ||\alpha||$. Moreover, due to (b) $\forall w \in W^c(\beta \in w$ iff $\models_w^{\mathcal{M}^c} \beta)$ holds. Hence, $\forall w' \in W^c(wR^c_{||\alpha||}w' \Rightarrow \models_{w'}^{\mathcal{M}^c} \beta)$. By $V_{\Box\!\rightarrow}$ it follows that $\models_w^{\mathcal{M}^c} \alpha \mathrel{\Box\!\!\rightarrow} \beta$.

"$\Leftarrow$": The proof is by contraposition. Let $\alpha \mathrel{\Box\!\!\rightarrow} \beta \notin w$ be the case. By Lemma 6.10 it follows that there exists a world $w' \in W^c$ such that $wR^c_{|\alpha|}w'$ and $\beta \notin w'$. By the induction hypothesis (a), $\mathrm{Def}_{|.|}$, and $\mathrm{Def}_{||.||}$ we get $|\alpha| = ||\alpha||$. Moreover, by (b) $|\beta| = ||\beta||$. Thus, it follows that there exists a world $w' \in W^c$ such that $wR^c_{||\alpha||}w'$ and $\not\models_{w'}^{\mathcal{M}^c} \beta$. By $V_{\Box\!\rightarrow}$ this implies that $\not\models_w^{\mathcal{M}^c} \alpha \mathrel{\Box\!\!\rightarrow} \beta$. □

Lemma 6.9 draws on Lemma 6.10 that in turn refers to Lemma 6.11.

Lemma 6.10. *For any canonical model $\mathcal{M}^c = \langle W^c, R^c, P^c, V^c\rangle$ of an extension* **L** *of System* **CK** *and all α,β and all $w \in W^c$ holds: $\alpha \mathrel{\Box\!\!\rightarrow} \beta \in w$ iff $\forall w' \in W^c(wR^c_{|\alpha|}w' \Rightarrow \beta \in w')$.*

Proof. "$\Rightarrow$": The proof is by contraposition. Suppose that for arbitrary formulas α, β and an arbitrary $w \in M^c$ it is not the case that $\forall w' \in W^c(wR^c_{|\alpha|}w' \Rightarrow$

$\beta \in w'$). Then, there exists a $w' \in W^c$ such that $wR^c_{|\alpha|}w'$ and $\beta \notin w'$. By Def$_{R^c}$ we obtain $\alpha \mathbin{\Box\!\!\rightarrow} \beta \notin w$.

"$\Leftarrow$": The proof is by contraposition. Suppose that for arbitrary α, β and an arbitrary world $w \in W^c$ it is the case that $\alpha \mathbin{\Box\!\!\rightarrow} \beta \notin w$, which implies by Lemma 6.5.b that $\neg(\alpha \mathbin{\Box\!\!\rightarrow} \beta) \in w$. The Lindenbaum Lemma, Lemma 6.11, and Def$_{W^c}$ give us that there exists a world $w' \in W^c$ such that $\{\gamma | \alpha \mathbin{\Box\!\!\rightarrow} \gamma \in w\} \cup \{\neg\beta\} \subseteq w'$. Since $\{\gamma | \alpha \mathbin{\Box\!\!\rightarrow} \gamma \in w\} \subseteq w'$, it holds that $\forall\gamma(\alpha \mathbin{\Box\!\!\rightarrow} \gamma \in w \Rightarrow \gamma \in w')$. Hence, it follows by Def$_{R^c}$ that $wR^c_{|\alpha|}w'$. Since $\{\neg\beta\} \subseteq w'$, we have $\neg\beta \in w'$. By Lemma 6.5.c it is, hence, the case that $\beta \notin w'$. Thus, there exists a world $w' \in W^c$ such that $wR^c_{|\alpha|}w'$ and $\beta \notin w'$. □

Lemma 6.11. *Let* **L** *be an extension of System* **CK**. *Then, if* Γ *is a* **L**-*consistent set of formulas such that* $\neg(\alpha \mathbin{\Box\!\!\rightarrow} \beta) \in \Gamma$, *then* $\{\gamma | \alpha \mathbin{\Box\!\!\rightarrow} \gamma \in \Gamma\} \cup \{\neg\beta\}$ *is* **L**-*consistent.*

Proof. The proof is by contraposition. Let $\{\gamma | \alpha \mathbin{\Box\!\!\rightarrow} \gamma \in \Gamma\} \cup \{\neg\beta\}$ be **L**-inconsistent. Then, there exists a $n \in \mathbb{N}_0$ such that $\{\gamma_1, \ldots, \gamma_n, \neg\beta\}$ is **L**-inconsistent for $\gamma_1, \ldots, \gamma_n \in \{\gamma | \alpha \mathbin{\Box\!\!\rightarrow} \gamma \in \Gamma\}$. (In the case of $n = 0$ the set $\{\neg\beta\}$ is **L**-inconsistent.) It follows that $\vdash_{\mathbf{L}} \neg(\gamma_1 \wedge \ldots \wedge \gamma_n \wedge \neg\beta)$. This implies that $\vdash_{\mathbf{L}} \gamma_1 \wedge \ldots \wedge \gamma_n \rightarrow \beta$ holds. Logic **L** is by assumption an extension of **CK**. Hence, RW, LT, and AND are theorems of **L**. By Lemma 4.10 the rule RCK is, then, admissible. When applied to the former inference step, it follows that $\vdash_{\mathbf{L}} (\alpha \mathbin{\Box\!\!\rightarrow} \gamma_1) \wedge \ldots \wedge (\alpha \mathbin{\Box\!\!\rightarrow} \gamma_n) \rightarrow (\alpha \mathbin{\Box\!\!\rightarrow} \beta)$. This implies in turn that $\vdash_{\mathbf{L}} \neg((\alpha \mathbin{\Box\!\!\rightarrow} \gamma_1) \wedge \ldots \wedge (\alpha \mathbin{\Box\!\!\rightarrow} \gamma_n) \wedge \neg(\alpha \mathbin{\Box\!\!\rightarrow} \beta))$. Hence, the set $\{\alpha \mathbin{\Box\!\!\rightarrow} \gamma_1, \ldots, \alpha \mathbin{\Box\!\!\rightarrow} \gamma_n\} \cup \{\neg(\alpha \mathbin{\Box\!\!\rightarrow} \beta)\}$ is **L**-inconsistent. Since it holds that $\gamma_1, \ldots, \gamma_n \in \{\gamma | \alpha \mathbin{\Box\!\!\rightarrow} \gamma \in \Gamma\}$, it follows that $\{\alpha \mathbin{\Box\!\!\rightarrow} \gamma_1, \ldots, \alpha \mathbin{\Box\!\!\rightarrow} \gamma_n\} \subseteq \Gamma$. This implies that $\Gamma \cup \{\neg(\alpha \mathbin{\Box\!\!\rightarrow} \beta)\}$ is **L**-inconsistent. □

6.4.3 Canonicity Proofs for Individual Principles

I will now give canonicity proofs (see Definition 6.1) for the principles in Tables 5.1 and 5.2 w.r.t. the corresponding frame conditions in Tables 5.3 and 5.4. The canonicity results establish the following for extensions **L** of System **CK**, as defined by Tables 5.1 and 5.2: If a given axiom schema from

Tables 5.1 and 5.2 is added to the logic **L**, then the frame of the canonical model of logic **L** satisfies the respective frame condition (see Section 6.1.2).

In the following proofs we often draw on $\text{Def}_{|.|}$, in particular for the following two equations concerning arbitrary formulas α and β: (a) $-|\alpha| = |\neg\alpha|$ and (b) $|\alpha| \cap |\beta| = |\alpha \wedge \beta|$. Equations (a) and (b) do not only draw on $\text{Def}_{|.|}$ but also on Lemma 6.5. For the sake of perspicuity we will not indicate these details in the individual canonicity proofs. By assumption all subsets of possible worlds referred to in the individual proofs are restricted to elements of P^c of the respective canonical model $\mathcal{M}^c = \langle W^c, R^c, P^c, V^c \rangle$.

System P

Axiom Schema Refl. Let $\mathbf{L}_{\text{Refl}}$ be an extension of System **CK** such that Refl holds and let $\mathcal{M}^c = \langle W^c, R^c, P^c, V^c \rangle$ be the canonical model of $\mathbf{L}_{\text{Refl}}$. Assume that Refl is not canonical. Then, there exist $X \in P^c$ and $w, w' \in W^c$ such that $wR^c_X w'$ and $w' \notin X$. By Def_{P^c} there is a formula α such that $X = |\alpha|$. Thus, $wR^c_{|\alpha|} w'$ and $w' \notin |\alpha|$. Due to Lemma 6.10 and $\text{Def}_{|.|}$ this implies $\alpha \mathrel{\Box\!\!\rightarrow} \alpha \notin w$. This contradicts the definition of $\mathcal{M}^c$ by Axiom Schema Refl, Lemma 6.5.a, and Lemma 6.5.c. □

Axiom Schema CM. Let $\mathbf{L}_{\text{CM}}$ be an extension of System **CK** such that CM holds. Moreover, let $\mathcal{M}^c = \langle W^c, R^c, P^c, V^c \rangle$ be the canonical model of $\mathbf{L}_{\text{CM}}$. Assume that CM is not canonical. Then, there exist $X, Y \in P^c$ and $w, w' \in W^c$ such that $\forall w''(wR^c_X w'' \Rightarrow w'' \in Y)$, $wR^c_{X\cap Y} w'$, and $\neg wR^c_X w'$. By Def_{P^c} there are formulas α and β such that $X = |\alpha|$ and $Y = |\beta|$. Hence, $\forall w''(wR^c_{|\alpha|} w'' \Rightarrow w'' \in |\beta|)$, $wR^c_{|\alpha|\cap|\beta|} w'$, and $\neg wR^c_{|\alpha|} w'$ hold. By $\text{Def}_{|.|}$, we get that $wR^c_{|\alpha\wedge\beta|} w'$. Since $\forall w''(wR^c_{|\alpha|} w'' \Rightarrow w'' \in |\beta|)$, Lemma 6.10 and $\text{Def}_{|.|}$ imply that $\alpha \mathrel{\Box\!\!\rightarrow} \beta \in w$. Due to Def_{R^c} and $\neg wR^c_{|\alpha|} w'$ there is a formula γ such that $\alpha \mathrel{\Box\!\!\rightarrow} \gamma \in w$ and $\gamma \notin w'$. Since $wR^c_{|\alpha\wedge\beta|} w'$, we obtain by Lemma 6.10 that $\alpha \wedge \beta \mathrel{\Box\!\!\rightarrow} \gamma \notin w$. As $\alpha \mathrel{\Box\!\!\rightarrow} \beta \in w$ and $\alpha \mathrel{\Box\!\!\rightarrow} \gamma \in w$, this contradicts the definition of $\mathcal{M}^c$ by Axiom Schema CM, Lemma 6.5.a, and Lemma 6.5.c. □

Axiom Schema CC. Let $\mathbf{L}_{\text{CC}}$ be an extension of System **CK** such that CC holds. Furthermore, let $\mathcal{M}^c = \langle W^c, R^c, P^c, V^c \rangle$ be the canonical model of $\mathbf{L}_{\text{CC}}$.

Assume that CC is not canonical. Then, there exist $X, Y \in P^c$ and $w, w' \in W^c$ such that $\forall w''(wR^c_X w'' \Rightarrow w'' \in Y)$, $wR^c_X w'$, and $\neg wR^c_{X\cap Y} w'$. By Def$_{P^c}$ there are formulas α and β such that $X = |\alpha|$ and $Y = |\beta|$. Hence, $\forall w''(wR^c_{|\alpha|} w'' \Rightarrow w'' \in |\beta|)$, $wR^c_{|\alpha|} w'$, and $\neg wR^c_{|\alpha|\cap|\beta|} w'$. Due to Def$_{|.|}$ this yields $\neg wR^c_{|\alpha\wedge\beta|} w'$. Since $\forall w''(wR^c_{|\alpha|} w'' \Rightarrow w'' \in |\beta|)$, Lemma 6.10 and Def$_{|.|}$ give us that $\alpha \,\Box\!\!\rightarrow \beta \in w$. Due to Def$_{R^c}$ and $\neg wR^c_{|\alpha\wedge\beta|} w'$ there is a formula γ such that $\alpha \wedge \beta \,\Box\!\!\rightarrow \gamma \in w$ and $\gamma \notin w'$. Since $wR^c_{|\alpha|} w'$, this implies by Lemma 6.10 that $\alpha \,\Box\!\!\rightarrow \gamma \notin w$. As $\alpha \,\Box\!\!\rightarrow \beta \in w$ and $\alpha \wedge \beta \,\Box\!\!\rightarrow \gamma \in w$ hold, this contradicts the definition of $\mathcal{M}^c$ by Axiom Schema CC, Lemma 6.5.a, and Lemma 6.5.c. □

Axiom Schema Loop. Let $\mathbf{L}_{\text{Loop}}$ be an extension of System **CK** such that Loop holds. In addition, let $\mathcal{M}^c = \langle W^c, R^c, P^c, V^c \rangle$ be the canonical model of $\mathbf{L}_{\text{Loop}}$. Assume that Loop is not canonical and that, hence, for some $k \in \mathbb{N}$ for $k \geq 2$, the following is the case: There exist $X_0, \ldots, X_k \in P^c$ and $w, w' \in W^c$ such that $\forall w''(wR^c_{X_0} w'' \Rightarrow w'' \in X_1), \ldots, \forall w''(wR^c_{X_{k-1}} w'' \Rightarrow w'' \in X_k)$, $\forall w''(wR^c_{X_k} w'' \Rightarrow w'' \in X_0)$, $wR^c_{X_0} w'$, and $w' \notin X_k$. By Def$_{P^c}$ there are formulas $\alpha_0, \ldots, \alpha_k$ such that $X_0 = |\alpha_0|, \ldots, X_k = |\alpha_k|$. We obtain, thus, $\forall w''(wR^c_{|\alpha_0|} w'' \Rightarrow w'' \in |\alpha_1|), \ldots, \forall w''(wR^c_{|\alpha_{k-1}|} w'' \Rightarrow w'' \in |\alpha_k|), \forall w''(wR^c_{|\alpha_k|} w'' \Rightarrow w'' \in |\alpha_0|)$, $wR^c_{|\alpha_0|} w'$, and $w' \notin |\alpha_k|$. The latter fact implies by Lemma 6.10 and Def$_{|.|}$ that $\alpha_0 \,\Box\!\!\rightarrow \alpha_1, \ldots, \alpha_{k-1} \,\Box\!\!\rightarrow \alpha_k, \alpha_k \,\Box\!\!\rightarrow \alpha_0 \in w$ and $\alpha_0 \,\Box\!\!\rightarrow \alpha_k \notin w$. This contradicts the definition of $\mathcal{M}^c$ by Axiom Schema Loop, Lemma 6.5.a, and Lemma 6.5.c. □

Axiom Schema Or. Let $\mathbf{L}_{\text{Or}}$ be an extension of System **CK** such that Or holds and let $\mathcal{M}^c = \langle W^c, R^c, P^c, V^c \rangle$ be the canonical model of $\mathbf{L}_{\text{Or}}$. Assume that Or is not canonical. Then, there exist $X, Y \in P^c$ and $w, w' \in W^c$ such that $wR^c_{X\cup Y} w'$ and both $\neg wR^c_X w'$ and $\neg wR^c_Y w'$. By Def$_{P^c}$ there are formulas α and β such that $X = |\alpha|$ and $Y = |\beta|$. Hence, $wR^c_{|\alpha|\cup|\beta|} w'$, $\neg wR^c_{|\alpha|} w'$, and $\neg wR^c_{|\beta|} w'$ result. Due to Def$_{|.|}$ we obtain $wR^c_{|\alpha\vee\beta|} w'$. Since $\neg wR^c_{|\alpha|} w'$ and $\neg wR^c_{|\beta|} w'$, we get by Def$_{R^c}$ that there are formulas γ and δ such that $\alpha \,\Box\!\!\rightarrow \gamma, \beta \,\Box\!\!\rightarrow \delta \in w$, and $\gamma, \delta \notin w'$. Hence, $\gamma \vee \delta \notin w'$ results by Lemma 6.5.d. In addition, rule RW of **CK** plus Lemma 6.5.a give us that $\alpha \,\Box\!\!\rightarrow \gamma \vee \delta \in w$ and $\beta \,\Box\!\!\rightarrow \gamma \vee \delta \in w$. Moreover, as $\gamma \vee \delta \notin w'$ and $wR^c_{|\alpha\vee\beta|} w'$, we get by Lemma 6.10 that $\alpha \vee \beta \,\Box\!\!\rightarrow$

$\gamma \vee \delta \notin w$. Since $\alpha \,\Box\!\!\rightarrow \gamma \vee \delta \in w$ and $\beta \,\Box\!\!\rightarrow \gamma \vee \delta \in w$, this contradicts the definition of $\mathcal{M}^c$ by Axiom Schema Or, Lemma 6.5.a, and Lemma 6.5.c. □

Axiom Schema S. Let $\mathbf{L}_{\mathrm{S}}$ be an extension of System **CK** such that S holds and let $\mathcal{M}^c = \langle W^c, R^c, P^c, V^c \rangle$ be the canonical model of $\mathbf{L}_{\mathrm{S}}$. Assume that S is not canonical. Then, there exist $X, Y \in P^c$ and $w, w' \in W^c$ such that $wR^c_X w'$, $w' \in Y$, and $\neg wR^c_{X \cap Y} w'$. By Def$_{P^c}$ there are formulas α and β such that $X = |\alpha|$ and $Y = |\beta|$. Hence, $wR^c_{|\alpha|} w'$, $w' \in |\beta|$, and $\neg wR^c_{|\alpha| \cap |\beta|} w'$. Due to Def$_{|.|}$ this yields $\neg wR^c_{|\alpha \wedge \beta|} w'$. Def$_{R^c}$ gives us that there is a formula γ such that $\alpha \wedge \beta \,\Box\!\!\rightarrow \gamma \in w$ and $\gamma \notin w'$. Since $w' \in |\beta|$, we get by Def$_{|.|}$ that $w' \notin |\beta \rightarrow \gamma|$. As $wR^c_{|\alpha|} w'$, Lemma 6.10 and Def$_{|.|}$ give us that $\alpha \,\Box\!\!\rightarrow (\beta \rightarrow \gamma) \notin w$. Since $\alpha \wedge \beta \,\Box\!\!\rightarrow \gamma \in w$, this contradicts the definition of $\mathcal{M}^c$ by Axiom Schema S, Lemma 6.5.a, and Lemma 6.5.c. □

Axiom Schema MOD. Let $\mathbf{L}_{\mathrm{MOD}}$ be an extension of System **CK** such that MOD holds and let $\mathcal{M}^c = \langle W^c, R^c, P^c, V^c \rangle$ be the canonical model of $\mathbf{L}_{\mathrm{MOD}}$. Assume that MOD is not canonical. Then, there exist $X, Y \in P^c$ and $w, w' \in W^c$ such that $\forall w''(wR^c_{-X} w'' \Rightarrow w'' \in X)$, $wR^c_Y w'$, and $w' \notin X$. By Def$_{P^c}$ there are formulas α and β such that $X = |\alpha|$ and $Y = |\beta|$. Hence, $\forall w''(wR^c_{-|\alpha|} w'' \Rightarrow w'' \in |\alpha|)$, $wR^c_{|\beta|} w'$, and $w' \notin |\alpha|$. By Def$_{|.|}$ we obtain $\forall w''(wR^c_{|\neg\alpha|} w'' \Rightarrow w'' \in |\alpha|)$. Hence, due to Lemma 6.10 and Def$_{|.|}$ we get $\neg\alpha \,\Box\!\!\rightarrow \alpha \in w$ and, since $wR^c_{|\beta|} w'$ and $w' \notin |\alpha|$, Lemma 6.10 gives us that $\beta \,\Box\!\!\rightarrow \alpha \notin w$. This contradicts the definition of $\mathcal{M}^c$ by Axiom Schema MOD, Lemma 6.5.a, and Lemma 6.5.c. □

Extensions of System P

Axiom Schema RM. Let $\mathbf{L}_{\mathrm{RM}}$ be an extension of System **CK** such that RM holds and let $\mathcal{M}^c = \langle W^c, R^c, P^c, V^c \rangle$ be the canonical model of $\mathbf{L}_{\mathrm{RM}}$. Assume that RM is not canonical. Then, there exist $X, Y \in P^c$ and $w, w', w'' \in W^c$ such that on the one hand $wR^c_X w''$ and $w'' \in Y$ and on the other that $wR^c_{X \cap Y} w'$ and $\neg wR^c_X w'$. By Def$_{P^c}$ there are formulas α and β such that $X = |\alpha|$ and $Y = |\beta|$. Hence, $wR^c_{|\alpha|} w''$, $w'' \in |\beta|$, $wR^c_{|\alpha| \cap |\beta|} w'$, and $\neg wR^c_{|\alpha|} w'$. By Def$_{|.|}$ we get $wR^c_{|\alpha \wedge \beta|} w'$. Hence, since $wR^c_{|\alpha|} w''$ and $w'' \in |\beta|$, Lemma 6.10, Def$_{|.|}$, and

$\text{Def}_{\diamond\!\!\rightarrow}$ imply that $\alpha \diamond\!\!\rightarrow \beta \in w$. Due to Def_{R^c} and $\neg wR^c_{|\alpha|}w'$ there is a formula γ such that $\alpha \Box\!\!\rightarrow \gamma \in w$ and $\gamma \notin w'$. Since $wR^c_{|\alpha\wedge\beta|}w'$ and $\gamma \notin w'$, it follows by Lemma 6.10 that $\alpha \wedge \beta \Box\!\!\rightarrow \gamma \notin w$. As $\alpha \diamond\!\!\rightarrow \beta \in w$ and $\alpha \Box\!\!\rightarrow \gamma \in w$, this contradicts the definition of $\mathcal{M}^c$ by Axiom Schema RM, Lemma 6.5.a, and Lemma 6.5.c. □

Axiom Schema CEM. Let $\mathbf{L}_{\text{CEM}}$ be an extension of System **CK** such that CEM holds. Moreover, let $\mathcal{M}^c = \langle W^c, R^c, P^c, V^c\rangle$ be the canonical model of $\mathbf{L}_{\text{CEM}}$. Assume that CEM is not canonical. Then, there exist $X \in P^c$ and $w, w', w'' \in W^c$ such that $wR^c_X w'$, $wR^c_X w''$, and $w'' \neq w'$. By Def_{P^c} there is a formula α such that $X = |\alpha|$. Thus, $wR^c_{|\alpha|}w'$ and $wR^c_{|\alpha|}w''$. Since $w'' \neq w'$, Def_{W^c} implies that there is a formula β such that $\beta \in w'$ and $\beta \notin w''$. Hence, as $wR^c_{|\alpha|}w''$, Lemma 6.10 implies that $\alpha \Box\!\!\rightarrow \beta \notin w$. Since $\beta \in w'$ it follows by Lemma 6.5.c that $\neg\beta \notin w'$. Hence, by Lemma 6.10 we have $\alpha \Box\!\!\rightarrow \neg\beta \notin w$. Since $\alpha \Box\!\!\rightarrow \beta \notin w$, this contradicts the definition of $\mathcal{M}^c$ by Axiom Schema CEM, Lemma 6.5.a, and Lemma 6.5.c.

□

Weak Probability Logic

Axiom Schema P-Cons. Let $\mathbf{L}_{\text{P-Cons}}$ be an extension of System **CK** such that P-Cons holds and let $\mathcal{M}^c = \langle W^c, R^c, P^c, V^c\rangle$ be the canonical model of $\mathbf{L}_{\text{P-Cons}}$. Assume that P-Cons is not canonical. Then, there exists a world $w \in W^c$ such that $\neg\exists w'(wR^c_{W^c}w')$. Since it holds by $\text{Def}_{|.|}$ and Def_{W^c} that $W^c = |\top|$, we get $\neg\exists w'(wR^c_{|\top|}w')$. Hence, it follows trivially by Lemma 6.10 that $\top \Box\!\!\rightarrow \bot \in w$. This contradicts the definition of $\mathcal{M}^c$ by Axiom Schema P-Cons, Lemma 6.5.a, and Lemma 6.5.c. □

Axiom Schema WOR. Let $\mathbf{L}_{\text{WOR}}$ be an extension of System **CK** such that WOR holds. Moreover, let $\mathcal{M}^c = \langle W^c, R^c, P^c, V^c\rangle$ be the canonical model of $\mathbf{L}_{\text{WOR}}$. Assume that WOR is not canonical. Then, there exist $X, Y \in P^c$ and $w, w' \in W^c$ such that $wR^c_X w'$, $\neg wR^c_{X\cap Y}w'$, and $\neg wR^c_{X\cap -Y}w'$. By Def_{P^c} there are formulas α and β such that $X = |\alpha|$ and $Y = |\beta|$. Thus, $wR^c_{|\alpha|}w'$, $\neg wR^c_{|\alpha|\cap|\beta|}w'$, and $\neg wR^c_{|\alpha|\cap-|\beta|}w'$. By $\text{Def}_{|.|}$ we get $\neg wR^c_{|\alpha\wedge\beta|}w'$ and $\neg wR^c_{|\alpha\wedge\neg\beta|}w'$. Due to

Def$_{R^c}$ it follows that there are formulas γ and δ such that $\alpha \wedge \beta \,\Box\!\!\rightarrow \gamma$, $\alpha \wedge \neg\beta \,\Box\!\!\rightarrow \delta \in w$, and $\gamma, \delta \notin w'$. Hence, $\gamma \vee \delta \notin w'$ results by Lemma 6.5.d. Then, rule RW plus Lemma 6.5.a give us that $\alpha \wedge \beta \,\Box\!\!\rightarrow \gamma \vee \delta \in w$ and $\alpha \wedge \neg\beta \,\Box\!\!\rightarrow \gamma \vee \delta \in w$. Moreover, as $\gamma \vee \delta \notin w'$ and $wR^c_{|\alpha|}w'$, we get by Lemma 6.10 that $\alpha \,\Box\!\!\rightarrow \gamma \vee \delta \notin w$. Since $\alpha \wedge \beta \,\Box\!\!\rightarrow \gamma \vee \delta \in w$ and $\alpha \wedge \neg\beta \,\Box\!\!\rightarrow \gamma \vee \delta \in w$, this contradicts the definition of $\mathcal{M}^c$ by Axiom Schema WOR, Lemma 6.5.a, and Lemma 6.5.c. □

Monotonic Systems

Axiom Schema Cut. Let $\mathbf{L}_{\text{Cut}}$ be an extension of System **CK** such that Cut holds. In addition let $\mathcal{M}^c = \langle W^c, R^c, P^c, V^c \rangle$ be the canonical model of $\mathbf{L}_{\text{Cut}}$. Assume that Cut is not canonical. Then, there exist $X, Y, Z \in P^c$ and $w, w' \in W^c$ such that $\forall w''(wR^c_Z w'' \Rightarrow w'' \in Y)$, $wR^c_{X\cap Z}w'$, and $\neg wR^c_{X\cap Y}w'$. By Def$_{P^c}$ there are formulas α, β, and γ such that $X = |\alpha|$, $Y = |\beta|$, and $Z = |\gamma|$. Thus, $\forall w''(wR^c_{|\gamma|}w'' \Rightarrow w'' \in |\beta|)$, $wR^c_{|\alpha|\cap|\gamma|}w'$, and $\neg wR^c_{|\alpha|\cap|\beta|}w'$. Def$_{|.|}$ gives us $wR^c_{|\alpha\wedge\gamma|}w'$ and $\neg wR^c_{|\alpha\wedge\beta|}w'$. Due to Def$_{R^c}$ there is a formula δ such that $\alpha \wedge \beta \,\Box\!\!\rightarrow \delta \in w$ and $\delta \notin w'$. As $wR^c_{|\alpha\wedge\gamma|}w'$, it follows by Lemma 6.10 that $\alpha \wedge \gamma \,\Box\!\!\rightarrow \delta \notin w$. Moreover, since $\forall w''(wR^c_{|\gamma|}w'' \Rightarrow w'' \in |\beta|)$, Lemma 6.10 implies by Def$_{|.|}$ that $\gamma \,\Box\!\!\rightarrow \beta \in w$. As $\alpha \wedge \beta \,\Box\!\!\rightarrow \delta \in w$ and $\alpha \wedge \gamma \,\Box\!\!\rightarrow \delta \notin w'$, this contradicts the definition of $\mathcal{M}^c$ by Axiom Schema Cut, Lemma 6.5.a, and Lemma 6.5.c. □

Axiom Schema Mon. Let $\mathbf{L}_{\text{Mon}}$ be an extension of System **CK** such that Mon holds and let $\mathcal{M}^c = \langle W^c, R^c, P^c, V^c \rangle$ be the canonical model of $\mathbf{L}_{\text{Mon}}$. Assume that Mon is not canonical. Then, there exist $X, Y \in P^c$ and $w, w' \in W^c$ such that $wR^c_{X\cap Y}w'$ and $\neg wR^c_X w'$. By Def$_{P^c}$ there are formulas α and β such that $X = |\alpha|$ and $Y = |\beta|$. Hence, $wR^c_{|\alpha|\cap|\beta|}w'$ and $\neg wR^c_{|\alpha|}w'$. By Def$_{|.|}$ we get $wR^c_{|\alpha\wedge\beta|}w'$. Since $\neg wR^c_{|\alpha|}w'$, due to Def$_{R^c}$ there is a formula γ such that $\alpha \,\Box\!\!\rightarrow \gamma \in w$ and $\gamma \notin w'$. Hence, by Lemma 6.10 follows that $\alpha \wedge \beta \,\Box\!\!\rightarrow \gamma \notin w$. As $\alpha \,\Box\!\!\rightarrow \gamma \in w$ holds, this contradicts the definition of $\mathcal{M}^c$ by Axiom Schema Mon, Lemma 6.5.a, and Lemma 6.5.c. □

Axiom Schema Trans. Let $\mathbf{L}_{\text{Trans}}$ be an extension of System **CK** such that Trans holds and let $\mathcal{M}^c = \langle W^c, R^c, P^c, V^c \rangle$ be the canonical model of $\mathbf{L}_{\text{Trans}}$.

Assume that Trans is not canonical. Then, there exist $X, Y \in P^c$ and $w, w' \in W^c$ such that $\forall w''(wR^c_X w'' \Rightarrow w'' \in Y)$, $wR^c_X w'$, and $\neg wR^c_Y w'$. By Def$_{P^c}$ there are formulas α and β such that $X = |\alpha|$ and $Y = |\beta|$. Thus, $\forall w''(wR^c_{|\alpha|} w'' \Rightarrow w'' \in |\beta|)$, $wR^c_{|\alpha|} w'$, and $\neg wR^c_{|\beta|} w'$. Due to Def$_{R^c}$ there exists a formula γ such that $\beta \mathrel{\Box\!\!\rightarrow} \gamma \in w$ and $\gamma \notin w'$. Hence, since $wR^c_{|\alpha|} w'$, we get by Lemma 6.10 that $\alpha \mathrel{\Box\!\!\rightarrow} \gamma \notin w$. Moreover, since $\forall w''(wR^c_{|\alpha|} w'' \Rightarrow w'' \in |\beta|)$, it follows by Lemma 6.10 and Def$_{|.|}$ that $\alpha \mathrel{\Box\!\!\rightarrow} \beta \in w$. As $\beta \mathrel{\Box\!\!\rightarrow} \gamma \in w$ and $\alpha \mathrel{\Box\!\!\rightarrow} \gamma \notin w$, this contradicts the definition of $\mathcal{M}^c$ by Axiom Schema Trans, Lemma 6.5.a, and Lemma 6.5.c. □

Axiom Schema CP. Let $\mathbf{L}_{\mathrm{CP}}$ be an extension of System **CK** such that CP holds, and let $\mathcal{M}^c = \langle W^c, R^c, P^c, V^c \rangle$ be the canonical model of $\mathbf{L}_{\mathrm{CP}}$. Assume that CP is not canonical. Then, there exist $X, Y \in P^c$ and $w, w' \in W^c$ such that $\forall w''(wR^c_X w'' \Rightarrow w'' \in Y)$, $wR^c_{-Y} w'$, and $w' \notin -X$. By Def$_{P^c}$ there are formulas α and β such that $X = |\alpha|$ and $Y = |\beta|$. Hence, $\forall w''(wR^c_{|\alpha|} w'' \Rightarrow w'' \in |\beta|)$, $wR^c_{-|\beta|} w'$, and $w' \notin -|\alpha|$. Due to Def$_{|.|}$ it follows that $wR^c_{|\neg\beta|} w'$ and $w' \notin |\neg\alpha|$. Hence, as $wR^c_{|\neg\beta|} w'$ and $w' \notin |\neg\alpha|$, Lemma 6.10, Def$_{|.|}$, and Def$_{\Diamond\!\rightarrow}$ imply that $\neg\beta \mathrel{\Box\!\!\rightarrow} \neg\alpha \notin w$. Since $\forall w''(wR^c_{|\alpha|} w'' \Rightarrow w'' \in |\beta|)$, Lemma 6.10 gives us that $\alpha \mathrel{\Box\!\!\rightarrow} \beta \in w$. Since $\neg\beta \mathrel{\Box\!\!\rightarrow} \neg\alpha \notin w$, this contradicts the definition of $\mathcal{M}^c$ by Axiom Schema CP, Lemma 6.5.a, and Lemma 6.5.c. □

Bridge Principles

Axiom Schema MP. Let $\mathbf{L}_{\mathrm{MP}}$ be an extension of System **CK** such that MP holds. Moreover, let $\mathcal{M}^c = \langle W^c, R^c, P^c, V^c \rangle$ be the canonical model of $\mathbf{L}_{\mathrm{MP}}$. Assume that MP is not canonical. Then, there exist $X \in P^c$ and $w \in W^c$ such that $w \in X$ and $\neg wR^c_X w$. By Def$_{P^c}$ there is a formula α such that $X = |\alpha|$. Hence, we have $w \in |\alpha|$ and $\neg wR^c_{|\alpha|} w$. Due to Def$_{R^c}$, it follows that there is a formula β such that $\alpha \mathrel{\Box\!\!\rightarrow} \beta \in w$ and $\beta \notin w$. Hence, since $w \in |\alpha|$ and $\beta \notin w$, we get by Def$_{|.|}$ that $\alpha \rightarrow \beta \notin w$. As $\alpha \mathrel{\Box\!\!\rightarrow} \beta \in w$ holds, this contradicts the definition of $\mathcal{M}^c$ by Axiom Schema MP, Lemma 6.5.a, and Lemma 6.5.c. □

Axiom Schema CS. Let $\mathbf{L}_{\mathrm{CS}}$ be an extension of System **CK** such that CS holds and let $\mathcal{M}^c = \langle W^c, R^c, P^c, V^c \rangle$ be the canonical model of $\mathbf{L}_{\mathrm{CS}}$. Assume that CS is not canonical. Then, there exist $X \in P^c$ and $w, w' \in W^c$ such that $w \in X$, $wR^c_X w'$, and $w' \neq w$. By Def$_{P^c}$ there is a formula α such that $X = |\alpha|$. Hence, we get $w \in |\alpha|$ and $wR^c_{|\alpha|}w'$. Since $w' \neq w$, by Def$_{W^c}$ there is a formula β such that $\beta \in w$ and $\beta \notin w'$. As $w \in |\alpha|$, this implies by Def$_{|.|}$ that $\alpha \wedge \beta \in w$. Since $wR^c_{|\alpha|}w'$ and $\beta \notin w'$, Lemma 6.10 gives us that $\alpha \,\square\!\!\rightarrow \beta \notin w$. Since $\alpha \wedge \beta \in w$, this contradicts the definition of $\mathcal{M}^c$ by Axiom Schema CS, Lemma 6.5.a, and Lemma 6.5.c. □

Axiom Schema TR. Let $\mathbf{L}_{\text{TR}}$ be an extension of System **CK** such that TR holds. Moreover, let $\mathcal{M}^c = \langle W^c, R^c, P^c, V^c \rangle$ be the canonical model of $\mathbf{L}_{\text{TR}}$. Assume that TR is not canonical. Then, there exist $X \in P^c$ and $w \in W^c$ such that $\forall w'(wR^c_{-X}w' \Rightarrow w' \in X)$ and $w \notin X$. By Def$_{P^c}$ there is a formula α such that $X = |\alpha|$. Hence, $\forall w'(wR^c_{-|\alpha|}w' \Rightarrow w' \in |\alpha|)$ and $w \notin |\alpha|$. Due to Def$_{|.|}$ we get $\forall w'(wR^c_{|\neg\alpha|}w' \Rightarrow w' \in |\alpha|)$. Lemma 6.10 and Def$_{|.|}$ give us, then, that $\neg\alpha \,\square\!\!\rightarrow \alpha \in w$. Since $w \notin |\alpha|$, by Def$_{|.|}$ $\alpha \notin w$. As $\neg\alpha \,\square\!\!\rightarrow \alpha \in w$, this contradicts the definition of $\mathcal{M}^c$ by Axiom Schema TR, Lemma 6.5.a, and Lemma 6.5.c. □

Axiom Schema Det. Let $\mathbf{L}_{\text{Det}}$ be an extension of System **CK** for which Det holds. Furthermore, let $\mathcal{M}^c = \langle W^c, R^c, P^c, V^c \rangle$ be the canonical model of $\mathbf{L}_{\text{Det}}$. Assume that Det is not canonical. Then, there exists a possible world $w \in W^c$ such that $\neg wR^c_{W^c}w$. Since due to Def$_{|.|}$ and Def$_{W^c}$ $W^c = |\top|$, we get $\neg wR^c_{|\top|}w$. By Def$_{R^c}$ it follows that there is a formula α such $\top \,\square\!\!\rightarrow \alpha \in w$ and $\alpha \notin w$. This contradicts the definition of $\mathcal{M}^c$ by Axiom Schema Det, Lemma 6.5.a, and Lemma 6.5.c. □

Axiom Schema Cond. Let $\mathbf{L}_{\text{Cond}}$ be an extension of System **CK** for which Cond holds. In addition, let $\mathcal{M}^c = \langle W^c, R^c, P^c, V^c \rangle$ be the canonical model of $\mathbf{L}_{\text{Cond}}$. Assume that Cond is not canonical. Then, there exist $w, w' \in W^c$ such that $wR^c_{W^c}w'$ and $w' \neq w$. Since due to Def$_{|.|}$ $W^c = |\top|$, we get $wR^c_{|\top|}w'$. Moreover, as $w' \neq w$, by the Def$_{W^c}$ it follows that there is a formula α such that $\alpha \in w$ and $\alpha \notin w'$. Since $wR^c_{|\top|}w'$, we get by Lemma 6.10 that $\top \,\square\!\!\rightarrow \alpha \notin w$. Since $\alpha \in w$, this contradicts the definition of $\mathcal{M}^c$ by Axiom Schema Cond, Lemma 6.5.a, and Lemma 6.5.c. □

Axiom Schema VEQ. Let $\mathbf{L}_{\text{VEQ}}$ be an extension of System **CK** for which VEQ holds and let $\mathcal{M}^c = \langle W^c, R^c, P^c, V^c \rangle$ be the canonical model of $\mathbf{L}_{\text{VEQ}}$. Assume that VEQ is not canonical. Then, there exist $X \in P^c$ and $w, w' \in W^c$ such that $wR^c_X w'$ and $w' \neq w$. By Def_{P^c} there is a formula α such that $X = |\alpha|$. Hence, $wR^c_{|\alpha|}w'$. Moreover, as $w' \neq w$, by Def_{W^c} it follows that there is a formula β such that $\beta \in w$ and $\beta \notin w'$. Since $wR^c_{|\alpha|}w'$, we get by Lemma 6.10 that $\alpha \;\Box\!\!\rightarrow \beta \notin w$. As $\beta \in w$, this contradicts the definition of $\mathcal{M}^c$ by Axiom Schema VEQ, Lemma 6.5.a, and Lemma 6.5.c. □

Axiom Schema EFQ. Let $\mathbf{L}_{\text{EFQ}}$ be an extension of System **CK** for which EFQ holds. Moreover, let $\mathcal{M}^c = \langle W^c, R^c, P^c, V^c \rangle$ be the canonical model of $\mathbf{L}_{\text{EFQ}}$. Assume that EFQ is not canonical. Then, there exist $X \in P^c$ and $w, w' \in W^c$ such that $w \in -X$ and $wR^c_X w'$. By Def_{P^c} there is a formula β such that $X = |\beta|$. Thus, we have $w \in -|\beta|$ and $wR^c_{|\beta|}w'$. By $\text{Def}_{|.|}$, it follows that $w \in |\neg\beta|$. By Def_{W^c} $\perp \notin w'$. Since $wR^c_{|\beta|}w'$, we get by Lemma 6.10 that $\beta \;\Box\!\!\rightarrow \perp \notin w$. As $\neg\beta \in w$, this contradicts the definition of $\mathcal{M}^c$ by Axiom Schema EFQ, Lemma 6.5.a, and Lemma 6.5.c. □

Traditional Extensions

Axiom Schema D. Let $\mathbf{L}_{\text{D}}$ be an extension of System **CK** for which D holds. Let $\mathcal{M}^c = \langle W^c, R^c, P^c, V^c \rangle$ be the canonical model of $\mathbf{L}_{\text{D}}$. Assume that D is not canonical. Then, there exist $X \in P^c$ and $w \in W^c$ such that $\neg\exists w'(wR^c_X w')$. By Def_{P^c} there is a formula α such that $X = |\alpha|$. Hence, $\neg\exists w'(wR^c_{|\alpha|}w')$. By Lemma 6.10 it follows trivially that $\alpha \;\Box\!\!\rightarrow \beta \in w$ and that $\alpha \;\Box\!\!\rightarrow \neg\beta \in w$. Thus, by Lemma 6.5.a and $\text{Def}_{\Diamond\!\rightarrow}$ this implies that $\alpha \;\Diamond\!\!\rightarrow \beta \notin w$. Since $\alpha \;\Box\!\!\rightarrow \beta \in w$, this contradicts the definition of $\mathcal{M}^c$ by Axiom Schema D, Lemma 6.5.a, and Lemma 6.5.c. □

Axiom Schema T. Let $\mathbf{L}_{\text{T}}$ be an extension of System **CK** for which T holds. Let $\mathcal{M}^c = \langle W^c, R^c, P^c, V^c \rangle$ be the canonical model of $\mathbf{L}_{\text{T}}$. Assume that T is not canonical. Then, there exist $X \in P^c$ and $w \in W^c$ such that $\neg wR^c_X w$. By Def_{P^c} there is a formula α such that $X = |\alpha|$. Hence, it follows that $\neg wR^c_{|\alpha|}w$. By Def_{R^c} this implies that there is a formula β such that $\alpha \;\Box\!\!\rightarrow \beta \in w$, and $\beta \notin w$.

This contradicts the definition of $\mathcal{M}^c$ by Axiom Schema T, Lemma 6.5.a, and Lemma 6.5.c. □

Axiom Schema B. Let $\mathbf{L}_\mathrm{B}$ be an extension of System **CK** for which B holds. Let $\mathcal{M}^c = \langle W^c, R^c, P^c, V^c \rangle$ be the canonical model of $\mathbf{L}_\mathrm{B}$. Assume that B is not canonical. Then, there exist $X \in P^c$ and $w, w' \in W^c$ such that $wR^c_X w'$ and $\neg w' R^c_X w$. By Def$_{P^c}$ there is a formula α such that $X = |\alpha|$. Hence, we have $wR^c_{|\alpha|} w'$ and $\neg w' R^c_{|\alpha|} w$. Def$_{R^c}$ gives us that there is a formula β such that $\alpha \,\Box\!\!\rightarrow \beta \in w'$ and $\beta \notin w$. Hence, by Lemma 6.5.b it follows that $\neg\beta \in w$. Moreover, since $\alpha \,\Box\!\!\rightarrow \beta \in w'$, Principle RW, Lemma 6.5.a, and Lemma 6.5.c imply that $\neg(\alpha \,\Box\!\!\rightarrow \neg\neg\beta) \notin w'$. Lemma 6.5.c and Def$_{\Diamond\!\rightarrow}$ give us that $\alpha \Diamond\!\!\rightarrow \neg\beta \notin w'$. Since $wR^c_{|\alpha|} w'$, we get by Lemma 6.10 that $\alpha \,\Box\!\!\rightarrow (\alpha \Diamond\!\!\rightarrow \neg\beta) \notin w$. As $\neg\beta \in w$, this contradicts the definition of $\mathcal{M}^c$ by Axiom Schema B, Lemma 6.5.a, and Lemma 6.5.c. □

Axiom Schema 4. Let $\mathbf{L}_4$ be an extension of System **CK** such that 4 holds. Let $\mathcal{M}^c = \langle W^c, R^c, P^c, V^c \rangle$ be the canonical model of $\mathbf{L}_4$. Assume that 4 is not canonical. Then, there exist $X \in P^c$ and $w, w', w'' \in W^c$ such that $wR^c_X w'$, $w' R^c_X w''$, and $\neg wR^c_X w''$. By Def$_{P^c}$ there is a formula α such that $X = |\alpha|$. Thus, $wR^c_{|\alpha|} w'$, $w' R^c_{|\alpha|} w''$, and $\neg wR^c_{|\alpha|} w''$. Due to the definition of R^c it follows that there is a formula β such that $\alpha \,\Box\!\!\rightarrow \beta \in w$ and $\beta \notin w''$. Since $w' R^c_{|\alpha|} w''$, we get by Lemma 6.10 that $\alpha \,\Box\!\!\rightarrow \beta \notin w'$. Since $wR^c_{|\alpha|} w'$ this implies by Lemma 6.10 that $\alpha \,\Box\!\!\rightarrow (\alpha \,\Box\!\!\rightarrow \beta) \notin w$. Since $\alpha \,\Box\!\!\rightarrow \beta \in w$, this contradicts the definition of $\mathcal{M}^c$ by Axiom Schema 4, Lemma 6.5.a, and Lemma 6.5.c. □

Axiom Schema 5. Let $\mathbf{L}_5$ be an extension of System **CK** such that 5 holds. Let $\mathcal{M}^c = \langle W^c, R^c, P^c, V^c \rangle$ be the canonical model of $\mathbf{L}_5$. Assume that 5 is not canonical. Then, there exist $X \in P^c$ and $w, w', w'' \in W^c$ such that $wR^c_X w'$, $wR^c_X w''$, and $\neg w' R^c_X w''$. By Def$_{P^c}$ there is a formula α such that $X = |\alpha|$. Thus, $wR^c_{|\alpha|} w'$, $wR^c_{|\alpha|} w''$, and $\neg w' R^c_{|\alpha|} w''$. This implies that there is a formula β such that $\alpha \,\Box\!\!\rightarrow \beta \in w'$ and $\beta \notin w''$. It follows by Lemma 6.5.b that $\neg\beta \in w''$. Moreover, since $\alpha \,\Box\!\!\rightarrow \beta \in w'$, we have by RW and Lemma 6.5.a that $\alpha \,\Box\!\!\rightarrow \neg\neg\beta \in w'$. Hence, Lemma 6.5.c and Def$_{\Diamond\!\rightarrow}$ imply that $\alpha \Diamond\!\!\rightarrow \neg\beta \notin w'$. As $wR^c_{|\alpha|} w'$, it follows by Lemma 6.10 that $\alpha \,\Box\!\!\rightarrow (\alpha \Diamond\!\!\rightarrow \neg\beta) \notin w$. Since

$wR^c_{|\alpha|}w''$ and $\neg\beta \in w''$, Lemma 6.10, $\mathrm{Def}_{|.|}$, and $\mathrm{Def}_{\Diamond\!\!\rightarrow}$ imply that $\alpha \Diamond\!\!\rightarrow \neg\beta \in w$. As $\alpha \Box\!\!\rightarrow (\alpha \Diamond\!\!\rightarrow \neg\beta) \notin w$, this contradicts the definition of $\mathcal{M}^c$ by Axiom Schema 5, Lemma 6.5.a, and Lemma 6.5.c. □

Iteration Principles

Axiom Schema Ex. Let $\mathbf{L}_{\mathrm{Ex}}$ be an extension of System **CK** such that Ex holds. Let $\mathcal{M}^c = \langle W^c, R^c, P^c, V^c\rangle$ be the canonical model of $\mathbf{L}_{\mathrm{Ex}}$. Assume that Ex is not canonical. Then, there exist $X, Y \in P^c$ and $w, w', w'' \in W^c$ such that $wR^c_X w'$, $w'R^c_Y w''$, and $\neg wR^c_{X\cap Y} w''$. By Def_{P^c} there are formulas α, β such that $X = |\alpha|$ and $Y = |\beta|$. Hence, $wR^c_{|\alpha|}w'$, $w'R^c_{|\beta|}w''$, and $\neg wR^c_{|\alpha|\cap|\beta|}w''$. $\mathrm{Def}_{|.|}$, then, implies that $\neg wR^c_{|\alpha\wedge\beta|}w''$. By Def_{R^c} it follows that there is a formula γ such that $\alpha \wedge \beta \Box\!\!\rightarrow \gamma \in w$, and $\gamma \notin w''$. As $w'R^c_{|\beta|}w''$, Lemma 6.10 gives us that $\beta \Box\!\!\rightarrow \gamma \notin w'$. Moreover, since $wR^c_{|\alpha|}w'$, Lemma 6.10 also implies that $\alpha \Box\!\!\rightarrow (\beta \Box\!\!\rightarrow \gamma) \notin w$. As $\alpha \wedge \beta \Box\!\!\rightarrow \gamma \in w$, this contradicts the definition of $\mathcal{M}^c$ by Axiom Schema Ex, Lemma 6.5.a, and Lemma 6.5.c. □

Axiom Schema Im. Let $\mathbf{L}_{\mathrm{Im}}$ be an extension of System **CK** such that Im holds, and let $\mathcal{M}^c = \langle W^c, R^c, P^c, V^c\rangle$ be the canonical model of $\mathbf{L}_{\mathrm{Im}}$. Assume that Im is not canonical. Then, there exist $X, Y \in P^c$ and $w, w' \in W^c$ such that $wR^c_{X\cap Y}w'$ and $\neg\exists w''(wR^c_X w'' \wedge w''R^c_Y w')$. By Def_{P^c} there are formulas α, β such that $X = |\alpha|$ and $Y = |\beta|$. Thus, $wR^c_{|\alpha|\cap|\beta|}w'$ and $\neg\exists w''(wR^c_{|\alpha|}w'' \wedge w''R^c_{|\beta|}w')$. Moreover, due to $\mathrm{Def}_{|.|}$ we have $wR^c_{|\alpha\wedge\beta|}w'$. By Def_{W^c} there is at least a formula γ such that $\gamma \in w'$ and $\gamma \notin w''$ for any other world $w'' \in W^c$. Since $\neg\exists w''(wR^c_{|\alpha|}w'' \wedge w''R^c_{|\beta|}w')$ it holds for all worlds w'' such that $wR^c_{|\alpha|}w''$ that they cannot possibly see w' by $w''R^c_{|\beta|}w'$ and that, hence, $\beta \Box\!\!\rightarrow \neg\gamma \in w''$ holds due to Lemma 6.10 and Lemma 6.5.b, as w' is the only world in W^c such that $\gamma \in w'$. Since this holds for all possible worlds w'' such that $w''R^c_{|\beta|}w'$ we get $\alpha \Box\!\!\rightarrow (\beta \Box\!\!\rightarrow \neg\gamma) \in w$. As $wR^c_{|\alpha\wedge\beta|}w'$ and $\gamma \in w'$, it follows by Lemma 6.10 and Lemma 6.5.c that $\alpha \wedge \beta \Box\!\!\rightarrow \neg\gamma \notin w$. Since $\alpha \Box\!\!\rightarrow (\beta \Box\!\!\rightarrow \gamma) \in w$ is the case, this contradicts the definition of $\mathcal{M}^c$ by Axiom Schema Im, Lemma 6.5.a, and Lemma 6.5.c.

□

Chapter 7

Chellas-Segerberg Semantics for Indicative and Counterfactual Conditionals

In this chapter I will align the formal results described in Chapters 4–6 with the philosophical discussion in the foregoing Chapters 1–3. My main thesis in this chapter is that Chellas-Segerberg (CS) semantics is a plausible and general semantics for indicative and counterfactual conditionals. To argue for this point I will give a subjective and an objective interpretation of basic CS semantics, which can serve as a justification of this semantics, and I will describe how a range of conditional logic proof systems can be accounted for by CS semantics (by means of the results from Chapters 4–6). I will discuss in more detail the following points:

(i) I shall give a description of basic Chellas-Segerberg semantics plus an objective and a subjective interpretation of this semantics (Section 7.1). The objective interpretation represents a generalization of the Kripke semantics, while the subjective interpretation is a modified Ramsey test interpretation in the sense of Section 3.3 (see in particular Section 3.3.8), but does not rely on a conditional consistency criterion, as described in Section 3.3.6 (cf. Section 3.3.7).

(ii) I will discuss individual conditional logic principles from Tables 5.1 and 5.2. These principles do not hold in the basic CS semantics and are, hence, in need of justification beyond the modified Ramsey test interpre-

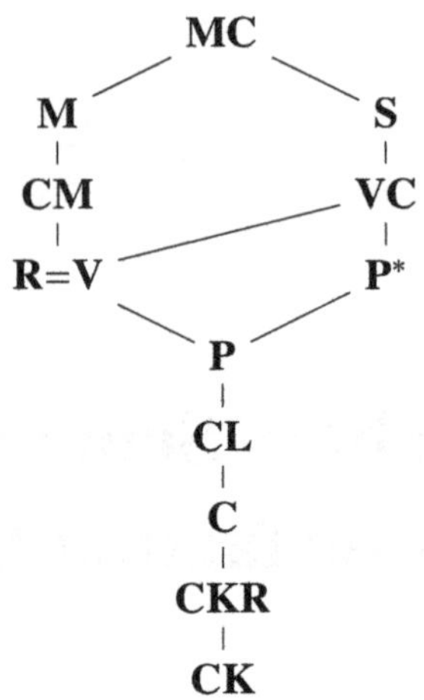

Figure 7.1
Systems Investigated in this chapter from the Lattice of Systems described by Tables 5.1 and 5.2

tation of the basic CS semantics. I argue in particular that CM ("Cautious Monotonicity"), CC ("Cautious Cut"), Or, and RM ("Rational Monotonicity") are plausible candidates for additional conditional logic principles. The discussion is not limited to these principles, but also includes the Principles Refl ("Reflexivity"), S, MOD ("Modality"), CEM ("Conditional Excluded Middle"), P-Cons ("Probabilistic Consistency"), WOR ("Weak Or"), Cut, Mon ("Monotonicity"), Trans ("Transitivity"), CP ("Contraposition"), MP ("Modus Ponens"), CS ("Conditional Sufficiency"), Det ("Detachment"), Cond ("Conditionalization"), VEQ ("Verum Ex Quodlibet"), EFQ ("Ex Falso Quodlibet"), and D (a conditional logic version of Axiom D from modal logic) (see Section 5.1 for a more detailed description of these principles.)

(iii) I will show that a range of systems for indicative and counterfactual conditionals can be described by means of our lattice of conditional logics defined by the axiom schemata in Tables 5.1 and 5.2 on the one hand

and frame conditions in Tables 5.3 and 5.4 on the other. See Figure 7.1 for an overview of the indicative and counterfactual conditional logic systems investigated in this chapter. System **CK** (see Sections 7.1 and 4.2.6) and System **MC** (Section 7.3.4) are the weakest and strongest systems, respectively. The edges in Figure 7.1 mark proper extensions of systems, where an edge between two systems indicates that the system placed "higher" in the hierarchy properly contains the system located "lower" in the hierarchy. System **CKR** (see Section 7.1) is the second-weakest system, which results from adding the Principle Refl ($\alpha \mathrel{\Box\!\!\rightarrow} \alpha$, see Table 5.1) to System **CK**. Full language versions of System **C** (Section 7.2.1), System **CL** (Section 7.2.2), System **P** (Section 7.2.3), and System **R** (Section 7.2.4) of Kraus et al. (1990) and Lehmann and Magidor (1992) follow. By a full language version of a system I mean a system which is formulated w.r.t. the full language $\mathcal{L}_{CL}$ rather than more restricted versions, such as $\mathcal{L}_{CL^-}$ and $\mathcal{L}_{rCL^*}$ (see Sections 3.5.3 and 4.2.1). Moreover, I will account for the counterfactual Conditional Logic System **V** (Section 7.2.5) of D. Lewis (1973b; see also Section 3.2) and describe full language versions of the two monotonic Systems **CM** and **M** (Section 7.2.6) defined in Kraus et al. (1990).

Then, I will focus on conditional logic systems with bridge principles: First, I will characterize Adams' (1965, 1966, 1977) original System **P*** (Section 7.3.1). Then, I will describe the counterfactual Conditional Logic System **VC** (Section 7.3.2) of D. Lewis (1973b; see also Section 3.2) and Stalnaker's (1968, Stalnaker & Thomason, 1970; see also Section 3.3.3) System **S** (Section 7.3.3). Furthermore, I will describe System **MC** for the full material collapse, in which all conditional formulas of the form $\alpha \mathrel{\Box\!\!\rightarrow} \beta$ are logically equivalent to material implications of the form $\alpha \rightarrow \beta$.

(iv) Finally, I will give a direct characterization of System **M** (the monotonic collapse without bridge principles, Section 7.2.6) and System **MC** (the monotonic collapse with bridge principles, Section 7.3.4) by means of frame conditions in terms of CS semantics (see Theorems 7.46 and 7.77, respectively). Both types of semantic characterizations are of interest for the investigation of conditional logics, insofar as they show which type of semantics corresponds to the limiting cases and, hence, should be avoided in conditional logic approaches.

Points (i) and (iv) are original contributions. Points (ii) and (iii) are partially original contributions, insofar as I will augment the discussion of conditional logic principles and provide object language proofs for a range of principles which are described in the literature – but for which only proof sketches are given.

The present discussion of conditional logics has some advantages over existing reviews of conditional logic such as Nejdl (1992), Kraus et al. (1990), and Lehmann and Magidor (1992). Unlike Nejdl (1992), I use the same semantic framework to discuss both weak systems of conditional logic (e.g., System **CK**) and strong systems (e.g., System **S**). Nejdl (1992) describes only System **CK** by means of Chellas models [frames] and then goes on to discuss stronger systems such as System **P** in terms of the relational Lewis frames (D. Lewis, 1973b, pp. 48–50; see Definitions 3.4–3.6). Moreover, in contrast to Kraus et al. (1990) and Lehmann and Magidor (1992) I will employ the full language $\mathcal{L}_{\mathrm{CL}}$, rather than a more restricted language, corresponding to $\mathcal{L}_{\mathrm{rrCL}}$ (cf. Section 3.5.3). My framework, hence, makes it possible to describe bridge principles (e.g., MP) and iterated and nested principles (e.g., Im). In the language of Kraus et al. (1990) and Lehmann and Magidor (1992) bridge principles and iterated and nested principles are not expressible. In addition, our semantic framework allows the characterization of much weaker systems than the semantic framework of Kraus et al. (1990) and Lehmann and Magidor (1992).

7.1 The Basic Chellas-Segerberg Systems (Systems CK and CKR)

Let us first clarify what I mean by basic CS semantics. I refer by that term to Chellas models and Segerberg models [and the respective notions of Chellas frames and Segerberg frames] as described in Sections 4.3.1 and 4.3.3 plus optionally the frame condition for Refl from Table 5.3. In proof-theoretic terms these semantics can be described by Systems **CK** and **CKR** (System **CK**+Refl), respectively.

Chellas models serve as basic units of our semantics. A Chellas model $\mathcal{M}_C$ has the form $\mathcal{M}_C = \langle W, R, V \rangle$, where W is a non-empty set of possible worlds and R an accessibility relation between elements $w, w' \in W$, which is additionally relativized to subsets of W. This accessibility relation is defined for all pairs of elements $w, w' \in W$ and subsets X of possible worlds in W. Since any formula α (of $\mathcal{L}_{\mathrm{CL}}$) is true in some subset X of W for any Chellas model $\mathcal{M}_C$, model $\mathcal{M}_C$ gives us for any formula α a natural relativization of R to formulas, namely by the subset of W, at which α is true (short: $||\alpha||$). A conditional $\alpha \mathrel{\Box\!\!\rightarrow} \beta$ is, then, determined to be true at world w in model $\mathcal{M}_C$ if and only if β is true at all worlds accessible from w by the accessibility relation $R_{||\alpha||}$.

In Chapters 4 and 6 I investigated Segerberg models in addition to Chellas models. Segerberg models differ from Chellas models insofar as an additional parameter P is employed. This parameter is used mainly for technical reasons. Both Segerberg models and Chellas models ensure that for all formulas α the accessibility relation $R_{||\alpha||}$ is defined. In addition, they agree on the truth conditions for modal formulas. Since these are the elements on which the following modified Ramsey test interpretation draws, our version of the Ramsey test is applicable to both Chellas models and Segerberg models.

7.1.1 Objective and Subjective Interpretations of Chellas-Segerberg Semantics

We saw in Section 3.4.3 that indicative conditionals are interpreted by taking one's current beliefs into account, while this element is in general missing when one is evaluating counterfactual conditionals. Hence, in order to provide an intuitively plausible interpretation of indicative and counterfactual conditionals for CS semantics, I have to be able to give both (a) an objective interpretation of CS semantics and (b) a subjective, agent-relative interpretation of this semantics.

I argue here that CS semantics can provide both (a) and (b). This is due to the fact that CS semantics is based on general structural conditions, which allow an interpretation of this semantics (i) in an objective way in line with traditional alethic Kripke semantics and (ii) subjectively, by means of a modified Ramsey test. Approach (i) gives us an objective interpretation of this semantics in terms of counterfactuals (requirement (a)), while approach (ii) allows for a subjective interpretation of indicative conditionals (requirement (b)). I will first describe (i) and then focus on (ii).

The Objective Alethic Interpretation

Let us now discuss how to interpret CS semantics in an objective way. I discussed in Section 3.4.3 that a standard way to describe alethic necessity is by means of Kripke models (cf. D. Lewis, 1973b, p. 5f; Schurz, 1997a, p. 17f). In Kripke semantics an accessibility relation R^* between possible worlds is employed (Hughes & Cresswell, 1996/2003, p. 37ff). This accessibility relation is, then, used to provide truth conditions of necessity formulas $\Box\alpha$ ("It is necessary that α") relative to possible worlds: A formula $\Box\alpha$ is true at a world w iff α is true at all worlds w' accessible from w by R^*. In alethic modal logic the accessibility relation R^* is interpreted in an objective way insofar as w stands in the relation R^* to w' if and only if w' is an alternative way the world might be w.r.t. the possible world w. I can add more flesh to this characterization by specifying in which sense R^* provides alternative ways the world might be by, for example, requiring that these alternatives are determined purely on logical grounds or by the physical laws of our world (see Section 3.4.3). The notion of alethic necessity is so broad that it covers all these more specific interpretations, as long as these amount to "necessary truth" rather than other types of necessity (see Section 3.4.3).

Observe here that the accessibility relation in CS semantics is not a two-place relation between possible worlds, but a three-place relation between pairs of possible worlds and sets of possible worlds. Thus, the accessibility relation R_X in CS semantics represents a generalization of the accessibility relation R^* in Kripke semantics, insofar as R is relativized to some proposition X.

I shall now give an objective interpretation of the accessibility relation R_X. Suppose that α is a formula such that in some Chellas model $\mathcal{M}_C = \langle W, R, V\rangle$ it holds that $X = ||\alpha||$. Then, all worlds $w' \in W$, which are accessible from a world $w \in W$ by R_X represent worlds that are compatible with the proposition $||\alpha||$ from the perspective of world w. Our interpretation of compatibility here is to be understood in a purely alethic way (cf. Section 3.4.3): Hence, I analyze $R_{||\alpha||}$ in such a way that those worlds and only those worlds are accessible by $R_{||\alpha||}$, which are compatible with the proposition $||\alpha||$ on the basis of logical principles alone or alternatively on the basis of the physical laws at the world w.

In order to allow for an adequate interpretation of CS semantics in terms of alethic modality, one should not be able to arrive from a world w by $R_{||\alpha||}$ at a world w' such that $\neg\alpha$ is true at w'. Since this might be the case for some Chellas models, one has to require that any Chellas model $\mathcal{M}_C = \langle W, R, V\rangle$, which is to be interpreted in an alethic way, has to satisfy the following property ($X, X_0, X_1, \ldots, Y, Z, \ldots$ range over sets of possible worlds):

C_{Refl} $\forall X \forall w, w'(wR_X w' \Rightarrow w' \in X)$

Due to the respective correspondence proof in Chapter 5, the following formula is valid in all Chellas models for which frame condition C_{Refl} holds:

Refl $\alpha \mathrel{\Box\!\!\rightarrow} \alpha$

Let us now define the result proof-theoretic system, which results from adding Refl to System **CK**:

Definition 7.1. *Logic* **CKR** *is the smallest logic containing* **CK**+*Refl.*

System **CK** is specified in Definition 4.1 as LLE+RW+AND+LT. Under certain conditions one would not like Refl to be valid in a model, for example if one interprets $R_{||\alpha||}$ in a deontic way. In this case $R_{||\alpha||}$ holds between w and $w' \in W$ if and only if w' represents a state of affairs, which is compatible with a certain pre-specified normative code, given that α holds. I will not investigate deontic interpretations of CS semantics here any further.

Let us now see in which sense the accessibility relation R_X represents a generalization of the accessibility relation R^* in Kripke semantics. I do so by reconstructing an absolute accessibility relation R^* on the basis of Chellas models $\mathcal{M}_C = \langle W, R, V \rangle$. One can achieve this by requiring (a) that wR^*w' holds (the possible world w stands in relation R^* to the possible world w') iff $\exists X(wR_Xw')$ (cf. Segerberg, 1989, p. 161). It follows from this definition that R_X for some $X \subseteq W$ is always a special case of R^*, since for any wR_Xw' we obtain wR^*w'. Interestingly, the alternative characterization of wR^*w' as $\forall X(wR_Xw')$ renders R^* void, in case I also endorse C_{Refl}: In order that wR^*w' holds it must be the case that wR_Xw' for all $X \subseteq W$, even for $X = \emptyset$. By C_{Refl} we get that for no worlds $w, w' \in W$ it is the case that $wR_{\emptyset}w'$.

The Modified Ramsey Test Interpretation

I will now give a modified Ramsey test interpretation of CS semantics. This interpretation differs in many respects from the objective interpretation described in the previous section. It is important to note here that I only give a Ramsey test interpretation – as well as the objective-alethic interpretation – for *basic* CS semantics, namely the class of all Chellas [Segerberg] models and the class of all Chellas [Segerberg] models for which C_{Refl} (see previous section) holds. I argue that additional conditions on Chellas [Segerberg] models are not justified by a Ramsey test interpretation, but must derive their intuitive import from other sources. I will, hence, discuss the intuitions of a range of important principles separately in Sections 7.2 and 7.3. I will thereby not draw – as Stalnaker (1968) does – on the notion of similarity of possible worlds (see Section 3.3.5) but maintain a pure Ramsey test interpretation of CS semantics.

I shall now repeat the properties our Ramsey test interpretation has to satisfy. For the sake of perspicuity I quote again the central features of the Ramsey test as described by Stalnaker (1968; cf. also Section 3.3):

> "add the antecedent (hypothetically) to your stock of knowledge (or be-

> liefs), and then consider whether or not the consequent is true. Your belief about the conditional should be the same as your hypothetical belief, under this condition, about the consequent" (Stalnaker, 1968, p. 101)

We saw in Section 3.3.8 that stocks of beliefs – contrary to what Stalnaker (1968) argues – correspond to sets of possible worlds rather than single possible worlds. In a Chellas model $\mathcal{M}_C = \langle W, R, V \rangle$ the accessibility relation R determines a set of possible worlds for each possible world $w \in W$ and any subset X of W so that these worlds are accessible from w by the proposition X. Suppose that there is a formula α such that $X = ||\alpha||^{\mathcal{M}_C}$. Then, $R_{||\alpha||}$ determines for a given world $w \in W$ the set of worlds, which result from revising our stock of beliefs by the proposition $||\alpha||$.

Since a Ramsey test interpretation is in general to be understood in subjective terms (see Section 3.3.8), it seems sensible to presuppose that the accessibility relation R in Chellas models $\mathcal{M}_C = \langle W, R, V \rangle$ has to be read in a subjective, person-relative way in line with epistemic or doxastic logics rather than objective alethic modal logics. In epistemic and doxastic logics a modal necessity operator $\Box$ is used. A necessity formula $\Box\alpha$ is, then, interpreted as "agent A knows that α" (Fagin, Halpern, Moses, & Vardi, 2003, p. 19; Meyer & van der Hoek, 1995, p. 7) and "agent A believes that α" (Meyer & van der Hoek, 1995, p. 68f), respectively. In standard systems of epistemic [doxastic] logic a Kripke semantics (see previous section) is used in such a way that a formula $\Box\alpha$ is true at a world w iff formula α is true at all epistemic [doxastic] alternatives to world w (Meyer & van der Hoek, 1995, p. 8). Two epistemic [doxastic] alternatives w' and w'' of a possible world w (for an agent A) are interpreted in such a way that A cannot, on the basis of her knowledge [beliefs], determine which of both worlds w' and w'' holds (Fagin et al., 2003, p. 20).

There is one essential difference between epistemic and doxastic logics on the one hand and CS semantics on the other hand. In epistemic and doxastic logics, the modal operator $\Box\alpha$ is introduced in order to be able to distinguish in a single framework state of affairs, which an agent knows [believes] and states of affairs that obtain factually (cf. Levesque, 1990, p. 269). In this type of semantics, formulas which do not contain a modal operator are understood in an objective way, whereas formulas of the form $\Box\alpha$ are interpreted subjectively.

In CS semantics the two-place modal operator $\Box\!\!\rightarrow$ is interpreted as a conditional operator, which allows for the representation of natural language conditionals. If we use the same interpretation for CS semantics as in epistemic or doxastic logic, then conditional formulas $\alpha \Box\!\!\rightarrow \beta$ are to be understood in a subjective way (as states of affairs known or believed respectively), while non-conditional formulas, such as $p \wedge q$, are interpreted as matters of fact. I argue that it is in general hardly plausible that conditional and non-conditional formulas have such a different cognitive status; this would contradict common sense assumptions and also seems from a philosophical perspective rather arbitrary (see Section 3.5.5).

A possible way out of this problem is the interpretation of a specific type of conditional formula, namely conditional formulas of the form $\top \Box\!\!\rightarrow \alpha$ – where $\top$ abbreviates the tautology $p \vee \neg p$ – as beliefs in a proposition α. This approach, however, implies that forced to reject that α and $\top \Box\!\!\rightarrow \alpha$ represent the same proposition, as I will show now.

If I were to accept that α and $\top \Box\!\!\rightarrow \alpha$ represent the same proposition, I would be required to accept the following two principles unconditionally:

Det $(\top \Box\!\!\rightarrow \alpha) \rightarrow \alpha$
Cond $\alpha \rightarrow (\top \Box\!\!\rightarrow \alpha)$

Here 'Det' and 'Cond' abbreviate 'Detachment' and 'Conditionalization', respectively. As we saw in Chapter 5, the Principles Det and Cond correspond to the following frame conditions in CS semantics:

C_{Det} $\forall w(wR_W w)$
C_{Cond} $\forall w(wR_W w' \Rightarrow w' = w)$

Conditions C_{Det} and C_{Cond} give us conjointly that each world w in a Chellas model $\mathcal{M}_C = \langle W, R, V\rangle$ can only see itself via $R_{\|\top\|}$. Hence, whenever one accepts that α is true exactly iff $\top \Box\!\!\rightarrow \alpha$ is the interpretation of $\top \Box\!\!\rightarrow \alpha$ as an agent's belief in α, implies that one believes proposition α if and only if α is objectively true. Note that unwarranted conditions of beliefs result not only from Principles Det and Cond but also from other bridge principles, as discussed in Section 3.4.2 (see also Table 5.1).

A general problem of an epistemic or a doxastic interpretation of CS semantics lies in the fact that specific worlds can plausibly be interpreted as describing states of affairs that factually obtain rather than stocks of beliefs (see Section 3.3.8). To circumvent this problem, I need not accept single worlds as interpreted units for encoding beliefs, but can instead use sets of possible worlds for that purpose. To ensure a doxastic/epistemic interpretation of conditionals in terms of the Ramsey test, one has to make sure that one can have beliefs that correspond to sets of possible worlds rather than single possible worlds, both prior to and after hypothetically putting the antecedent to one's stock of beliefs.

The basic idea of my Ramsey interpretation of CS semantics is the following: One can interpret the set of possible worlds Y, as given by the $R_{||\alpha||}$ for a possible world w in W, as the set of possible worlds, which describe one's belief after putting the antecedent α hypothetically to one's stock of beliefs. It is ensured that this interpretation always works, since any Chellas model $\mathcal{M}_C = \langle W, R, V \rangle$ specifies R_X between pairs of possible worlds, for arbitrary sets of possible worlds X, hence also for arbitrary propositions $||\alpha||$.

This proposal has to be modified for the following reason: The accessibility relation gives one the set of possible worlds Y, which are accessible by $R_{||\alpha||}$, for each *single* possible world and not for sets of possible worlds. The solution for this problem goes as follows: We can, given a set of possible worlds $X \subseteq W$ as a starting point, specify the set of possible worlds which are accessible from the set X, by requiring that any world accessible in this way must be also accessible from some world in X, formally: $XR^{+}_{Y}w'$ iff $\exists w \in X(wR_{Y}w')$. All formulas true in the set X are regarded true by the agent, prior to putting any antecedent hypothetically to one's stock of beliefs. The set X is, hence, the "ontological analogue" to an initial stock of beliefs (cf. Sections 3.3.5 and 3.3.8). Hypothetically putting a proposition

$||\alpha||$ to one's initial stock of beliefs results, then, in a new stock of (hypothetical) beliefs represented by a further set of possible worlds $Y \subseteq W$. In this approach the set Y is determined by the accessibility relation $R_{||\alpha||}$ in such a way that for every $w' \in Y$ every world $w \in X$ stands in the relation $R_{||\alpha||}$. This condition naturally gives us that a conditional $\alpha \mathrel{\Box\!\!\rightarrow} \beta$ is true w.r.t. the set of possible worlds X iff β is true at every world $w \in Y$.

Moreover, it seems reasonable to endorse a version of the Ramsey test which requires that any stock of (hypothetical) beliefs A that is updated by a proposition $||\alpha||$, must always arrive at possible worlds at which α is also true. In such an interpretation of CS semantics the formula $\alpha \mathrel{\Box\!\!\rightarrow} \alpha$ (Refl, see previous section) is always true. In order to ensure that this condition is always satisfied I can require that any Chellas (Segerberg) model interpreted in such a way has to satisfy condition C_{Refl} (see previous section).

There are certain interpretations of the Ramsey test which do not make Refl valid. In such an interpretation it is possible that hypothetically adding $||\alpha||$ to one's stock of knowledge does not lead to the acceptance of α. In such a case I presuppose α, but end up – on the basis of my background knowledge – rejecting α rather than accepting it. Such a scenario can arise if I, for example, endorse the following strong conditional consistency requirement (criterion (B) from Sections 3.1 and 3.3.6):

CNC $\neg((\alpha \mathrel{\Box\!\!\rightarrow} \beta) \land (\alpha \mathrel{\Box\!\!\rightarrow} \neg\beta))$

'CNC' abbreviates 'Conditional Consistency Criterion' and I discussed this principle (and a weakened version) in detail in Sections 3.3.6, 3.6.3, and 3.8. We saw that, for example, Bennett (2003, p. 84) argues that CNC is valid for all conditionals. Suppose one wants to find out whether the formula $\bot \mathrel{\Box\!\!\rightarrow} \alpha$ holds for a stock of (hypothetical) beliefs described by a subset X of W in a Chellas model $\mathcal{M}_C = \langle W, R, V\rangle$. One then puts $||\bot||$ to one's stock of beliefs. (For any Chellas [Segerberg] model it is the case that $||\bot|| = \varnothing$.) Due to CNC one is not allowed to arrive at an absurd belief, according to which one believes α and $\neg\alpha$. This means that one has to reject the assumption $\bot$. Hence, even if one presupposes $\bot$, one is allowed only to arrive at mutually consistent formulas.

It can thus be seen as an advantage of CS semantics that Refl is not valid in every Chellas model, since both CNC and Refl contradict each other given minimal preconditions (see Section 3.6.3). Due to that fact, one can also describe a conditional logic with the unrestricted conditional consistency requirement CNC. Principle D (a generalization of Principle D from Kripke semantics) from Table 5.2 is in fact equivalent to CNC given $\text{Def}_{\Diamond\!\!\rightarrow}$ (see Theorem 3.24). Hence, the frame condition for Principle D (C_{D}) is also a frame condition for CNC.

One can alternatively also use the following weaker conditional consistency criterion (criterion (A); cf. Sections 3.1 and 3.3.6, see also Table 5.1):

P-Cons $\neg(\top \Box\!\!\rightarrow \bot)$

This principle is, for example, discussed by Hawthorne (1996, pp. 194–196) and Schurz (1998, p. 88 and p. 90; see Section 3.3.6). P-Cons is consistent with Refl given System **CK**. Hence, one can apply $C_{\text{P-Cons}}$ in combination with C_{Refl}. This principle is much weaker than CNC and ensures that the consequent must be consistent (cf. Section 3.3.6) only in case the antecedent is a logically true formula.

7.1.2 Alternative Axiomatizations of System CKR

In the previous section I gave a definition of System **CKR** in terms of **CK** +Refl (see Definition 7.1). Observe that there is a more parsimonious axiomatization available, namely the one given in Theorem 7.2. This is due to the fact that the following Principle LT (see Table 4.1) is redundant given Refl and RW:

LT $\alpha \Box\!\!\rightarrow \top$

I shall henceforth indicate derivability of principles by $\Rightarrow$. In addition, axiom schemata and rules of **CK**, which I rely on in an derivation are indicated by parentheses.

Let us now state the following theorem:

Theorem 7.2. *System* **CKR** *can be axiomatized as LLE+RW+AND+Refl.*

Proof. By Definition 4.1 and Lemma 7.3 we have Theorem 7.2. □

Lemma 7.3. *(RW)+Refl ⇒ LT*

Proof.

1. $\alpha \,\Box\!\!\rightarrow \alpha$ Refl
2. $\alpha \,\Box\!\!\rightarrow \top$ 1, RW

□

One can describe the System **CKR** alternatively on the basis of the following rule:

SC if $\vdash \alpha \rightarrow \beta$, then $\vdash \alpha \,\Box\!\!\rightarrow \beta$

'SC' abbreviates for 'Supraclassicality' (e.g., Schurz, 1998, p. 84). This rules states that any logically true material implication $\alpha \rightarrow \beta$ results in the corresponding conditional $\alpha \,\Box\!\!\rightarrow \beta$ being true. Please note that SC is not an instance of the rule RCK from Section 4.2.7. The main difference is that in RCK – unlike in SC – the material implication '$\rightarrow$' is not replaced by conditional operator '$\Box\!\!\rightarrow$'. That SC is not an instance of RCK can also be seen from the fact that Refl is derivable from LLE+RW+SC (see Lemma 7.5) but not from LLE+RW+RCK. Moreover, Rule SC is not, as Schurz (1996, Footnote 3, p. 201) argues, entailed by LLE and RW. For this to hold SC would be required to be a theorem of **CK**, which it is clearly not. Let us now give the alternative axiomatization of System **CKR** on the basis of Principle SC:

Theorem 7.4. *System* **CKR** *can be axiomatized by LLE+RW+AND+SC.*

Proof. By Definition 4.1 and Lemma 7.5 we obtain Theorem 7.4. □

Lemma 7.5. *(LLE+RW) ⇒ (Refl ⇔ SC)*

Proof. By Lemmata 7.6 and 7.7 we get Lemma 7.5. □

Lemma 7.6. *(LLE+RW)+Refl* $\Rightarrow$ *SC*

Proof.

1.	$\vdash \alpha \rightarrow \beta$	given
2.	$\vdash \alpha \wedge \beta \leftrightarrow \alpha$	1, p.c.
3.	$\alpha \wedge \beta \mathrel{\Box\!\!\rightarrow} \alpha \wedge \beta$	Refl
4.	$\alpha \mathrel{\Box\!\!\rightarrow} \alpha \wedge \beta$	3, 2, LLE
5	$\alpha \mathrel{\Box\!\!\rightarrow} \beta$	4, RW

□

Lemma 7.7. *SC* $\Rightarrow$ *Refl*

Proof.

1.	$\vdash \alpha \rightarrow \alpha$	p.c.
2.	$\alpha \mathrel{\Box\!\!\rightarrow} \alpha$	1, SC

□

7.2 Conditional Logics without Bridge Principles

In this section I will discuss the following conditional logic systems: System **C** (Section 7.2.1), System **CL** (Section 7.2.2), System **P** (Section 7.2.3), System **R** (Section 7.2.4), System **V** (Section 7.2.5), System **CM** (Section 7.2.6), and System **M** (Section 7.2.6). Versions of Systems **C**, **CL**, **P**, **CM**, and **M** can be found in Kraus et al. (1990). Moreover, Lehmann and Magidor (1992) investigate versions of Systems **P** and **R**. I also discussed Systems **P** and **R** in Section 3.6.1. Observe that in Section 3.6.1 I used the expression '$\mathbf{P}^{+}$' to refer to System **R**.

Moreover, the systems described in Kraus et al. (1990) and Lehmann and Magidor (1992) are not formulated in the full language (see the introductory part to this chapter) while all of our formulations pertain to the full language $\mathcal{L}_{\mathrm{CL}}$. System **V** is in that sense exceptional, since it was already formulated by D. Lewis in the full language $\mathcal{L}_{\mathrm{CL}}$ (D. Lewis, 1973b, p. 132f, pp. 120–123).

7.2.1 System C

In Section 7.2 I will focus exclusively on systems without bridge principles. This type of system can be described by presupposing – opposed to systems with bridge principles – that no fixed relation between conditional and non-conditional formulas holds (see Section 4.2.1). I argued in Chapter 1 that any conditional logic approach aims to render Principles S1′–S5′ non-valid. Interestingly, some of the Principles S1′–S5′ are bridge principles, namely S1′ (EFQ) and S2′ (VEQ). It seems intuitive that bridge principles such as S1′ and S2′ cannot be derived in any system without bridge principles. The question whether bridge principles are derivable in systems, which can be axiomatized without reference to bridge principles is far from trivial and investigated in depth in Schurz (1997a). Schurz's (1997a) formal investigation of the Is-Ought problem shows for a range of multi-modal systems that no bridge principles are provable when those systems can be axiomatized without reference to bridge principles (see Schurz, 1997a, Definition 8 [Special Hume Thesis], p. 72; Theorem 3, p. 118; Theorem 4, p. 121; Theorem 5, p. 124). Schurz (1997a, p. 173f), furthermore, conjectures that an extension of his results also holds for two-place operators, such as the conditional operator $\Box\!\!\rightarrow$. Although I am not aware of a proof of this result, I will not discuss this issue here. Rather the above assumption is intended to only motivate our approach to discuss conditional logics with bridge principles and without bridge principles in separate sections.

I will also postpose the discussion of EFQ and VEQ to Section 7.3 and discuss here only principles which are not bridge principles, such as Principles S3′ (Trans, "Transitivity"), S4′ (Mon, "Monotonicity"), and S5′ (CP, "Contraposition") (see also Table 5.1):

Trans $(\alpha \Box\!\!\rightarrow \beta) \wedge (\beta \Box\!\!\rightarrow \gamma) \rightarrow (\alpha \Box\!\!\rightarrow \gamma)$
Mon $(\alpha \Box\!\!\rightarrow \gamma) \rightarrow (\alpha \wedge \beta \Box\!\!\rightarrow \gamma)$
CP $(\alpha \Box\!\!\rightarrow \beta) \rightarrow (\neg\beta \Box\!\!\rightarrow \neg\alpha)$

Mon is the central principle here. This is due to the fact that Mon is a theorem of both **CK**+CP and **CK**+Ref+Trans (see Lemmata 7.45 and 7.51, respectively; cf. also Section 1.2). So, any extension of **CK**+Refl which is non-monotonic, viz. in which Mon is not a theorem, has neither Trans nor CP as theorems. Note that Refl is often regarded a cornerstone for any conditional logic (see Section 7.1.1).

Let us now focus on System **C**, which can be defined the following way:

Definition 7.8. *Logic* **C** *is the smallest logic containing* **CK**+*Refl*+*CM*+*CC*.

Specific to this system are the following two principles (see Table 5.1):

CM $(\alpha \mathrel{\Box\!\!\rightarrow} \gamma) \wedge (\alpha \mathrel{\Box\!\!\rightarrow} \beta) \rightarrow (\alpha \wedge \beta \mathrel{\Box\!\!\rightarrow} \gamma)$
CC $(\alpha \wedge \beta \mathrel{\Box\!\!\rightarrow} \gamma) \wedge (\alpha \mathrel{\Box\!\!\rightarrow} \beta) \rightarrow (\alpha \mathrel{\Box\!\!\rightarrow} \gamma)$

'CM' and 'CC' stand for 'Cautious Monotonicity' and 'Cautious Cut', respectively. The names 'Cautious Monotonicity' and 'Cautious Cut' are motivated by the fact that CM and CC represent weakened versions of the Principles Mon and Cut (see Section 7.2.6 and Table 5.1), respectively:

Cut $(\alpha \wedge \beta \mathrel{\Box\!\!\rightarrow} \delta) \wedge (\gamma \mathrel{\Box\!\!\rightarrow} \beta) \rightarrow (\alpha \wedge \gamma \mathrel{\Box\!\!\rightarrow} \delta)$

Interestingly CC can also be seen as a weakened version of Trans (see above). Since the relationship between CC and Trans is somewhat more perspicuous, I shall henceforth focus on the relationship between CC and Trans rather than the relationship between CC and Cut.

Principles CM and CC correspond to Mon and Trans, respectively, in the following sense: Mon allows one to conclude $\alpha \wedge \beta \mathrel{\Box\!\!\rightarrow} \gamma$ from $\alpha \mathrel{\Box\!\!\rightarrow} \gamma$, whereas CM licenses to infer $\alpha \wedge \beta \mathrel{\Box\!\!\rightarrow} \gamma$ from $\alpha \mathrel{\Box\!\!\rightarrow} \gamma$ and $\alpha \mathrel{\Box\!\!\rightarrow} \beta$. Thus, CM is essentially a restricted version of Mon (hence the name 'Cautious Monotonicity'). Furthermore, Trans allows one to infer $\alpha \mathrel{\Box\!\!\rightarrow} \gamma$ from $\alpha \mathrel{\Box\!\!\rightarrow} \beta$ and $\beta \mathrel{\Box\!\!\rightarrow} \gamma$ while CC needs $\alpha \mathrel{\Box\!\!\rightarrow} \beta$ and $\alpha \wedge \beta \mathrel{\Box\!\!\rightarrow} \gamma$ in order for $\alpha \mathrel{\Box\!\!\rightarrow} \gamma$ to be deducible. The difference between both principles is that in the case of CC the second antecedent requires $\alpha \wedge \beta \mathrel{\Box\!\!\rightarrow} \gamma$ to hold rather than only $\alpha \mathrel{\Box\!\!\rightarrow} \gamma$. In all counter-examples to Mon (E4, E4′, and E4″) and Trans (E3, E3′, and E3″), discussed in Sections 1.2, 1.3, and 1.4.3, the antecedents of Mon and Trans are satisfied, but not the stricter ones of CM and CC.

This argumentation shows that one may block counter-intuitive inferences by employing only CM and CC instead of Mon and Trans, respectively. This insight alone, however, does not establish that CM and CC are rationally justified principles. Further support for this thesis is given by the fact that Principles CM and CC (as well as the Principles Or and RM ("Rational Monotonicity") discussed in Sections 7.2.3 and 7.2.4) respectively) are probabilistically justified. They translate into probabilistically valid inference schemata in Systems **P** (Schurz, 2005b), $\mathbf{P}^*$ (Adams, 1965, 1966, 1977), and $\mathbf{P}^+$ (Adams, 1986; Schurz, 1998; Bamber, 1994; see Section 3.6.1) for both a validity criterion of infinitesimal probability (see Definition 3.18) and a validity criterion with a high, but non-infinitesimal probability (see Theorem 3.23). Principles CM and CC (and Or and RM) are, hence, statistically reliable.

In addition Principles CM and CC are also in line with one's general intuitions regarding conditionals. To provide support for this thesis, I will draw on Kraus et al. (1990, p. 178f) and for that purpose adopt their examples for the present discussion. Let us begin with Principle CM. Assume for that purpose that you believe the two following propositions:

(a1) If Fireball takes part in the race, then it will normally win.

(b1) If Fireball takes part in the race, then normally Thunderhead takes place in the race.

Are you, then, not rationally forced to believe the following as well?

(c1) If Fireball takes part in the race and Thunderhead takes place in the race, then normally Fireball wins the race?

The inference from (a1) and (b1) to (c1) is statistically reliable (see above). Moreover, it seems also quite natural to accept (c1) given that one accepts (a1) and (b1). Furthermore, if one does not accept (c1) then it seems natural to reject (a1). The inference from (a1) and (b1) to (c1) is a natural language representation of the above Principle CM. A parallel argument can also be made for Principle CC. Assume that you believe following two propositions:

(a2) If Fireball takes part in the race and Thunderhead takes part in the race, then Fireball will normally win.
(b2) If Fireball takes part in the race, then normally Thunderhead takes part in the race.

Would believing (a2) and (b2) not rationally commit one to also believe the following proposition?

(c2) If Fireball takes part in the race, then normally Fireball wins the race.

Again, the inference from (a2) and (b2) to (c2) is statistically reliable (in the above sense). Moreover, it seems natural to reject (a2) (or (b2)) if one rejects (c2).

Let us now discuss the semantic representation of CM and CC in CS semantics. I list here the non-trivial frame conditions for CM and CC and the non-trivial frame conditions for Mon and Trans (where X and Y refer to subsets of possible worlds; see also Table 5.3):

C_{CM} $\forall X, Y \forall w (\forall w' (wR_X w' \Rightarrow w' \in Y) \Rightarrow \forall w' (wR_{X \cap Y} w' \Rightarrow wR_X w'))$
C_{CC} $\forall X, Y \forall w (\forall w' (wR_X w' \Rightarrow w' \in Y) \Rightarrow \forall w' (wR_X w' \Rightarrow wR_{X \cap Y} w'))$
C_{Mon} $\forall X, Y \forall w, w' (wR_{X \cap Y} w' \Rightarrow wR_X w')$
C_{Trans} $\forall X, Y \forall w (\forall w' (wR_X w' \Rightarrow w' \in Y) \Rightarrow \forall w' (wR_X w' \Rightarrow wR_Y w'))$

The frame conditions C_{CM}, C_{CC}, and C_{Trans} contain preconditions, which require worlds, which are accessible via an accessibility relation R_X, to be members of some set Y. I will call this type of frame condition ‘relativized’. C_{Mon} is not relativized in that sense: It rather represents a closure condition, which does not depend on accessible worlds being members of some set and, thus, can be applied in this sense non-conditionally. It is the relativization of C_{CM} that gives CM its intuitive import compared to Mon.

The case is different for C_{Trans}. Provided that C_{Refl} holds for all subsets of possible worlds X specified by the given Chellas frame $\mathcal{F}_C$ – namely that it is the case that $\forall X \forall w, w' (wR_X w' \Rightarrow w' \in X)$ – C_{Trans} implies C_{Mon}. Hence, the precondition C_{Trans} becomes in the presence of C_{Refl} “ineffective”. This can be seen if one replaces variables 'X' and 'Y' in C_{Trans} by '$X \cap Y$' and 'X', respectively. Then, the following frame condition results:

(1) $\forall X, Y \forall w(\forall w'(wR_{X\cap Y}w' \Rightarrow w' \in X) \Rightarrow \forall w'(wR_{X\cap Y}w' \Rightarrow wR_X w'))$

Principle Refl guarantees that the precondition of (1), namely $\forall X, Y \forall w, w'$ $(wR_{X\cap Y}w' \Rightarrow w' \in X)$, holds in all cases, and thus, C_{Trans} implies in the presence of C_{Ref} frame condition C_{Mon}. From a proof-theoretic point of view this is mirrored by the fact that **CK**+Refl+Trans implies Mon (see Lemma 7.51).

Let us now turn to the proof theory of System **C**. I will provide two alternative and more parsimonious axiomatizations of System **C** compared to Definition 7.8, as described by Theorems 7.9 and 7.11. These alternative characterizations are important, since they might be invoked for the description of extensions of System **C**, such as System **CL** (Definition 7.14), System **P** (Definition 7.15), and System **R** (Definition 7.27). Let us now describe these alternative axiomatizations of System **C**:

Theorem 7.9. *System* **C** *can be axiomatized by LLE+RW+Refl+CM+CC. (cf. Kraus et al., 1990, p. 176).*

Proof. System **C** is by Definition 7.8 **CK**+Refl+CM+CC. Since **CK** is due to Definition 4.1 LLE+RW+AND+LT, System **C** is LLE+RW+AND+LT +Refl+CM+CC. To show that **C** is LLE+RW+Refl+CM+CC, we, hence, have to prove that (a) LT and (b) AND are theorems of LLE+RW+Refl+CM +CC. Lemmata 7.3 and 7.10 give us that LT is derivable from RW+Refl and that AND is a theorem of RW+Refl+CM+CC, respectively. Hence, (a) and (b) follow. □

Lemma 7.10. *(RW)+Refl+CM+CC ⇒ AND (Kraus et al., 1990, p. 179)*

Proof.

1.	$\alpha \mathrel{\Box\!\!\rightarrow} \beta$	given
2.	$\alpha \mathrel{\Box\!\!\rightarrow} \gamma$	given
3.	$\alpha \wedge \beta \mathrel{\Box\!\!\rightarrow} \gamma$	2, 1, CM
4.	$\alpha \wedge \beta \wedge \gamma \mathrel{\Box\!\!\rightarrow} \alpha \wedge \beta \wedge \gamma$	Refl
5.	$\alpha \wedge \beta \wedge \gamma \mathrel{\Box\!\!\rightarrow} \beta \wedge \gamma$	4, RW

6.	$\alpha \wedge \beta \mathrel{\Box\!\!\rightarrow} \beta \wedge \gamma$	5, 3, CC
7.	$\alpha \mathrel{\Box\!\!\rightarrow} \beta \wedge \gamma$	6, 1, CC

□

Let us now prove the second alternative axiomatization:

Theorem 7.11. *System* **C** *can be axiomatized by LLE+SC+CM+CC.*

Proof. System **C** is by Theorem 7.9 LLE+RW+Refl+CM+CC. To prove that **C** is LLE+SC+CM+CC, we have to show that (a) RW and (b) Refl are theorems of LLE+SC+CM+CC and that (c) SC is derivable from LLE+RW+Refl+CM+CC. Lemmata 7.12 and 7.7 give us that RW is a theorem of SC+CC and that Refl can be obtained from SC, respectively. Thus, (a) and (b) follow. Moreover, by Lemma 7.6 SC is a theorem of LLE+RW+Refl. Hence, (c) is the case. □

Lemma 7.12. *SC+CC $\Rightarrow$ RW (cf. Adams, 1966, p. 280f)*

Proof.

1.	$\alpha \mathrel{\Box\!\!\rightarrow} \beta$	given
2.	$\vdash \beta \rightarrow \gamma$	given
3.	$\vdash \alpha \wedge \beta \rightarrow \gamma$	2, p.c.
4.	$\alpha \wedge \beta \mathrel{\Box\!\!\rightarrow} \gamma$	3, SC
5.	$\alpha \mathrel{\Box\!\!\rightarrow} \gamma$	4, 1, CC

□

Principle LLE is included in our axiomatization of System **C** in Theorem 7.11. Sometimes it is argued that the stronger System **P** (Geffner & Pearl, 1994, p. 71; Schurz, 2005b, p. 42; Schurz, 1998, p. 84) and System $\mathbf{P}_{\epsilon}$ (Schurz, 1994, p. 251) can be axiomatized by inference schema corresponding to SC+CM+CC+Or and SC+CM+CC+Or+RCNC, respectively, hence *without* reference to LLE. (For a discussion of the Principles Or and RCNC see Sections 7.2.3 and 3.6.3, respectively.) In their proof of LLE Geffner and Pearl (1994, Theorem 2, p. 72) refer explicitly only to Principles SC, CM, and CC (which are all theorems of System **C**). There is one important difference between our and Geffner and Pearl's treatment of conditionals: Geffner and Pearl (1994) conceptualize antecedents of conditionals as sets of formulas, where elements of "antecedent sets" are interpreted as "unordered conjunctions" of formulas (p. 70f). Since in Geffner and Pearl's account both $\{\alpha,\beta\} \boxright \gamma$ and $\{\beta,\alpha\} \boxright \gamma$ represent the "same" conditional, Geffner and Pearl (1994), hence, presuppose that the following principle for conditionals holds:

ECA $(\alpha \wedge \beta \boxright \gamma) \rightarrow (\beta \wedge \alpha \boxright \gamma)$

Here 'ECA' stands for 'Exchangability of Conjuncts in the Antecedent'. Moreover, given ECA, LLE is derivable from CM, CC, and SC, as the following lemma shows:

Lemma 7.13. *ECA+SC+CC+CM ⇒ LLE (cf. Geffner & Pearl, 1994, Theorem 2, p. 72)*

Proof.

1.	$\alpha \boxright \gamma$	given
2.	$\vdash \alpha \leftrightarrow \beta$	given
3.	$\vdash \alpha \rightarrow \beta$	2, p.c.
4.	$\alpha \boxright \beta$	3, SC
5.	$\alpha \wedge \beta \boxright \gamma$	1, 4, CM
6.	$\beta \wedge \alpha \boxright \gamma$	5, ECA
7.	$\vdash \beta \rightarrow \alpha$	2, p.c.
8.	$\beta \boxright \alpha$	7, SC
9.	$\beta \boxright \gamma$	6, 8, CC

□

Principle ECA is essential for Lemma 7.13 to go through. I am, moreover, not aware of any alternative proof of Lemma 7.13 which does not rely on an at least implicit version of ECA.[1] I hence conclude that the above axiomatizations of System **P** (Geffner & Pearl, 1994, p. 71; Schurz, 2005b, p. 42; Schurz, 1998, p. 84) and System $\mathbf{P}_\epsilon$ (Schurz, 1994, p. 251) have to

[1] For example, in Schurz (1994, 2005b) no alternative proof or reference to an alternative proof of Lemma 7.13 without ECA is provided.

either include ECA as a further axiom schema or conceptualize antecedents of conditionals as sets of formulas rather than single formulas (in the way described above).

7.2.2 System CL

System **CL** is System **C** plus the following Principle Loop (see Table 5.1):

Loop $(\alpha_0 \,\Box\!\!\rightarrow \alpha_1) \wedge \ldots \wedge (\alpha_{k-1} \,\Box\!\!\rightarrow \alpha_k) \wedge (\alpha_k \,\Box\!\!\rightarrow \alpha_0) \rightarrow (\alpha_0 \,\Box\!\!\rightarrow \alpha_k)$ $(k \geq 2)$

Hence, System **CL** can be defined accordingly:

Definition 7.14. *Logic* **CL** *is the smallest logic containing* **CK**+*Refl*+*CM*+*CC*+*Loop (cf. Kraus et al., 1990, p. 187).*

This principle has been first suggested by Kraus et al. (1990, p. 187). The name 'Loop' is also appropriate from a semantical perspective. The non-trivial frame condition for Principle Loop is the following (see Table 5.3):

C_{Loop} $\forall X_0,\ldots,X_k \forall w(\forall w'(wR_{X_0}w' \Rightarrow w' \in X_1) \wedge \ldots \wedge \forall w'(wR_{X_{k-1}}w' \Rightarrow w' \in X_k) \wedge \forall w'(wR_{X_k}w' \Rightarrow w' \in X_0) \Rightarrow \forall w'(wR_{X_0}w' \Rightarrow w' \in X_k))$ $(k \geq 2)$

The frame condition can be interpreted in the following sense: If for a given number of sets the accessibility relations relativized to sets and the worlds accessible form a loop (by being members in the respective sets and being relativized to the respective sets), then all worlds accessibly by the first set are members of the last set.

7.2.3 System P

System **P** is defined the following way:

Definition 7.15. *Logic* **P** *is the smallest logic containing* **C**+*Or* *(Kraus et al., 1990, p. 190).*

The main difference between System **P** on the one hand and System **CL** and **C** on the other hand is, hence, the fact that System **P** contains the additional Principle Or (see Table 5.1):

Or $(\alpha \mathrel{\Box\!\!\rightarrow} \gamma) \wedge (\beta \mathrel{\Box\!\!\rightarrow} \gamma) \rightarrow (\alpha \vee \beta \mathrel{\Box\!\!\rightarrow} \gamma)$

Principle Or describes probabilistically valid inferences and, thus, is probabilistically justified (cf. Section 7.2.1; see also Section 3.6). Moreover, to demonstrate the intuitive support for Principle Or, let us consider the following inference, which corresponds to Principle Or (Kraus et al., 1990, p. 190):

(a3) If Cathy attends the party then normally the evening will be great.
(b3) If John attends the party then normally the evening will be great.
(c3) Therefore: If Cathy or John attends the party, then normally the evening will be great.

Given that (a3) and (b3) hold, it is seem reasonable to infer (c3). Example (a3)-(c3) is one of many Or inferences, which provide intuitive support for the Principle Or. A further motivation for including Or is that this type of inference is unproblematic insofar as it does not render, given further reasonable assumptions, the Principles Mon, Trans, and CP valid. This is due to the complementary relation of both Principles Mon and Or. Let us repeat the non-trivial frame condition corresponding to Or (see Table 5.3):

C_{Or} $\forall X, Y \forall w, w' (wR_{X \cup Y}w' \Rightarrow wR_X w' \vee wR_Y w')$

The frame condition for Or implies that for any worlds w, w' in any Chellas frame $\mathcal{F}_C = \langle W, R \rangle$ and subsets X, Y of W that if $wR_X w'$, then either $wR_{X \cap Y}w'$ or $wR_{X \cap -Y}w'$ (see also Lemma 7.25 and the frame condition C_{WOR} from Table 5.3). The Principle Mon on the other hand gives one for arbitrary worlds w, w' in $\mathcal{F}_C$ and X, Y being subsets of W that, if $wR_{X \cap Y}w'$ then $wR_X w'$. Both frame conditions for Mon and Or are from a technical perspective a type of closure condition, which is not relativized, as CM and CC are (see Section 7.2.1). The complementary relation of Or and Mon can also be seen from a proof-theoretic perspective: While System **CK**+Refl+Mon+Or allows for an axiomatization of the monotonic collapse without bridge principles (System **M**; see Definition 7.40), the following holds: (a) System **P** contains Or but does not collapse into **M** (moreover Mon is not valid in System **P**) and (b) System **CM** contains Mon, but not Or.

Furthermore, Principle Or implies – given Refl – Principle CC (see Lemma 7.16). Hence, the axiomatization given in Definition 7.15 suffices.

Lemma 7.16. *(LLE+RW+AND)+Refl+Or ⇒ CC*

Proof.

1.	$\alpha \wedge \beta \mathrel{\Box\!\!\rightarrow} \gamma$	given
2.	$\alpha \mathrel{\Box\!\!\rightarrow} \beta$	given
3.	$\alpha \wedge \neg\beta \mathrel{\Box\!\!\rightarrow} \alpha \wedge \neg\beta$	Refl
4.	$\alpha \wedge \beta \mathrel{\Box\!\!\rightarrow} \gamma \vee (\alpha \wedge \neg\beta)$	1, RW
5.	$\alpha \wedge \neg\beta \mathrel{\Box\!\!\rightarrow} \gamma \vee (\alpha \wedge \neg\beta)$	3, RW
6.	$(\alpha \wedge \beta) \vee (\alpha \wedge \neg\beta) \mathrel{\Box\!\!\rightarrow} \gamma \vee (\alpha \wedge \neg\beta)$	4, 5, Or
7.	$\alpha \mathrel{\Box\!\!\rightarrow} \gamma \vee (\alpha \wedge \neg\beta)$	6, LLE
8.	$\alpha \mathrel{\Box\!\!\rightarrow} (\gamma \vee (\alpha \wedge \neg\beta)) \wedge \beta$	7, 2, AND
9.	$\alpha \mathrel{\Box\!\!\rightarrow} \gamma$	8, RW

□

Let me now give two alternative axiomatizations of System **P** (cf. Definition 3.19):

Theorem 7.17. *System* **P** *can be axiomatized as LLE+RW+AND+Refl+CM +Or.*

Proof. By Definition 7.15 System **P** is **C**+Or. Hence, since by Theorem 7.9 System **C** is LLE+RW+Refl+CM+CC, System **P** is LLE+RW+Refl+CM+CC+Or. As by Lemma 7.16 Principle CC is derivable from LLE+RW+AND +Refl+Or, System **P** is LLE+RW+AND+Refl+CM+Or. □

Theorem 7.18. *System* **P** *can be axiomatized as LLE+SC+CM+CC+Or.*

Proof. By Theorem 7.17 System **P** is LLE+RW+AND+Refl+CM+Or. As by Lemma 7.16 Principle CC is derivable from LLE+RW+AND+Refl+Or, System **P** is LLE+RW+AND+Refl+CM+CC+Or. Due to Lemma 7.10 AND is a theorem of RW+Refl+CM+CC. Hence, System **P** is LLE+RW+Refl+ CM+CC+Or. As Lemma 7.5 gives us that the Principles Refl and SC are logically equivalent given LLE+RW, System **P** is LLE+RW+ SC+CM+CC+Or.

Moreover, since RW can be derived from SC+CC by Lemma 7.12, it follows that System **P** is LLE+SC+CM+CC+Or. □

Our axiomatization of System **P** in Theorem 7.18 differs from the axiomatizations given in Schurz (1998, p. 84), Schurz (2005b, p. 42), and Geffner and Pearl (1994, p. 71) insofar it includes the Principle LLE (see comments to Theorem 7.11).

Furthermore, Principle Or is closely related to the following two principles (see Table 5.1):

S $(\alpha \wedge \beta \; \square\!\!\rightarrow \gamma) \rightarrow (\alpha \; \square\!\!\rightarrow (\beta \rightarrow \gamma))$
WOR $(\alpha \wedge \beta \; \square\!\!\rightarrow \gamma) \wedge (\alpha \wedge \neg\beta \; \square\!\!\rightarrow \gamma) \rightarrow (\alpha \; \square\!\!\rightarrow \gamma)$

The name 'S' is due to Kraus et al. (1990, p. 191) and Lehmann and Magidor (1992, p. 6). Unfortunately, neither Kraus et al. (1990) nor Lehmann and Magidor (1992) indicate what the label 'S' stands for. Kraus et al. (1990, p.191) also investigate WOR ("Weak Or", see Hawthorne & Makinson, 2007, p. 252). In Kraus et al.'s terminology the Principle WOR is called 'D' (p. 191).

Principle WOR is an axiom schema of the probabilistic threshold logic **O** of Hawthorne and Makinson (2007, p. 252; see also Section 3.6.2), which is strictly weaker than System **P** (Hawthorne & Makinson, 2007, p. 253). The main difference between System **O** and **P** is from a proof-theoretic perspective that AND is a theorem of **P**, but not of System **O** (Hawthorne & Makinson, 2007, p. 259; see also Section 3.6.1). Lemmata 7.19 and 7.25 give us that both Principles S and WOR are logically equivalent to Or provided **CK**+Refl. In contrast, the equivalence between WOR and Or does not hold given the rules and axioms of System **O** (cf. Hawthorne & Makinson, 2007, p. 253). Let us now establish Lemmata 7.19 and 7.25:

Lemma 7.19. *(LLE+RW+AND)+Refl ⇒ (Or ⇔ S)*

Proof. By Lemmata 7.20 and 7.21 we obtain Lemma 7.19. □

Lemma 7.20. *(LLE+RW+AND)+S ⇒ Or*

Proof.

1.	$\alpha \Box\!\!\rightarrow \gamma$	given
2.	$\beta \Box\!\!\rightarrow \gamma$	given
3.	$(\alpha \vee \beta) \wedge (\alpha \vee \neg\beta) \Box\!\!\rightarrow \gamma$	1, LLE
4.	$(\alpha \vee \beta) \Box\!\!\rightarrow ((\alpha \vee \neg\beta) \rightarrow \gamma)$	3, S
5.	$(\alpha \vee \beta) \wedge (\neg\alpha \vee \beta) \Box\!\!\rightarrow \gamma$	2, LLE
6.	$(\alpha \vee \beta) \Box\!\!\rightarrow ((\neg\alpha \vee \beta) \rightarrow \gamma)$	5, S
7.	$(\alpha \vee \beta) \Box\!\!\rightarrow ((\alpha \vee \neg\beta) \rightarrow \gamma) \wedge ((\neg\alpha \vee \beta) \rightarrow \gamma)$	4, 6, AND
8.	$(\alpha \vee \beta) \Box\!\!\rightarrow \gamma$	7, RW

□

Lemma 7.21. *(LLE+RW)+Refl+Or ⇒ S (Kraus et al., 1990, p. 191)*

Proof.

1.	$\alpha \wedge \beta \Box\!\!\rightarrow \gamma$	given
2.	$\alpha \wedge \beta \Box\!\!\rightarrow (\beta \rightarrow \gamma)$	1, RW
3.	$\alpha \wedge \neg\beta \Box\!\!\rightarrow \alpha \wedge \neg\beta$	Refl
4.	$\alpha \wedge \neg\beta \Box\!\!\rightarrow (\beta \rightarrow \gamma)$	3, RW
5.	$(\alpha \wedge \beta) \vee (\alpha \wedge \neg\beta) \Box\!\!\rightarrow (\beta \rightarrow \gamma)$	2, 4, Or
6.	$\alpha \Box\!\!\rightarrow (\beta \rightarrow \gamma)$	5, LLE

□

The foregoing Lemma 7.19 and the following Lemma 7.22 do not only draw on the Principle Refl (which is a theorem of System **O**), but also on the Principle AND (which is not a theorem of System **O**) to prove the logical equivalence of S and Or on the one hand and WOR and Or on the other.

Lemma 7.22. *(RW+AND)+Refl ⇒ (S ⇔ WOR)*

Proof. By Lemmata 7.23 and 7.24 we get Lemma 7.22. □

Lemma 7.23. *(RW+AND)+S* $\Rightarrow$ *WOR*

Proof.

1.	$\alpha \wedge \beta \mathrel{\Box\!\!\rightarrow} \gamma$	given
2.	$\alpha \wedge \neg\beta \mathrel{\Box\!\!\rightarrow} \gamma$	given
3.	$\alpha \mathrel{\Box\!\!\rightarrow} (\beta \rightarrow \gamma)$	1, S
4.	$\alpha \mathrel{\Box\!\!\rightarrow} (\neg\beta \rightarrow \gamma)$	2, S
5.	$\alpha \mathrel{\Box\!\!\rightarrow} (\beta \rightarrow \gamma) \wedge (\neg\beta \rightarrow \gamma)$	3, 4, AND
6.	$\alpha \mathrel{\Box\!\!\rightarrow} \gamma$	5, RW

□

Lemma 7.24. *(RW)+Refl+WOR* $\Rightarrow$ *S*

Proof.

1.	$\alpha \wedge \beta \mathrel{\Box\!\!\rightarrow} \gamma$	given
2.	$\alpha \wedge \neg\beta \mathrel{\Box\!\!\rightarrow} \alpha \wedge \neg\beta$	Refl
3.	$\alpha \wedge \neg\beta \mathrel{\Box\!\!\rightarrow} (\beta \rightarrow \gamma)$	2, RW
4.	$\alpha \wedge \beta \mathrel{\Box\!\!\rightarrow} (\beta \rightarrow \gamma)$	1, RW
5.	$\alpha \mathrel{\Box\!\!\rightarrow} (\beta \rightarrow \gamma)$	4, 3, WOR

□

We also obtain the following lemma:

Lemma 7.25. *(LLE+RW+AND)+Refl* $\Rightarrow$ (*Or* $\Leftrightarrow$ *WOR*) *(Kraus et al., 1990, p. 191)*

Proof. By Lemmata 7.19 and 7.22 Theorem 7.25 results. □

Let us, finally, discuss the following Principle MOD ("Modality", see Table 5.1):

MOD $(\neg\alpha \mathrel{\Box\!\!\rightarrow} \alpha) \rightarrow (\beta \mathrel{\Box\!\!\rightarrow} \alpha)$

The name 'MOD' is taken from Nute and Cross (2001, p. 10) and Nute (1980, p. 52). Interestingly, Nute and Cross (2001) and Nute (1980) do not indicate what 'MOD' abbreviates. Nute and Cross' use of the name suggests that 'MOD' is a proxy for 'Modality'. (I will henceforth read 'MOD' in that sense).

The Principle MOD can be interpreted in the following way: Provided that α is the case even if $\neg\alpha$ holds, then α holds, no matter what the case is. In this reading the formula $\neg\alpha \;\Box\!\!\rightarrow \alpha$ expresses that α is necessary, in the sense that in any case α holds. Nute (1980, p. 52), accordingly, defines $\neg\alpha \;\Box\!\!\rightarrow \alpha$ as $\Box\alpha$, where $\Box$ is a one-place modal necessity operator (see also Nute, 1980, p. 16 and Section 7.1.1). The reading of $\neg\alpha \;\Box\!\!\rightarrow \alpha$ as $\Box\alpha$ (α being necessary) is also suggested by an analogous interpretation of universal quantifiers in f.o.l. One can interpret $\forall x(\neg Fx \rightarrow Fx)$ the following way: Given that (an arbitrary) x is not is not a F, x is a F. This is just to say in f.o.l. that Fx holds for any x. It is, hence, not surprising that $\forall x(\neg Fx \rightarrow Fx)$ is f.o.l.-equivalent to $\forall xFx$.

In System **CK** and in CS semantics the case of MOD is not perfectly analogous to the interpretation of universal quantifies in f.o.l. In order to establish an interpretation of $\neg\alpha \;\Box\!\!\rightarrow \alpha$ as a modal necessity in line with Kripke semantics (Hughes & Cresswell, 1996/2003, p. 38f), one has to show that the truth of $\neg\alpha \;\Box\!\!\rightarrow \alpha$ at a world w in a Chellas model $\langle W, R, V\rangle$ guarantees that α is true at all worlds accessible by some accessibility relation R_X from w, where $X \subseteq W$. In other words, for $\Box\alpha$ the following Kripke truth-conditions should hold in $\mathcal{M}_C$, where wR^*w' is the case iff $\exists X(wR_Xw')$ (see Section 7.1.1): $\Box\alpha$ is true at a world $w \in W$ iff α is true at all worlds w' such that wR^*w'. Principle MOD now corresponds to the following non-trivial frame condition (see Table 5.3):

$$C_{\text{MOD}} \quad \forall X, Y \forall w, w'((wR_{-X}w' \Rightarrow w' \in X) \Rightarrow \forall w'(wR_Yw' \Rightarrow w' \in X))$$

The frame condition C_{MOD} does not give us that result, because the condition $\forall X \forall w'(wR_{-X}w' \Rightarrow w' \in X)$ allows that there are even formulas α and worlds w, w' in a Chellas models $\langle W, R, V\rangle$ such that $wR_{||\neg\alpha||}w'$ and $w' \in ||\alpha||$. One might think that the addition of the Principle Refl ($\alpha \;\Box\!\!\rightarrow \alpha$) makes it possible to interpret $\neg\alpha \;\Box\!\!\rightarrow \alpha$ as $\Box\alpha$. Let us, for that purpose, consider the non-trivial frame condition C_{Refl} (see Section 7.1.1 and Table 5.3):

C_{Refl} $\forall X \forall w, w'(wR_X w' \Rightarrow w' \in X)$

The frame condition C_{Refl} and $\forall X \forall w'(wR_{-X}w' \Rightarrow w' \in X)$ do not guarantee that $\neg\alpha \,\Box\!\!\rightarrow \alpha$ can be interpreted as $\Box\alpha$, since the truth of $\neg\alpha \,\Box\!\!\rightarrow \alpha$ at a world w in a Chellas model $\mathcal{M}_C = \langle W, R, V\rangle$ implies – given C_{Refl} – only that there are no worlds w' in $\mathcal{M}_C$ such that $wR_{\|\neg\alpha\|}w'$. There might still be worlds w' which are not attainable by $R_{\|\neg\alpha\|}$ but via some other accessibility relation $R_{\|\beta\|}$ and at which, hence, $\neg\alpha$ might be true. Thus, $\forall X \forall w'(wR_{-X}w' \Rightarrow w' \in X)$ and C_{Refl} do not make sure that formula α is true at all worlds attainable by the accessibility relation R^*. An alternative formulation of MOD investigated by Segerberg (1989, Principle #2, p. 158) is the following:

MOD′ $(\neg\alpha \,\Box\!\!\rightarrow \bot) \rightarrow (\beta \,\Box\!\!\rightarrow \alpha)$

MOD is logically equivalent to MOD′ given **CK**+Refl. The non-trivial frame condition corresponding to MOD′ is the following (cf. Segerberg 1989, p. 163):

C_{MOD} $\forall X, Y \forall w(\neg\exists w'(wR_{-X}w') \Rightarrow \forall w'(wR_Y w' \Rightarrow w' \in X))$

The truth of the formula $\neg\alpha \,\Box\!\!\rightarrow \bot$ at a world w in a Chellas model $\mathcal{M}_C = \langle W, R, V\rangle$ gives us again only that there is no world w' in $\mathcal{M}_C$ such that $wR_{\|\neg\alpha\|}w'$. It does not guarantee that α is true at all worlds attainable by accessibility R^*, as defined above. One can alternatively define $\Box\alpha$ directly the following way (based on a Chellas model $\mathcal{M} = \langle W, R, V\rangle$, cf. Segerberg, 1989, p. 161; cf. Section 7.1.1):

$V_\Box$ $\models_w^{\mathcal{M}_C} \Box\alpha$ iff $\forall w' \forall X \subseteq W(wR_X w' \Rightarrow \models_{w'}^{\mathcal{M}_C} \alpha)$

Observe that $V_\Box$ involves a second-order quantification over subsets of the set of possible worlds W. Open problems regarding the notion of necessity include the following: (a) Can modal necessity ($\Box$) also be defined without second-order quantification? (b) What is the weakest conditional logic system such that $(\neg\alpha \,\Box\!\!\rightarrow \alpha) \leftrightarrow \Box\alpha$ holds? Due to space and time limitations I have to leave these problems for further investigations.

Let me expand upon why I include the discussion of MOD in this section rather than other sections. The Principle MOD is derivable in System **P**, as the following lemma shows:

Lemma 7.26. *(LLE+RW+AND)+Refl+CM+Or ⇒ MOD*

Proof.

1.	$\neg\alpha \,\Box\!\!\rightarrow \alpha$	given
2.	$\neg\alpha \,\Box\!\!\rightarrow \neg\alpha$	Refl
3.	$\neg\alpha \,\Box\!\!\rightarrow \alpha \wedge \neg\alpha$	1,2, AND
4.	$\neg\alpha \,\Box\!\!\rightarrow \beta$	3, RW
5.	$\neg\alpha \wedge \beta \,\Box\!\!\rightarrow \alpha$	1, 4, CM
6.	$\alpha \wedge \beta \,\Box\!\!\rightarrow \alpha \wedge \beta$	Refl
7.	$\alpha \wedge \beta \,\Box\!\!\rightarrow \alpha$	6, RW
8.	$(\neg\alpha \wedge \beta) \vee (\alpha \wedge \beta) \,\Box\!\!\rightarrow \alpha$	5, 7, Or
9.	$\beta \,\Box\!\!\rightarrow \alpha$	8, LLE

□

7.2.4 System R

System **R** differs from System **P** insofar as the former, but not the latter, includes the following additional Principle RM ("Rational Monotonicity"; see Table 5.1; cf. Section 2.2.6):

RM $(\alpha \,\Box\!\!\rightarrow \gamma) \wedge (\alpha \,\Diamond\!\!\rightarrow \beta) \rightarrow (\alpha \wedge \beta \,\Box\!\!\rightarrow \gamma)$

The formula $\alpha \,\Diamond\!\!\rightarrow \beta$ abbreviates $\neg(\alpha \,\Box\!\!\rightarrow \neg\beta)$ (see $\text{Def}_{\Diamond\rightarrow}$, Section 4.2.1). Principle RM differs from Principle CM, insofar as it requires $\alpha \,\Diamond\!\!\rightarrow \beta$ instead of $\alpha \,\Box\!\!\rightarrow \beta$. Let me postpone the discussion of Principle RM and, first, describe System **R**:

Definition 7.27. *Logic* **R** *is the smallest logic containing* **P**+*RM (cf. Lehmann & Magidor, 1992, p. 19).*

Theorem 7.28. *Logic* **R** *can be axiomatized by LLE+RW+Refl+CM+RM +Or.*

Proof. By Definition 7.27 and Theorem 7.17 System **R** is LLE+RW+Refl+ CM+RM+Or. □

Let us give here a further axiomatization of System **R**:

Theorem 7.29. *System* **R** *can be axiomatized by LLE+SC+CM+RM+CC+Or (cf. Schurz, 1998, p. 84).*

Proof. By Definition 7.27 and Theorem 7.18 System **R** is LLE+SC+CM+RM +CC+Or. □

Theorem 7.29 gives us that System $\mathbf{P}^+$ by Schurz (1998, cf. Adams, 1986; see also Section 3.6.1) and System **R** concur, since Schurz (1998, p. 84) effectively axiomatizes System $\mathbf{P}^+$ as LLE+SC+CM+RM+CC+Or. [2]

Let us now focus on Principle RM. The Principle RM corresponds to following frame condition (see Table 5.3):

$$C_{\mathrm{RM}} \quad \forall X, Y \forall w(\exists w'(wR_X w' \wedge w' \in Y) \Rightarrow \forall w'(wR_{X \cap Y} w' \Rightarrow wR_X w'))$$

Principle RM represents – as CM – a restricted version of Principle Mon (see Section 7.2.1). In case of the corresponding frame condition C_{RM} an existential claim is required to be satisfied rather than a universal one as in C_{RM}. One can give reasons for accepting RM analogously to CM. Consider the following argument:

(a4) If Fireball takes part in the race, then it will normally win.
(b4) The following is not the case: If Fireball takes part in the race, then normally Thunderhead does not take part in the race.
(c4) Therefore: If Fireball takes part in the race and Thunderhead takes part in the race, then normally Fireball wins the race.

[2] I included here additionally LLE, since Schurz's (1998) axiomatization seems to implicitly presuppose that this principle holds (see Section 7.2.1).

Again Principle RM – and thus the inference from (a4) and (b4) to (c4) – is probabilistically justified based on infinitesimal and non-infinitesimal probabilistic validity criteria (see Schurz, 2005b, pp. 42–45;[3] cf. Section 7.2.1; see also Section 3.6.1). Moreover, analogously as in the case for CM, Principle RM seems to describe rationally justified inferences (cf. Lehmann & Magidor, 1992, p. 18f). Moreover, although RM might look like a default inference, it is not, since it neither depends on non-derivability conditions nor on consistency conditions (cf. Section 2.2.6). The discussion of Kraus et al. (1990, p. 198) and Lehmann and Magidor (1992, p. 18f) is somewhat misleading in that respect.

Schurz (2012, p. 552) suggested that RM is in fact stronger than Principle CM. However, Principle CM is not a theorem of **CK**+Refl+CC+RM+Or as the following lemma shows:

Lemma 7.30. *Given* **CK**+*Refl+RM+CC+Or the Principle CM is not derivable.*

Proof. To prove this lemma, I draw on the soundness result described in Chapter 6 for Chellas frames. We know by the correspondence results from Chapter 5 that, given C_{Refl}, C_{RM}, C_{CC}, and C_{Or} hold for a Chellas frame $\mathcal{F}_C = \langle W, R\rangle$, then Refl, RM, CC, and Or are valid on $\mathcal{F}_C$. Moreover, all **CK** theorems are valid on $\mathcal{F}_C$, including theorems derivable in **CK** on the basis of Refl, RM, CC, and Or. Hence, if for $\mathcal{F}_C$ the frame conditions C_{Refl}, C_{RM}, C_{CC}, and C_{Or} and are the case, but CM is not valid in $\mathcal{F}_C$, then CM is not a theorem of **CK**+Refl+RM+CC+Or. By the correspondence results C_{CM} holds for a frame $\mathcal{F}_C$ iff CM is valid on $\mathcal{F}_C$. Thus, it suffices to show that there is some Chellas frame $\mathcal{F}_C$ such that C_{Refl}, C_{RM}, C_{CC}, and C_{Or} hold, but C_{CM} does not.

Let $\mathcal{F}_C = \langle W, R\rangle$ be a Chellas frame such that $W = \{w_1, w_2\}$. Moreover, let R be $\{\langle w_1, w_2, \{w_2\}\rangle\}$. It is easy to see that C_{Refl}, C_{RM}, C_{CC}, and C_{Or} apply

[3] The Principle RM corresponds in Schurz's (2005b) terminology to the Principle WRM (p. 42, "Weak Rational Monotonicity").

to $\mathcal{F}_C$. C_{CM} does not hold for $\mathcal{F}_C$, since $\forall w_3(w_1 R_{\{w_1,w_2\}} w_3 \Rightarrow w_3 \in \{w_2\})$, $w_1 R_{\{w_1,w_2\}\cap\{w_2\}} w_2$, but not $w_1 R_{\{w_1,w_2\}} w_2$. □

Since by Lemma 7.30 Principle CM is not a theorem of **CK**+Refl+RM+CC +Or, it is, hence, needed for the axiomatization of System **R**. Observe that Schurz (2012, p.552f) discusses Principle RM in the context of System $\mathbf{P}^+$ (see Section 3.6), which is – as I proved in Theorem 7.29 – logically equivalent to System **R**.

7.2.5 D. Lewis' System V

The main aim of this section is to show that D. Lewis' (1973b) proof-theoretic System **V** is System **R** from the preceding section (see Definition 7.27). This alternative axiomatization of D. Lewis' (1973b) System **V** allows us, then, to describe System **V** in terms of principles from Table 5.1. As a result, one can account for D. Lewis' (1973b) System **V** purely in terms of CS semantics.

I will, then, show that the principles of System **R** are valid in all Lewis models (without centering conditions), which we discussed in Section 3.2 (see Definitions 3.1–3.3), and discuss translations from Lewis models into Chellas models. Both points contribute to a better understanding of the relation of CS semantics and the systems of spheres semantics of D. Lewis (1973b).

Let us now focus on the alternative axiomatization D. Lewis' (1973b) System **V**. D. Lewis does not directly give an axiomatization for his System **V** in terms of a conditional operator, but he does so for System **VC** (p. 132). System **VC** is defined as a specific System **V** plus the both centering Principles MP ("Modus Ponens") and CS ("Conjunctive Sufficiency") (see D. Lewis, 1973b, p. 132, p. 120f). The Principles MP and CS correspond – as we saw in Section 3.2 – to the frame conditions C_{MP} and C_{CS}, respectively. Furthermore, C_{MP} and C_{CS} are conjointly equivalent to the centering condition in Lewis models (see Section 3.2; cf. D. Lewis, 1973b, p. 123, p. 120f). Observe that System **V** is sound and complete w.r.t. the class of all Lewis models described in Definitions 3.2 and 3.3, which do not include the centering conditions C_{MP} and C_{CS} (cf. D. Lewis, 1973b, p. 123, p. 120ff).

D. Lewis (1973b) draws effectively on the following principles for the axiomatization of System **VC** (apart from MP and CM): [4]

LV $(\alpha \boxright \neg\beta) \vee ((\alpha \wedge \beta \boxright \gamma) \leftrightarrow (\alpha \boxright (\beta \rightarrow \gamma))$

LE Exchange of Logical Equivalents

Here 'LV' and 'LE' abbreviate D. Lewis' (1973b) specific axiom for System **V**' and 'Logical Equivalents', respectively. The rule LE is equivalent to conjointly both rules LLE and rule RLE (D. Lewis, 1973b, p. 132). I repeat the rules LLE and RLE from Section 4.2.6 for mnemonic reasons:

LLE if $\vdash \alpha \leftrightarrow \beta$ and $\alpha \boxright \gamma$, then $\beta \boxright \gamma$

RLE if $\vdash \beta \leftrightarrow \gamma$ and $\alpha \boxright \beta$ then $\alpha \boxright \gamma$

Moreover, for an easier formal treatment of System **V**, I divide LV into the following two principles by employing $\text{Def}_{\diamondright}$ (see Section 4.2.1):

LV′ $(\alpha \diamondright \beta) \wedge (\alpha \wedge \beta \boxright \gamma) \rightarrow (\alpha \boxright (\beta \rightarrow \gamma))$

LV″ $(\alpha \diamondright \beta) \wedge (\alpha \boxright (\beta \rightarrow \gamma)) \rightarrow (\alpha \wedge \beta \boxright \gamma)$.

It is easy to show that the conjunction of LV′ and LV″ is p.c. equivalent to LV given $\text{Def}_{\diamondright}$. D. Lewis, then, defines **V** indirectly (as described above) in the following way:

Definition 7.31. *Logic* **V** *is the smallest logic containing LE+RCK+Refl +MOD+LV (D. Lewis, 1973b, p. 132, p. 123, p. 120f).*

Since I will prove that System **V** is logically equivalent with an axiomatization which refers solely to principles in Table 5.1 (Theorem 7.32) – for which I identified non-trivial corresponding frame conditions (see Table 5.3) – I will not discuss non-trivial frame conditions for LV, LV′, or LV″. A further motivation not to pursue that goal stems from the fact that Principles

[4] I omit the definitions of non-primitive operators for the axiomatizations of Systems **V** and **VC** by D. Lewis (1973b), since I will not discuss them here.

LV and LV″ seem hardly intuitive (cf. D. Lewis, 1973b, p. 133) and are seldomly discussed in the conditional logic literature. Hence, I see no point in including frame conditions for LV′ and LV″, as Segerberg (1989, Principle #6, p. 159) does for LV″. Please note that I use the full language $\mathcal{L}_{CL}$ for the proof of the equivalence of System **R** and System **V**. In the restricted language $\mathcal{L}_{CL^-}$ (see Section 4.2.1) employed by Adams (1965, 1966, 1977) D. Lewis' (1973b) systems of spheres semantics is sound and complete with respect to System **P** rather than **R** (Adams, 1977, pp. 188–190), due to the limited expressiveness of $\mathcal{L}_{CL^-}$ (see Section 3.6.1). Let us now prove the equivalence of Systems **VC** and **R**:

Theorem 7.32. *For logics* **V** *and* **R** *holds:* **V**=**R**. *(Gärdenfors, 1979, p. 393)*

Proof. Since by Definition 7.27 and Theorem 7.17 System **R** is LLE+RW+AND+Refl+CM+RM+Or, we have to show that LE+RCK+Refl+MOD+LV is logically equivalent to LLE+RW+AND+Refl+CM+RM+Or. As LE is p.c.-equivalent to LLE+RLE and LV is LV′ +LV″, it follows that System **V** is LLE+RLE+RCK+Refl+MOD+ LV′+LV″. Lemma 4.6 gives us that LLE+RW+AND+LT is logically equivalent to RCK+LLE. Hence, System **V** is LLE+RLE +RW+AND+LT+Refl+MOD+LV′+LV″. Since by Lemmata 4.4 and 7.3 Principles RLE and LT are theorems of Principle RW and RW+Refl, respectively, it follows that System **V** is LLE+RW+AND+Refl+MOD+LV′ +LV″. It remains to be shown that (a) MOD, (b) LV′, and (c) LV″ are derivable from LLE+RW+AND+Refl+CM+RM+Or and that (d) CM, (e) RM, and (f) Or are theorems of LLE+RW+AND+Refl+MOD+LV′+LV″. Lemma 7.26 gives us that MOD is a theorem of LLE+RW+AND+ Refl+CM +Or. Hence, (a) follows. Moreover, Lemma 7.33 shows that LV′ is derivable from Principle S. Since by Lemma 7.21 Principle S is a theorem of LLE+RW +Refl+Or, we obtain (b). Lemma 7.34 implies that LV″ is derivable from RW+AND+Refl+RM. Hence (c) holds as well. Lemma 7.35 gives us that CM is derivable from LLE+RW+AND+Refl+RM+MOD. By Lemma 7.37 RM is derivable from RW+LV″. So, we can get CM and RM from LLE+RW +AND+Refl+MOD+LV″ and, thus, we obtain (d) and (e). Finally, as by

Lemma 7.36 Principle S is a theorem of RW+LV′ and Or is derivable from LLE+RW+AND+S by Lemma 7.20, it follows that Or is a theorem of LLE+RW+AND+LV′. So, (f) results. □

Lemma 7.33. *S ⇒ LV′*

Proof.

1.	$\alpha \diamond\!\!\rightarrow \beta$	given
2.	$\alpha \wedge \beta \,\Box\!\!\rightarrow \gamma$	given
3.	$\alpha \,\Box\!\!\rightarrow (\beta \rightarrow \gamma)$	2, S

□

Lemma 7.34. *(RW+AND)+Refl+RM ⇒ LV″*

Proof.

1.	$\alpha \diamond\!\!\rightarrow \beta$	given
2.	$\alpha \,\Box\!\!\rightarrow (\beta \rightarrow \gamma)$	given
3.	$\alpha \wedge \beta \,\Box\!\!\rightarrow (\beta \rightarrow \gamma)$	2, 1, RM
4.	$\alpha \wedge \beta \,\Box\!\!\rightarrow \alpha \wedge \beta$	Refl
5.	$\alpha \wedge \beta \,\Box\!\!\rightarrow (\beta \rightarrow \gamma) \wedge (\alpha \wedge \beta)$	3, 4, AND
6.	$\alpha \wedge \beta \,\Box\!\!\rightarrow \gamma$	5, RW

□

Lemma 7.35. *(LLE+RW+AND)+Refl+RM+MOD ⇒ CM*

Proof.

1.	$\alpha \,\Box\!\!\rightarrow \gamma$	given
2.	$\alpha \,\Box\!\!\rightarrow \beta$	given
3.	$\alpha \,\Box\!\!\rightarrow \neg\beta$	assumption 1, proof by cases
4.	$\alpha \,\Box\!\!\rightarrow \beta \wedge \neg\beta$	2, 3, AND
5.	$\alpha \,\Box\!\!\rightarrow \neg\alpha$	4, RW
6.	$\neg\neg\alpha \,\Box\!\!\rightarrow \neg\alpha$	5, LLE
7.	$\alpha \wedge \beta \,\Box\!\!\rightarrow \neg\alpha$	6, MOD

8.	$\alpha \wedge \beta \,\Box\!\!\rightarrow \alpha \wedge \beta$	Refl
9.	$\alpha \wedge \beta \,\Box\!\!\rightarrow \neg\alpha \wedge (\alpha \wedge \beta)$	7, 8, AND
10.	$\alpha \wedge \beta \,\Box\!\!\rightarrow \gamma$	9, RW
11.	$\neg(\alpha \,\Box\!\!\rightarrow \neg\beta)$	assumption 2, proof by cases
12.	$\alpha \,\Diamond\!\!\rightarrow \beta$	11, $\mathrm{Def}_{\Diamond\!\rightarrow}$
13.	$\alpha \wedge \beta \,\Box\!\!\rightarrow \gamma$	1, 12, RM
14.	$\alpha \wedge \beta \,\Box\!\!\rightarrow \gamma$	3–10,11–13, proof by cases

□

Lemma 7.36. *(RW)+LV′ ⇒ S*

Proof.

1.	$\alpha \wedge \beta \,\Box\!\!\rightarrow \gamma$	given
2.	$\alpha \,\Box\!\!\rightarrow \neg\beta$	assumption 1, proof by cases
3.	$\alpha \,\Box\!\!\rightarrow (\beta \rightarrow \gamma)$	2, RW
4.	$\neg(\alpha \,\Box\!\!\rightarrow \neg\beta)$	assumption 2, proof by cases
5.	$\alpha \,\Diamond\!\!\rightarrow \beta$	4, $\mathrm{Def}_{\Diamond\!\rightarrow}$
6.	$\alpha \,\Box\!\!\rightarrow (\beta \rightarrow \gamma)$	5, 1, LV′
7.	$\alpha \,\Box\!\!\rightarrow (\beta \rightarrow \gamma)$	2–3, 4–6, proof by cases

□

Lemma 7.37. *(RW)+LV″ ⇒ RM*

Proof.

1.	$\alpha \,\Diamond\!\!\rightarrow \beta$	given
2.	$\alpha \,\Box\!\!\rightarrow \gamma$	given
3.	$\alpha \,\Box\!\!\rightarrow (\beta \rightarrow \gamma)$	2, RW
4.	$\alpha \wedge \beta \,\Box\!\!\rightarrow \gamma$	1, 3, LV″

□

I shall now prove that the rules and axiom schemata and rules of System **R** hold in Lewis models (see Definitions 3.1–3.3). For that purpose I abbreviate truth and falsehood of a formula α at a possible world w in a Lewis model $\mathcal{M}_L = \langle W, R, V \rangle$ by $\models_w^{\mathcal{M}_L} \alpha$ and $\not\models_w^{\mathcal{M}_L} \alpha$, respectively. $\mathrm{V}_{\Box\!\rightarrow}$ in this proof refers to Definition 3.3 (Lewis models).

Theorem 7.38. *The rules and axioms of System* **R** *are valid in all Lewis models.*

Proof. We have to show that the axiom schemata and rules of System **R** hold for all worlds in all Lewis models. By Definition 7.27 and Theorem 7.17 System **R** is LLE+RW+AND+Refl+CM+RM+Or. It is easy to prove that LLE, RW, AND, and Refl are true in all Lewis models. Hence, we only prove that (I) CM, (II) Or, and (III) RM are valid in the class of all Lewis models.

(I) We show that CM, that is $(\alpha \mathrel{\Box\!\!\rightarrow} \gamma) \wedge (\alpha \mathrel{\Box\!\!\rightarrow} \beta) \rightarrow (\alpha \wedge \beta \mathrel{\Box\!\!\rightarrow} \gamma)$, is true at an arbitrary world w in an arbitrary Lewis model $\mathcal{M}_L = \langle W, \$, V \rangle$. Suppose that (a) $\models_w^{\mathcal{M}_L} \alpha \mathrel{\Box\!\!\rightarrow} \gamma$ and (b) $\models_w^{\mathcal{M}_L} \alpha \mathrel{\Box\!\!\rightarrow} \beta$. By (a) and $V_{\Box\!\rightarrow}$ there is either (a1) no α-permitting sphere in $\$_w$ or (a2) there is an α-permitting sphere $S \in \$_w$ such that for all worlds $w' \in S$ it is the case that $\models_{w'}^{\mathcal{M}_L} \alpha \rightarrow \gamma$. In case of (a1), it follows by the definition of α-permitting spheres that there is also no $\alpha \wedge \beta$-permitting sphere in $\$_w$. Hence, by $V_{\Box\!\rightarrow}$ it holds trivially that $\models_w^{\mathcal{M}_L} \alpha \wedge \beta \mathrel{\Box\!\!\rightarrow} \gamma$. Suppose (a2). By (b) it follows that either (b1) no α-permitting sphere exists in $\$_w$ or (b2) there is an α-permitting sphere $S' \in \$_w$ such that for all worlds $w' \in S'$ it is the case that $\models_{w'}^{\mathcal{M}_L} \alpha \rightarrow \beta$. Case (b1) is contradicted by assumption (a2). Hence, (b2) holds. Moreover, by the inclusion requirement b.ii of Definition 3.1 either (A) $S \subseteq S'$ or (B) $S' \subseteq S$. Suppose that (A) is the case. Then, since by (b2) for all worlds $w' \in S'$ it is the case that $\models_{w'}^{\mathcal{M}_L} \alpha \rightarrow \beta$, it follows for all worlds $w'' \in S$ that $\models_{w''}^{\mathcal{M}_L} \alpha \rightarrow \beta$. As by (a2) S is an α-permitting sphere, there is a world $w''' \in S$ such that $\models_{w'''}^{\mathcal{M}_L} \alpha$. As for all worlds $w' \in S$ it is due to (b2) the case that $\models_{w'}^{\mathcal{M}_L} \alpha \rightarrow \beta$, we obtain $\models_{w'''}^{\mathcal{M}_L} \alpha \wedge \beta$. Hence, S is an $\alpha \wedge \beta$-permitting sphere. By (a2) follows that $\models_{w'}^{\mathcal{M}_L} \alpha \rightarrow \gamma$ for all worlds $w' \in S$. As a result, also $\models_{w'}^{\mathcal{M}_L} \alpha \wedge \beta \rightarrow \gamma$ is the case for all worlds $w' \in S$. Thus, since S is an $\alpha \wedge \beta$-permitting sphere, condition $V_{\Box\!\rightarrow}$ implies that $\models_w^{\mathcal{M}_L} \alpha \wedge \beta \mathrel{\Box\!\!\rightarrow} \gamma$. Case (B) is analogous to case (A).

(II) We now show that Or, that is $(\alpha \mathrel{\Box\!\!\rightarrow} \gamma) \wedge (\beta \mathrel{\Box\!\!\rightarrow} \gamma) \rightarrow (\alpha \vee \beta \mathrel{\Box\!\!\rightarrow} \gamma)$, is true at all worlds in all Lewis models. Suppose that (a) $\models_w^{\mathcal{M}_L} \alpha \mathrel{\Box\!\!\rightarrow} \gamma$ and (b) $\models_w^{\mathcal{M}_L} \beta \mathrel{\Box\!\!\rightarrow} \gamma$ hold for an arbitrary world w in an arbitrary Lewis model $\mathcal{M}_L =$

$\langle W,\$,V\rangle$. By (a) and $V_{\Box\!\!\rightarrow}$ there is either (a1) no α-permitting sphere in $\$_w$ or (a2) there is an α-permitting sphere $S \in \$_w$ such that for all worlds $w' \in S$ it is the case that $\models_{w'}^{\mathcal{M}_L} \alpha \rightarrow \gamma$. Moreover, by (b) and $V_{\Box\!\!\rightarrow}$ there is either (b1) no β-permitting sphere in $\$_w$ or (b2) there is a β-permitting sphere $S' \in \$_w$ such that for all worlds $w' \in S'$ it is the case that $\models_{w'}^{\mathcal{M}_L} \beta \rightarrow \gamma$. There are four possible cases: *(i)* (a1) and (b1), *(ii)* (a1) and (b2), *(iii)* (a2) and (b1), and *(iv)* (a2) and (b2). *Case (i)*: Suppose (a1) and (b1). As there is no α-permitting sphere and no β-permitting sphere in $\$_w$, there is no $\alpha \vee \beta$-permitting sphere in $\$_w$ either. Thus, by $V_{\Box\!\!\rightarrow}$ it follows trivially that $\models_{w}^{\mathcal{M}_L} \alpha \vee \beta \mathrel{\Box\!\!\rightarrow} \gamma$. *Case (ii)*: Let (a1) and (b2) be the case. Since by (a1) there is no α-permitting sphere in $\$_w$, it holds that for all worlds $w' \in S'$ that $\models_{w'}^{\mathcal{M}_L} \neg\alpha$. Hence, it follows that $\models_{w'}^{\mathcal{M}_L} \alpha \rightarrow \gamma$ is the case for all worlds $w' \in S'$. Moreover, by (b2) the sphere S' in $\$_w$ is such that for all $w' \in S'$ holds that $\models_{w'}^{\mathcal{M}_L} \beta \rightarrow \gamma$. It follows that for all worlds $w' \in S'$ it is the case that $\models_{w'}^{\mathcal{M}_L} \alpha \vee \beta \rightarrow \gamma$. Since S' is by (b2) a β-permitting sphere and any β-permitting sphere is also an $\alpha \vee \beta$-permitting sphere, we get by $V_{\Box\!\!\rightarrow}$ that $\models_{w}^{\mathcal{M}_L} \alpha \vee \beta \mathrel{\Box\!\!\rightarrow} \gamma$. *Case (iii)*: Analogous to case (ii). *Case (iv)*: Suppose that (a2) and (b2) hold. Then, by (a2) and (b2) there are spheres S and $S' \in \$_w$, such that for all $w' \in S$ and $w'' \in S'$ it is the case that $\models_{w'}^{\mathcal{M}_L} \alpha \rightarrow \gamma$ and $\models_{w''}^{\mathcal{M}_L} \beta \rightarrow \gamma$, respectively. By the inclusion requirement b.ii of Definition 3.1 it follows that either (A) $S' \subseteq S$ or (B) $S \subseteq S'$. Suppose that (A) is the case. Then, since for all worlds $w' \in S$ it is the case that $\models_{w'}^{\mathcal{M}_L} \alpha \rightarrow \gamma$, for all worlds $w'' \in S'$ holds that $\models_{w''}^{\mathcal{M}_L} \alpha \rightarrow \gamma$. As by (b2) it is the case that $\models_{w''}^{\mathcal{M}_L} \beta \rightarrow \gamma$ for all worlds $w'' \in S'$, we get $\models_{w''}^{\mathcal{M}_L} \alpha \vee \beta \rightarrow \gamma$ for all worlds w'' in S'. Since S' is by (b2) a β-permitting sphere, it is also an $\alpha \vee \beta$-permitting sphere. Hence, by $V_{\Box\!\!\rightarrow}$ it follows that $\models_{w}^{\mathcal{M}_L} \alpha \vee \beta \mathrel{\Box\!\!\rightarrow} \gamma$. Case (B) is analogous to case (A).

(III) We shall now show that Principle RM, namely $(\alpha \mathrel{\Box\!\!\rightarrow} \gamma) \wedge (\alpha \mathrel{\Diamond\!\!\rightarrow} \beta) \rightarrow (\alpha \wedge \beta \mathrel{\Box\!\!\rightarrow} \gamma)$, is valid in the class of all Lewis models. Suppose (a) $\models_{w}^{\mathcal{M}_L} \alpha \mathrel{\Box\!\!\rightarrow} \gamma$ and (b) $\models_{w}^{\mathcal{M}_L} \alpha \mathrel{\Diamond\!\!\rightarrow} \beta$ for an arbitrary world w in an arbitrary Lewis model $\mathcal{M}_L = \langle W,\$,V\rangle$. By (a) and $V_{\Box\!\!\rightarrow}$ there is either (a1) no α-permitting sphere in $\$_w$ or (a2) there is an α-permitting sphere $S \in \$_w$ such that for all world $w' \in S$

holds that $\models^{\mathcal{M}_L}_{w'} \alpha \to \gamma$. In case of (a1), it follows that there is also no $\alpha \wedge \beta$-permitting sphere in $\$_w$ by the definition of α-permitting spheres. Hence, due to $V_{\Box\!\!\to}$ this implies trivially that $\models^{\mathcal{M}_L}_{w} \alpha \wedge \beta \Box\!\!\to \gamma$. Suppose that (a2) is the case. By (b) and $\text{Def}_{\Diamond\!\!\to}$ (see Section 4.2.1) follows that $\not\models^{\mathcal{M}_L}_{w} \alpha \Box\!\!\to \neg\beta$. On the basis of Definition $V_{\Box\!\!\to}$ this implies that there is an α-permitting sphere in $\$_w$ and that for all spheres $S' \in \$_w$ there is a world $w' \in S'$, such that $\not\models^{\mathcal{M}_L}_{w'} \alpha \to \neg\beta$. Hence, each sphere $S' \in \$_w$ contains a world w', such that $\models^{\mathcal{M}_L}_{w'} \alpha \wedge \beta$. Thus, every sphere in $\$_w$ is an $\alpha \wedge \beta$-permitting sphere, including sphere S. By (a2) for all worlds $w' \in S$ it is the case that $\models^{\mathcal{M}_L}_{w'} \alpha \to \gamma$. Hence, for all $w' \in S$ it holds that $\models^{\mathcal{M}_L}_{w'} \alpha \wedge \beta \to \gamma$. Since S is an $\alpha \wedge \beta$-permitting sphere, we get by $V_{\Box\!\!\to}$ that $\models^{\mathcal{M}_L}_{w} \alpha \wedge \beta \Box\!\!\to \gamma$. □

Finally, let us discuss translations from Lewis models into Chellas models. I am only interested here in translations from Lewis models $\mathcal{M}_L = \langle W, \$, V \rangle$ into Chellas models $\mathcal{M}_C = \langle W, R, V \rangle$, which guarantee that for all formulas α in language $\mathcal{L}_{\text{CL}}$ and all worlds $w \in W$ it is the case that $\models^{\mathcal{M}_L}_{w} \alpha$ iff $\models^{\mathcal{M}_C}_{w} \alpha$. I will call this type of translation 'point-to-point translation'. My specification of the point-to-point translation draws on the fact that the parameter V in Lewis models $\mathcal{M}_L = \langle W, \$, V \rangle$ and Chellas models $\mathcal{M}_C = \langle W, R, V \rangle$ is defined only for atomic propositions (see Definitions 3.2 and 4.12, respectively). The truth of arbitrary formulas is specified separately in so-called extension of Lewis models (see Definition 3.3) and Chellas models (see Definition 4.13).

One might suspect that one can provide a point-to-point translation from a Lewis model $\mathcal{M}_L = \langle W, \$, V \rangle$ into Chellas models $\mathcal{M}_C = \langle W, R, V \rangle$ the following way: Specify the accessibility relation R for $w, w' \in W$ and $X \subseteq W$ in such a way that wR_Xw' obtains iff $w' \in \text{minS}^w_X$, where minS^w_X is defined as $\{w' \in W \mid \exists S \in \$_w(w' \in S \cap X \wedge \neg\exists S' \in \$_w(w' \in S' \cap X \wedge S' \subset S'))\}$. Here minS^w_X is the minimal sphere $S \in \$_w$, such that $X \cap S$ is non-empty, in case such a sphere exists. In all other cases minS^w_X is empty. I discussed in Section 3.2 minimal α-spheres in a systems of spheres $\$_w$ rather than minimal spheres w.r.t. propositions $X \subseteq W$. A minimal α-sphere in the sense of Section 3.2 concurs with a minimal X-sphere (w.r.t. w) minS^w_X in a Lewis model $\mathcal{M}_L =$

$\langle W,\$,V\rangle$ iff $X = ||\alpha||^{\mathcal{M}_L}$. Here the set $||\alpha||^{\mathcal{M}_L}$ is defined as $\{w \mid V(\alpha,w) = 1\}$ for model $\mathcal{M}_L$ (see Section 4.3).

There are two reasons why there might be no minimal α-sphere in a system of spheres $\$_w$ in a Lewis model $\mathcal{M}_L = \langle W,\$,V\rangle$. First, no α-permitting sphere might exist, viz. there is no sphere S in $\$_w$ such that it contains a world w' at which α is true. Second, no minimal α-sphere might exist in the system of spheres $\$_w$ due to the presence of infinite sequences of α-permitting spheres in $\$_w$ (see Section 3.2).

The infinite sequences of α-permitting spheres are a problem in the above point-to-point translations from Lewis models into Chellas models. The truth condition $V_{\Box\!\!\rightarrow}$ for conditional formulas $\alpha \Box\!\!\rightarrow \beta$ in Definition 4.13 guarantees that if for a world w in a Chellas model $\mathcal{M}_C = \langle W,R,V\rangle$ there are no worlds $w' \in W$ such that $wR_{||\alpha||^{\mathcal{M}_C}}w'$, then $\alpha \Box\!\!\rightarrow \beta$ is trivially true at w. Hence, the above translation would render all conditional formulas $\alpha \Box\!\!\rightarrow \beta$ true at a world w iff there is an infinite descending chain of α-permitting spheres in $\$_w$ in the respective Lewis model $\mathcal{M}_L = \langle W,\$,V\rangle$. The presence of an infinite descending chain of α-permitting spheres in $\$_w$ in a Lewis model $\mathcal{M}_L = \langle W,\$,V\rangle$ for $w \in W$, however, does not guarantee that $\alpha \Box\!\!\rightarrow \beta$ is true at world w: In this infinite sequence of α-permitting spheres each sphere be such that there exists a world w' in each such sphere at which $\alpha \wedge \neg\beta$ is true. In such a case the truth-condition $V_{\Box\!\!\rightarrow}$ in Definition 3.3 renders the conditional $\alpha \Box\!\!\rightarrow \beta$ false at w, while the above translation procedure would make $\alpha \Box\!\!\rightarrow \beta$ true at w. Although there remains much to be said on the relationship of Lewis models [frames] on the one hand and Chellas models [frames] on the other, I am not able pursue this issue further here.

7.2.6 Monotonic Systems without Bridge Principles (Systems CM and M)

In this section I will discuss the two monotonic Systems **CM** and **M**, which can be defined as follows:

Definition 7.39. *Logic* **CM** *is the smallest logic containing* **CK**+*Refl*+*CC*

+*Mon*.

Definition 7.40. *Logic* **M** *is the smallest logic containing* **CK**+*Refl*+*Mon*+*Or*.

System **M** is the monotonic collapse without bridge principles. System **CM**, despite having Mon as a theorem, is not the full monotonic collapse, since Or – as can easily be seen – is not valid in Chellas frames for **CM** (see also Kraus et al., 1990, p. 201). Definitions 7.39 and 7.40 deviate from the definitions of **CM** and **M** given in Kraus et al. (1990, p. 200f) and Kraus et al. (1990, p. 202), respectively, insofar as they are more parsimonious. Kraus et al. (1990, p. 200f) characterize **CM** as all rules and axiom schemata for System **C** (i.e., LLE+RW+AND+Refl+CM+CC; see Theorem 7.9) plus the following rule:

RMon if $\vdash \alpha \rightarrow \beta$ and $\beta \,\Box\!\!\rightarrow \gamma$, then $\alpha \,\Box\!\!\rightarrow \gamma$

One can easily see that – given LLE – RMon is p.c. equivalent to Mon from Table 5.1. Observe for that purpose that $\vdash \alpha \rightarrow \beta$ implies $\vdash \alpha \leftrightarrow (\alpha \wedge \beta)$. In addition, System **C** contains the Principle CM, which can be derived trivially from Mon given the rules of **CK**.

Kraus et al. (1990, p. 202) axiomatize System **M** as System **C** (i.e., LLE+RW+AND+Refl+CM+CC; see Theorem 7.9) plus CP. Kraus et al.'s (1990) axiomatization is redundant, since LLE+RW+AND+CP implies Mon (Lemma 7.45) and Or (Lemma 7.44). In the presence of LLE+RW+AND+Refl Principles Mon and Or give one Principles CM (without proof) and CC (Lemma 7.16), respectively. Hence, System **M** can be alternatively axiomatized as **CK**+Refl+CP. Let us now prove Theorem 7.41, which shows that our axiomatization of System **M** (**CK**+Refl+Mon+Or) and **CK**+Refl+CP are logically equivalent:

Theorem 7.41. *System* **M** *can be axiomatized by* **CK**+*Refl*+*CP (cf. Kraus et al., 1990, p. 202).*

Proof. By Definition 7.40 System **M** is **CK**+Refl+Mon+Or. We have to show that **CK**+Refl+CP is **CK**+Refl+Mon+Or. Lemma 7.42 gives us that

CK+Refl $\Rightarrow$ (Mon+Or $\Leftrightarrow$ CP) holds. It follows that **CK**+Refl+CP=**CK** +Refl+Mon+Or. □

Lemma 7.42. *(LLE+RW+AND)+Refl ⇒ (Mon+Or ⇔ CP)*

Proof. By Lemmata 7.43, 7.44, and 7.45 we get Lemma 7.42. □

Lemma 7.43. *(LLE+RW+AND)+Refl+Mon+Or ⇒ CP*

Proof.

1.	$\alpha \mathrel{\Box\!\!\rightarrow} \beta$	given
2.	$\alpha \wedge \neg\beta \mathrel{\Box\!\!\rightarrow} \beta$	Mon
3.	$\alpha \wedge \neg\beta \mathrel{\Box\!\!\rightarrow} \alpha \wedge \neg\beta$	Refl
4.	$\alpha \wedge \neg\beta \mathrel{\Box\!\!\rightarrow} \beta \wedge (\alpha \wedge \neg\beta)$	2, 3, AND
5.	$\alpha \wedge \neg\beta \mathrel{\Box\!\!\rightarrow} \neg\alpha$	4, RW
6.	$\neg\alpha \mathrel{\Box\!\!\rightarrow} \neg\alpha$	Refl
7.	$\neg\alpha \wedge \neg\beta \mathrel{\Box\!\!\rightarrow} \neg\alpha$	6, Mon
8.	$(\alpha \wedge \neg\beta) \vee (\neg\alpha \wedge \neg\beta) \mathrel{\Box\!\!\rightarrow} \neg\alpha$	5, 7, Or
9.	$\neg\beta \mathrel{\Box\!\!\rightarrow} \neg\alpha$	8, LLE

□

Lemma 7.44. *(LLE+RW+AND)+CP ⇒ Or (Kraus et al., 1990, p. 202)*

Proof.

1.	$\alpha \mathrel{\Box\!\!\rightarrow} \gamma$	given
2.	$\beta \mathrel{\Box\!\!\rightarrow} \gamma$	given
3.	$\neg\gamma \mathrel{\Box\!\!\rightarrow} \neg\alpha$	1, CP
4.	$\neg\gamma \mathrel{\Box\!\!\rightarrow} \neg\beta$	2, CP
5.	$\neg\gamma \mathrel{\Box\!\!\rightarrow} \neg\alpha \wedge \neg\beta$	3, 4, AND
6.	$\neg(\neg\alpha \wedge \neg\beta) \mathrel{\Box\!\!\rightarrow} \neg\neg\gamma$	5, CP
7.	$\alpha \vee \beta \mathrel{\Box\!\!\rightarrow} \neg\neg\gamma$	6, LLE
8.	$\alpha \vee \beta \mathrel{\Box\!\!\rightarrow} \gamma$	7, RW

□

Lemma 7.45. *(LLE+RW)+CP ⇒ Mon (Kraus et al., 1990, p. 180f)*

Proof.

1.	$\alpha \mathrel{\Box\!\!\rightarrow} \gamma$	given
2.	$\neg\gamma \mathrel{\Box\!\!\rightarrow} \neg\alpha$	1, CP
3.	$\neg\gamma \mathrel{\Box\!\!\rightarrow} \neg\alpha \vee \neg\beta$	2, RW
4.	$\neg(\neg\alpha \vee \neg\beta) \mathrel{\Box\!\!\rightarrow} \neg\neg\gamma$	3, CP
5.	$\alpha \wedge \beta \mathrel{\Box\!\!\rightarrow} \neg\neg\gamma$	4, LLE
6.	$\alpha \wedge \beta \mathrel{\Box\!\!\rightarrow} \gamma$	5, RW

□

Model Theory

Let us now focus on the semantic characterization of System **M**. This system can, as Theorem 7.46 shows, be described by the following frame condition:

$$C_{\mathrm{M}} \quad \forall X \forall w, w'(wR_X w' \Leftrightarrow \exists Y(wR_Y w') \wedge w' \in X)$$

Applied to syntactically representable sets of possible worlds, frame condition C_{M} gives the following intuitive reading: If a world w can see a world w' in a Chellas model $\mathcal{M}_C = \langle W, R, V\rangle$ by R_Y with Y being a subset of W, then, for all formulas α holds: α is true at w' (i.e., $w' \in ||\alpha||$) if and only if it is the case that $wR_{||\alpha||}w'$. Let us now turn to Theorem 7.46:

Theorem 7.46. *The class of Chellas frames for System* **M** *can be characterized by the frame condition* C_{M}.

Proof. By Definition 7.40, logic **M** is **CK**+Refl+Mon+Or. Hence, all theorems of **M** are valid on a Chellas frame $\mathcal{F}_C = \langle W, R\rangle$ iff C_{Refl}, C_{Mon}, and C_{Or} hold for $\mathcal{F}_C$. Hence, we have to show that C_{Refl}, C_{Mon}, and C_{Or} imply C_{M}, and vice versa. Let us first focus on (1) $C_{\mathrm{Refl}}+C_{\mathrm{Mon}}+C_{\mathrm{Or}} \Rightarrow C_{\mathrm{M}}$ and then show (2) $C_{\mathrm{M}} \Rightarrow C_{\mathrm{Refl}}+C_{\mathrm{Mon}}+C_{\mathrm{Or}}$. I will abbreviate the left-to-right and the right-to-left direction of C_{M} by '$C_{\mathrm{M}}^{\Rightarrow}$' and '$C_{\mathrm{M}}^{\Leftarrow}$', respectively.

(1) $C_{\mathrm{Refl}}+C_{\mathrm{Mon}}+C_{\mathrm{Or}} \Rightarrow C_{\mathrm{M}}^{\Rightarrow}$: Suppose that worlds w, w' are arbitrary worlds in a Chellas frame $\mathcal{F}_C = \langle W, R\rangle$ and X is an arbitrary subset of W, such that

wR_Xw'. It trivially holds that there is a subset Y of W, such that wR_Yw', namely X. Moreover, by C_{Refl} follows that $w' \in X$.

$C_{\text{Refl}}+C_{\text{Mon}}+C_{\text{Or}} \Rightarrow C_{\text{M}}^{\Leftarrow}$: Suppose that worlds w, w' are arbitrary worlds in a Chellas frame $\mathcal{F}_C = \langle W,R\rangle$ and let X, Y be arbitrary subsets of W, such that wR_Yw' and $w' \in X$. By C_{Or} follows that either $wR_{X\cap Y}w'$ or $wR_{-X\cap Y}w'$. $wR_{-X\cap Y}w'$ cannot be the case, since otherwise C_{Refl} implies that $w' \in -X \cap Y$, contradicting $w' \in X$. Hence, $wR_{X\cap Y}w'$ is the case. From this observation follows by C_{Mon} that wR_Xw'.

(2) $C_{\text{M}} \Rightarrow C_{\text{Refl}}$: Suppose that worlds w, w' are arbitrary worlds in a Chellas frame $\mathcal{F}_C = \langle W,R\rangle$ and let X be an arbitrary subset of W, such that wR_Xw'. Then, by C_{M}, it follows that $\exists Y(wR_Yw') \wedge w' \in X$. Hence, $w' \in X$ is the case.

$C_{\text{M}} \Rightarrow C_{\text{Mon}}$: Suppose that worlds w, w' are arbitrary worlds in a Chellas frame $\mathcal{F}_C = \langle W,R\rangle$ and let X, Y be arbitrary subsets of W, such that $wR_{X\cap Y}w'$. By C_{M} follows that $w' \in X \cap Y$. So $w' \in X$. Since there is a subset Z, namely $Z = X \cap Y$, such that wR_Zw', this implies by C_{M} that wR_Xw'.

$C_{\text{M}} \Rightarrow C_{\text{Or}}$: Suppose that w, w' are arbitrary worlds in a Chellas frame $\mathcal{F}_C = \langle W,R\rangle$ and let X, Y be arbitrary subsets of W, such $wR_{X\cup Y}w'$. By C_{M} it follows that $w' \in X \cup Y$. So, either $w' \in X$ or $w' \in Y$ is the case. Since $\exists Z(wR_Zw')$, namely for $Z = X \cup Y$, C_{M} implies that either wR_Xw' or wR_Yw'. □

Let us now focus on the Principle CEM ("Conditional Excluded Middle", see Table 5.1):

CEM $(\alpha \mathrel{\Box\!\!\rightarrow} \beta) \vee (\alpha \mathrel{\Box\!\!\rightarrow} \neg\beta)$

CEM is the characteristic principle of Stalnaker's system (see Sections 3.3.6 and 7.3.3). It is somewhat surprising that it is not derivable in System **M**, since it is not a bridge principle. Let us for the sake of perspicuity repeat the non-trivial frame condition corresponding to CEM (see Table 5.3) here:

C_{CEM} $\forall X \forall w,w',w''(wR_Xw' \wedge wR_Xw'' \Rightarrow w'' = w')$

The frame condition C_{CEM} requires that each accessibility relation R_X in a Chellas frame $\mathcal{F}_C = \langle W,R\rangle$ is functional in the sense that any world $w \in W$

can access at most one world $w' \in W$ via R_X. In an ordinary Chellas frame $\mathcal{F}_C = \langle W, R \rangle$ the accessibility relation R_X need not be unique, i.e., there may exist worlds w, w', and w'' and a subset of possible worlds X of W, such that wR_Xw', wR_Xw'', but $w'' \neq w'$. Lemma 7.47 gives us the corresponding formal result.

Lemma 7.47. *In System* **M** *the Principle CEM is not derivable.*

Proof. To prove Lemma 7.47, we construct a Chellas frame $\mathcal{F}_C = \langle W, R \rangle$, such that C_{M}, but not C_{CEM} holds (cf. Lemma 7.30).

Let $\mathcal{F}_C = \langle W, R \rangle$ be a Chellas frame, such that $W = \{w_1, w_2, w_3\}$. Moreover, let R be $\{\langle w_1, w_2, \{w_2\}\rangle, \langle w_1, w_2, \{w_1, w_2\}\rangle, \langle w_1, w_2, \{w_2, w_3\}\rangle, \langle w_1, w_2, \{w_1, w_2, w_3\}\rangle, \langle w_1, w_3, \{w_3\}\rangle, \langle w_1, w_3, \{w_1, w_3\}\rangle, \langle w_1, w_3, \{w_2, w_3\}\rangle, \langle w_1, w_3, \{w_1, w_2, w_3\}\rangle\}$. It is easy to check that C_{M} is valid for $\mathcal{F}_C$. C_{CEM} does not hold for $\mathcal{F}_C$, since, for example, $wR_{\{w_2,w_3\}}w_2$ and $wR_{\{w_2,w_3\}}w_3$, but $w_2 \neq w_3$. □

The proof of Lemma 7.47 also shows that CEM is not valid in the semantics for System **M**. Principle CEM is derivable in System **MC** (see Lemma 7.83), as described by Definition 7.76.

Proof Theory

Let us now turn to a further proof-theoretic discussion of Systems **M** and **CM**. We already observed that both Systems **M** and **CM** are monotonic in the sense that Mon is a theorem of both systems. Only System **M** represents the monotonic collapse without bridge principles. Since I identify monotonic systems with systems in which Mon is derivable, one can construct weaker monotonic systems in that sense, for example **CK**+Mon and **CK**+Refl+Mon.

For the proof-theoretic discussion of Systems **M** and **CM** I focus on Principles Trans (Principle S3′ from Chapter 1) and Cut (see Table 5.1). We encountered Principles Trans and Cut in Section 7.2.1, but I repeat both principles for the sake of perspicuity:

Trans $(\alpha \mathrel{\Box\!\!\rightarrow} \beta) \wedge (\beta \mathrel{\Box\!\!\rightarrow} \gamma) \rightarrow (\alpha \mathrel{\Box\!\!\rightarrow} \gamma)$
Cut $(\alpha \wedge \beta \mathrel{\Box\!\!\rightarrow} \delta) \wedge (\gamma \mathrel{\Box\!\!\rightarrow} \beta) \rightarrow (\alpha \wedge \gamma \mathrel{\Box\!\!\rightarrow} \delta)$

Principle Cut does not correspond to "Cut" in Kraus et al. (1990, p. 177) and Lehmann and Magidor (1992, p. 6), the latter of which corresponds to our notion of Cautious Cut, denoted by 'CC'. Our notion of Cut translates into Principle (9) in Lehmann and Magidor (1992, p. 6).

In Table 5.1 I classified the Principles Cut, Mon, Trans, and CP under the label 'Monotonic Systems'. I gave a justification for this terminology only for Mon (which is trivially derivable in a monotonic system) and CP, which implies Mon (see Lemma 7.45) given System **CK**. Hence, it remains to be shown that Cut and Trans are monotonic in the above sense, namely that Mon is derivable from both Trans and CP. I do so in Lemma 7.51 and Lemma 7.56, respectively. The proofs for Lemma 7.51 and Lemma 7.56 draw on **CKR** (= **CK**+Refl) rather than only on **CK**. As Refl is regarded as a centerpiece of almost all conditional logics (see Section 7.1.1), one should arguably categorize both Principles Trans and Cut also as monotonic principles. One can even prove a stronger result based on Lemmata 7.51 and 7.56, namely that System **CM** can alternatively be axiomatized as **CK**+Refl+Trans+CC (see Theorem 7.48) and **CK**+Refl+Cut (see Theorem 7.52).

Theorem 7.48. *System* **CM** *can be axiomatized by* **CK**+*Refl+Trans+CC.*

Proof. By Definition 7.39 holds that **CM**=**CK**+Refl+Mon+CC. Lemma 7.49 establishes, then, that Mon and Trans can be derived from each other given **CK**+Refl+CC. □

Lemma 7.49. *(RW)+Refl+CC* $\Rightarrow$ *(Trans* $\Leftrightarrow$ *Mon)*

Proof. By Lemmata 7.50 and 7.51 we obtain Lemma 7.49. □

Lemma 7.50. *Mon+CC* $\Rightarrow$ *Trans*

Proof.

1. $\alpha \mathrel{\Box\!\!\rightarrow} \beta$ given
2. $\beta \mathrel{\Box\!\!\rightarrow} \gamma$ given

3.	$\alpha \wedge \beta \;\square\!\!\rightarrow \gamma$	2, Mon
4.	$\alpha \;\square\!\!\rightarrow \gamma$	3, 1, CC

□

Lemma 7.51. *(RW)+Refl+Trans ⇒ Mon*

Proof.

1.	$\alpha \;\square\!\!\rightarrow \gamma$	given
2.	$\alpha \wedge \beta \;\square\!\!\rightarrow \alpha \wedge \beta$	Refl
3.	$\alpha \wedge \beta \;\square\!\!\rightarrow \alpha$	2, RW
4.	$\alpha \wedge \beta \;\square\!\!\rightarrow \gamma$	3, 1, Trans

□

Let us now focus on the second alternative axiomatization of System **CM**:

Theorem 7.52. *System* **CM** *can be axiomatized by* **CK**+*Refl*+*Cut.*

Proof. By Definition 7.39 holds that **CM**=**CK**+Refl+Mon+CC. Lemma 7.53 gives us that Mon and CC on the one hand and Cut on the other are derivable from each other given **CK**+Refl. □

Lemma 7.53. *(LLE+RW)+Refl ⇒ (Mon+CC ⇔ Cut)*

Proof. By Lemmata 7.54–7.56 we get Lemma 7.53 □

Lemma 7.54. *(LLE)+Mon+CC ⇒ Cut*

Proof.

1.	$\alpha \wedge \beta \;\square\!\!\rightarrow \delta$	given
2.	$\gamma \;\square\!\!\rightarrow \beta$	given
3.	$(\alpha \wedge \beta) \wedge \gamma \;\square\!\!\rightarrow \delta$	1, Mon
4.	$(\alpha \wedge \gamma) \wedge \beta \;\square\!\!\rightarrow \delta$	3, LLE
5.	$\gamma \wedge \alpha \;\square\!\!\rightarrow \beta$	2, Mon
6.	$\alpha \wedge \gamma \;\square\!\!\rightarrow \beta$	5, LLE
7.	$\alpha \wedge \gamma \;\square\!\!\rightarrow \delta$	4, 6, CC

□

Lemma 7.55. *(LLE)+Cut* $\Rightarrow$ *CC*

Proof.

1.	$\alpha \wedge \beta \,\Box\!\!\rightarrow \gamma$	given
2.	$\alpha \,\Box\!\!\rightarrow \beta$	given
3.	$\alpha \wedge \alpha \,\Box\!\!\rightarrow \gamma$	1, 2, Cut
4.	$\alpha \,\Box\!\!\rightarrow \gamma$	3, LLE

$\Box$

Lemma 7.56. *(LLE+RW)+Refl+Cut* $\Rightarrow$ *Mon (cf. Lehmann & Magidor, 1992, p. 6)*

Proof.

1.	$\alpha \,\Box\!\!\rightarrow \gamma$	given
2.	$\alpha \wedge \top \,\Box\!\!\rightarrow \gamma$	1, LLE
3.	$\beta \,\Box\!\!\rightarrow \beta$	Refl
4.	$\beta \,\Box\!\!\rightarrow \top$	3, RW
5.	$\alpha \wedge \beta \,\Box\!\!\rightarrow \gamma$	2, 4, Cut

$\Box$

7.3 Conditional Logics with Bridge Principles

In this section I will discuss conditional logic systems, which contain bridge principles in the sense of Section 4.2.1. For that purpose I will explicitly address here the Principles MP, CS, Det, Cond, VEQ, and EFQ from Tables 5.1 and 5.2. In addition, I give an account of Adams' (1965, 1966, 1977) original System **P*** (see also Section 3.6.1), D. Lewis' (1973b) System **VC** (see also Section 3.2) and Stalnaker's System **S** (1968; Stalnaker & Thomason, 1970; see also Section 3.3.3) in terms of CS semantics. I will finally describe a frame condition for CS semantics which characterizes System **MC**, the monotonic collapse system with bridge principles.

7.3.1 Adams' System P*

Before we discuss Adams' (1965, 1966, 1977) original System **P***, let me repeat Principles MP ("Modus Ponens"), CS ("Conditional Sufficiency"),

Det ("Detachment"), and Cond ("Conditionalization") from Table 5.1 (we discussed Principles MP and CS already in Sections 3.2 and 3.4.2):

MP $(\alpha \mathrel{\Box\!\!\rightarrow} \beta) \rightarrow (\alpha \rightarrow \beta)$
CS $\alpha \wedge \beta \rightarrow (\alpha \mathrel{\Box\!\!\rightarrow} \beta)$
Det $(\top \mathrel{\Box\!\!\rightarrow} \alpha) \rightarrow \alpha$
Cond $\alpha \rightarrow (\top \mathrel{\Box\!\!\rightarrow} \alpha)$

The Principles MP, CS, Det, and Cond are bridge principles in the sense of our definition of bridge principles in Section 4.2.1. Bridge principles presuppose a fixed relationship between conditional and non-conditional facts (see Section 3.4.2). Non-trivial frame conditions corresponding to the Principles MP, CS, Det, and Cond can be found in Table 5.3.

The Principles Det and Cond are not as weak as they might seem at first. In fact, Det and Cond are logically equivalent to MP and CS, given the rules and axiom schemata of System **P** of Kraus et al. (1990; see Section 7.2.3), as Theorem 7.63 shows. To establish this result we first prove Lemmata 7.57 and 7.60:

Lemma 7.57. *(LLE)+S $\Rightarrow$ (MP $\Leftrightarrow$ Det)*

Proof. Lemmata 7.58 and 7.59 give us Lemma 7.57. □

Lemma 7.58. *(LLE)+S+Det $\Rightarrow$ MP (Segerberg, 1989, Theorem 1.2.x, p. 159; cf. Adams, 1998, p. 157f)*

Proof.

1.	$\alpha \mathrel{\Box\!\!\rightarrow} \beta$	given
2.	$\top \wedge \alpha \mathrel{\Box\!\!\rightarrow} \beta$	1, LLE
3.	$\top \mathrel{\Box\!\!\rightarrow} (\alpha \rightarrow \beta)$	2, S
4.	$\alpha \rightarrow \beta$	3, Det

□

Lemma 7.59. *MP ⇒ Det*

Proof.

1.	$\top \mathrel{\Box\!\!\to} \alpha$	given
2.	$(\top \mathrel{\Box\!\!\to} \alpha) \to (\top \to \alpha)$	MP
3.	α	2, 1, p.c.

□

Principles Cond and CS are logically equivalent given **CK**+CM:

Lemma 7.60. *(LLE+RW)+CM ⇒ (CS ⇔ Cond)*

Proof. By Lemmata 7.61 and 7.62 Lemma 7.60 results. □

Lemma 7.61. *(LLE+RW)+CM+Cond ⇒ CS (cf. Adams, 1998, p. 157)*

Proof.

1.	$\alpha \wedge \beta$	given
2.	$\top \mathrel{\Box\!\!\to} \alpha \wedge \beta$	1, Cond
3.	$\top \mathrel{\Box\!\!\to} \alpha$	2, RW
4.	$\top \wedge \alpha \mathrel{\Box\!\!\to} \alpha \wedge \beta$	2, 3, CM
5.	$\alpha \mathrel{\Box\!\!\to} \alpha \wedge \beta$	4, LLE
6.	$\alpha \mathrel{\Box\!\!\to} \beta$	5, RW

□

Lemma 7.62. *CS ⇒ Cond*

Proof.

1.	α	given
2.	$\top \wedge \alpha$	1, p.c.
3.	$\top \mathrel{\Box\!\!\to} \alpha$	2, CS

□

We can now prove the following theorem:

Theorem 7.63. *Given System* **P**, *(a) MP and Det and (b) CS and Cond are logically equivalent.*

Proof. By Definition 7.15 it is the case that **P**=**CK**+Refl+CM+Or. Since **CK**=LLE+RW+AND+LT and by Lemma 7.19 **CK**+Refl+Or implies S, we obtain (a) by Lemma 7.57. Moreover, (b) is the case due to Definition 7.15 and Lemma 7.60. □

Let us now connect the present discussion to the earlier points I made in Chapter 1 and 3 on bridge principles for indicative and counterfactual conditionals. I argued in Section 3.4 that the Bridge Principles MP and CS are not warranted for a logic of indicative conditional logics. We saw in particular that Principle MP is problematic, although it seems to be prima facie intuitive. One reason for rejecting MP is the observation that MP does not allow us to adequately describe normic conditionals, such as 'Fishes are normally cold-blooded' (see Sections 1.4 and 3.4.2). Since Det and Cond are logically equivalent to MP and E, respectively, on the basis of the rules and axiom schemata of System **P** (see Theorem 7.63), our arguments extend to systems which contain Det and Cond and are at least as strong as System **P**.

What about counterfactual conditionals? Let us first repeat the criteria for counterfactual conditionals discussed in Section 3.4.1. We saw in these sections that there are at least two approaches: In the first approach (approach A) the difference between indicative and counterfactual conditionals is located in the mood of the conditional sentence. The presence of the subjunctive mood (and the absence of the indicative mood) indicate(s) that a conditional is a counterfactual. In the second approach (approach B) – which I endorse here – counterfactuals are conceptualized as being "counter to the facts". W.r.t. approach (B) one can further distinguish between between (B1) "genuine" counterfactuals and (B2) tentative counterfactuals. Counterfactuals of the first type are conditionals whose antecedent is false, whereas counterfactuals of type (B2) presuppose that their antecedents are merely improbable.

In Section 3.4.1 we saw that given an interpretation of counterfactuals in terms of (B1), Principles MP and CS are warranted. On the other hand it is not clear that the interpretation of counterfactuals in terms of (B2) makes MP and CS valid principles. Observe that our arguments extend again to Principles Det and Cond for systems which endorse the rules and axiom schemata of System **P**.

Let us now focus on the axiomatization of Adams' (1965, 1966, 1977) original System **P** (which we call 'System **P***'). Adams' System **P*** differs from Kraus et al.'s (1990) System **P** insofar as it additionally contains the Bridge Principle PC2 (see Adams, 1966, p. 277; see also Adams, 1965, p. 189), which can be formulated in a logically-equivalent way by the conjunction of both Det and Cond. Since Kraus et al. (1990) do not employ bridge principles in their axiomatization of their System **P**, arguably both versions of System **P** differ. I will, henceforth, call Adams' original system 'System **P***'. Adams' (1965, 1966, 1977) axiomatization of System **P***, then, draws in addition on the following two axiom schemata (e.g., Adams, 1966, p. 277; Adams, 1965, p. 189):

ED $(\alpha \vee \beta \mathrel{\Box\!\!\rightarrow} \gamma) \wedge (\beta \mathrel{\Box\!\!\rightarrow} \neg\gamma) \rightarrow (\alpha \mathrel{\Box\!\!\rightarrow} \gamma)$

RW′ $(\alpha \mathrel{\Box\!\!\rightarrow} \beta \wedge \gamma) \rightarrow (\alpha \mathrel{\Box\!\!\rightarrow} \beta)$

'ED' stands for 'elimination of disjunctions' and RW′ represents a variation of Right Weakening (my own terminology). Principle RW′ is logically equivalent to Principle CW ("consequence weakening") described in Section 4.2.6 – as one can easily prove – provided RLE holds (see also Section 4.2.6). Let us now turn to Adams' axiomatization of System **P***. Adams (1965, 1966, 1977) defines System **P***, then, in the following way:

Definition 7.64. *Adams' (1965, 1966, 1977) probabilistic conditional logic* **P*** *is the smallest logic containing LLE+RW′+AND+SC+ED+CC+Or+Det +Cond (Adams, 1965, p. 189; Adams, 1966, p. 277).*

We will see that System **P*** is in fact System **P** of Kraus et al. (1990) and Lehmann and Magidor (1992) plus the Bridge Principles MP and CS (Theorem 7.69). In order to make the proof more perspicuous I shall first prove that **P*** without the Bridge Principles Det and Cond is logically equivalent to **P** (Theorem 7.65).

Theorem 7.65. *System* **P** *(Kraus et al., 1990; Lehmann & Magidor, 1992) is LLE+RW′+AND+SC+ED+CC+Or.*

Proof. By Theorem 7.18 System **P** can be axiomatized by LLE+SC+CM+CC+Or. So, it remains to be shown that (a) RW′, (b) AND, and (c) ED are derivable from LLE+SC+CM+CC+Or and that (d) CM is a theorem of LLE+RW′+AND+SC+ED+CC+Or.

RW is derivable from SC+CC by Lemma 7.12. Since by Lemma 7.66 RW′ is derivable from RW, we obtain (a). Lemma 7.10 gives us that AND is a theorem of RW+Refl+CM+CC. By Lemma 7.7 it holds that Refl can be derived from SC and, hence, that AND is a theorem of RW+SC+CM+CC. As RW is derivable from SC+CC by Lemma 7.12, (b) results. Lemma 7.67 shows that ED is a theorem of LLE+RW+AND+SC+CM+Or. Since RW can be derived from SC+CC by Lemma 7.12 and AND can be obtained from LLE+SC+CM+CC+Or by (b), we get that ED is a theorem of LLE+SC +CM+CC+Or as well. Finally, Lemma 7.68 shows that CM can be derived from LLE+RW+AND+SC+ED. Since Lemma 7.12 shows that RW can be obtained from SC+CC, (d) follows.

□

Lemma 7.66. *RW ⇒ RW′*

Proof.

1.	$\alpha \mathrel{\Box\!\!\rightarrow} \beta \wedge \gamma$	given
2.	$\alpha \mathrel{\Box\!\!\rightarrow} \beta$	1, RW

□

Lemma 7.67. *(LLE+RW+AND)+SC+CM+Or ⇒ ED*

Proof.

1.	$\alpha \vee \beta \mathrel{\Box\!\!\rightarrow} \gamma$	given
2.	$\beta \mathrel{\Box\!\!\rightarrow} \neg\gamma$	given
3.	$\beta \mathrel{\Box\!\!\rightarrow} \neg\gamma \vee \alpha$	2, RW
4.	$\vdash \alpha \rightarrow \neg\gamma \vee \alpha$	p.c.
5.	$\alpha \mathrel{\Box\!\!\rightarrow} \neg\gamma \vee \alpha$	4, SC
6.	$\alpha \vee \beta \mathrel{\Box\!\!\rightarrow} \neg\gamma \vee \alpha$	5, 3, Or
7.	$\alpha \vee \beta \mathrel{\Box\!\!\rightarrow} \gamma \wedge (\neg\gamma \vee \alpha)$	1, 6, AND
8.	$\alpha \vee \beta \mathrel{\Box\!\!\rightarrow} \alpha$	7, RW

9. $(\alpha \vee \beta) \wedge \alpha \; \Box\!\!\rightarrow \gamma$ — 1, 8, CM
10. $\alpha \; \Box\!\!\rightarrow \gamma$ — 9, LLE

□

Lemma 7.68. *(LLE+RW+AND)+SC+ED ⇒ CM*

Proof.

1. $\alpha \; \Box\!\!\rightarrow \gamma$ — given
2. $\alpha \; \Box\!\!\rightarrow \beta$ — given
3. $\alpha \; \Box\!\!\rightarrow \beta \wedge \gamma$ — 2, 1, AND
4. $(\alpha \wedge \beta) \vee (\alpha \wedge \neg\beta) \; \Box\!\!\rightarrow \beta \wedge \gamma$ — 3, LLE
5. $\vdash \alpha \wedge \neg\beta \rightarrow \neg(\beta \wedge \gamma)$ — p.c.
6. $\alpha \wedge \neg\beta \; \Box\!\!\rightarrow \neg(\beta \wedge \gamma)$ — 5, SC
7. $\alpha \wedge \beta \; \Box\!\!\rightarrow \beta \wedge \gamma$ — 4, 6, ED
8. $\alpha \wedge \beta \; \Box\!\!\rightarrow \gamma$ — 7, RW

□

Theorem 7.69. *System* **P*** *(Adams, 1965, 1966, 1977) is* **P**+*MP+CS, where System* **P** *is defined as in Definition 7.2.3.*

Proof. By Definition 7.64 System **P*** is LLE+RW′+AND+SC+ED+CC+Or +Det+Cond. Theorem 7.65 gives us that System **P** is LLE+RW′+AND+SC +ED+CC+Or. Thus, System **P*** is **P**+Det+Cond. Theorem 7.63 implies that Det and MP on the one hand and Cond and CS are logically equivalent given the rules and axiom schemata of System **P**. Hence, it follows that System **P*** is **P**+MP+CS. □

7.3.2 D. Lewis' System VC

In Section 3.2 I discussed D. Lewis' (1973b) systems of spheres semantics (see Definitions 3.1–3.3). Apart from this semantics description, D. Lewis (1973b) also provides a proof-theoretic characterization, System **VC**, of his systems of spheres. This proof-theoretic System **VC** is sound and complete w.r.t. Lewis models as characterized by Definitions 3.1–3.3 with the centering conditions described in Section 3.2. In this section I will provide an alternative characterization of D. Lewis' (1973b) proof-theoretic System **VC** in terms of CS semantics, which allows one to represent D. Lewis' proof-theoretic system in terms of principles from Tables 5.1 and 5.2.

D. Lewis' (1973b) interprets his conditional System **VC** primarily as a system for counterfactual conditionals. The objective interpretation of CS semantics described in Section 7.1.1 can thus be used as an alternative to D. Lewis' own similarity ranking of possible worlds for counterfactuals (Section 3.2). Let us now characterize D. Lewis' (1973b) System **VC**:

Definition 7.70. *D. Lewis' counterfactual System* **VC** *is the smallest logic containing V+MP+CS (D. Lewis, 1973b, p. 132).*

Theorem 7.71. VC=R+*MP+CS*

Proof. Definitions 7.27, 7.70, and Theorem 7.32 give us that **V=R**. □

Theorem 7.71 holds if one uses the full language $\mathcal{L}_{\mathrm{CL}}$. If one restricts the language to $\mathcal{L}_{\mathrm{CL}^-}$ (see Section 4.2.1), D. Lewis' (1973b) System **VC** becomes logically equivalent to System **P*** (see Adams, 1977, pp. 188–190; see Section 3.6.1).

7.3.3 Stalnaker's System S

In this section I will discuss the conditional logic system of Stalnaker and Thomason (Stalnaker, 1968; Stalnaker & Thomason, 1970). I described the basic semantic notion of the Stalnaker and Thomason's system, namely Stalnaker models (see Definition 3.7). In Sections 3.3.2–3.3.6 I furthermore discussed the intuitions underlying Stalnaker and Thomason's conditional logic and its relation to the proposed conditional consistency criterion in the Ramsey test (cf. Sections 3.3.1 and 3.3.2).

In this section I will focus on the proof-theoretic side of Stalnaker and Thomason's conditional logic, which I will call 'System **S**'. I will, then, provide an alternative axiomatization in terms of System **CK** (Definition 4.1) plus principles from Tables 5.1 and 5.2. By these means, one can account for Stalnaker and Thomason's system in terms of CS semantics.

My description of System **S** does not directly refer to Stalnaker (1968) nor to Stalnaker and Thomason (1970), but draws on D. Lewis' (1973b, p. 132f) proof-theoretic axiomatization of System **S**. This is due to the fact that the axiomatization of System **S** by D. Lewis (1973b) is more perspicuous than the axiomatizations by Stalnaker (1968) and Stalnaker and Thomason (1970). I will – unlike Stalnaker and Thomason (1970) – only focus here on the propositional part of System **S**. Let us now describe the characteristic principle of System **S**, namely CEM (Table 5.1; see also Section 7.2.6):

CEM $(\alpha \mathrel{\Box\!\!\rightarrow} \beta) \vee (\alpha \mathrel{\Box\!\!\rightarrow} \neg\beta)$

'CEM' stands for 'Conditional Excluded Middle'. D. Lewis (1973b, p. 79 and p. 133). D. Lewis (1973b), then, defines System **S** as follows:

Definition 7.72. *Stalnaker's (1968) and Stalnaker and Thomason's (1970) indicative/counterfactual Conditional Logic System* **S** *is System* **V**+*MP*+*CS*+ *CEM (D. Lewis, 1973b, p. 132f).*

Theorem 7.73. **S**=**P**+*MP*+*CEM*

Proof. By Definition 7.72 System **S** is **V**+MP+CS+CEM. Theorem 7.32 gives us that **V**=**R**. As System **R** is **P**+RM by Definition 7.27, it follows that System **S** is **P**+RM+MP+CS+CEM, where System **P** is LLE+RW+AND+ Refl+CM+Or by Theorem 7.17. Since by Lemma 7.74 CM and CEM imply RM, this yields that **S** is **P**+MP+CS+CEM. Lemma 7.75 gives us that CS is derivable from MP+CEM. Hence, it follows that **S** = **P**+MP+CEM. □

We prove now the remaining lemmata, on which Theorem 7.73 draws:

Lemma 7.74. *CM+CEM* $\Rightarrow$ *RM*

Proof.

1.	$\alpha \mathrel{\Box\!\!\rightarrow} \gamma$	given
2.	$\alpha \mathrel{\Diamond\!\!\rightarrow} \beta$	given

3. $\neg(\alpha \mathrel{\Box\!\!\rightarrow} \neg\beta)$ 2, Def$_{\Diamond\!\rightarrow}$
4. $\alpha \mathrel{\Box\!\!\rightarrow} \beta$ 3, CEM
5. $\alpha \wedge \beta \mathrel{\Box\!\!\rightarrow} \gamma$ 1,4, CM

□

Lemma 7.75. *MP+CEM ⇒ CS*

Proof.

1. $\alpha \wedge \beta$ given
2. $\neg(\alpha \rightarrow \neg\beta)$ 1, p.c.
3. $\neg(\alpha \mathrel{\Box\!\!\rightarrow} \neg\beta)$ 2, MP
4. $\alpha \mathrel{\Box\!\!\rightarrow} \beta$ 3, CEM

□

7.3.4 The Material Collapse System MC

The material collapse System **MC** is defined the following way:

Definition 7.76. *The material collapse System* **MC** *is the smallest logic containing* **CK**+*MP*+*VEQ*+*EFQ*.

We will give a model-theoretic characterization of System **MC** and then discuss the proof-theoretic basis of **MC**.

Model Theory

In this section we will give a semantic characterization of the full material collapse System **MC** in terms of CS semantics (Theorem 7.77). We draw for that purpose on following frame condition:

C_{MC} $\forall X \forall w, w'(wR_X w' \Leftrightarrow w' = w \wedge w \in X)$

Frame condition C_{MC} has – applied to syntactically representable sets – the following intuitive reading: If a formula α is true at a world w in a Chellas model, then w sees only itself via accessibility relation $R_{\|\alpha\|}$. If α is false at w , w cannot see any world via $R_{\|\alpha\|}$. Let us now turn to Theorem 7.77:

Theorem 7.77. *The class of Chellas frames for System* **MC** *can be characterized by the frame condition* C_{MC}.

Proof. By Definition 7.76, logic **MC** is **CK** plus Principles MP, VEQ, and EFQ. Hence, all theorems of **MC** are valid on a Chellas frame $\mathcal{F}_C = \langle W, R\rangle$ iff C_{MP}, C_{VEQ}, and C_{EFQ} hold for $\mathcal{F}_C$. Hence, we have to show that C_{MP}, C_{VEQ}, and C_{EFQ} imply C_{MC}, and vice versa. We first prove (1) $C_{\mathrm{MP}}+C_{\mathrm{VEQ}}+C_{\mathrm{EFQ}} \Rightarrow C_{\mathrm{MC}}$ and, then, show (2) $C_{\mathrm{MC}} \Rightarrow C_{\mathrm{MP}}+C_{\mathrm{VEQ}}+C_{\mathrm{EFQ}}$. I abbreviate the left-to-right and the right-to-left direction of C_{MC} by '$C^{\Rightarrow}_{\mathrm{MC}}$' and '$C^{\Leftarrow}_{\mathrm{MC}}$', respectively.

(1) $C_{\mathrm{MP}}+C_{\mathrm{VEQ}}+C_{\mathrm{EFQ}} \Rightarrow C^{\Rightarrow}_{\mathrm{MC}}$: Let w, w' be arbitrary worlds in a Chellas frame $\mathcal{F}_C = \langle W, R\rangle$ and let X be an arbitrary subset of W, such that wR_Xw'. By C_{VEQ} it follows that $w' = w$. Moreover, since there is a world w'' – namely w' – such that wR_Xw'', we obtain by C_{EFQ} that $w \notin -X$ and, hence, $w \in X$ is the case.

$C_{\mathrm{MP}}+C_{\mathrm{VEQ}}+C_{\mathrm{EFQ}} \Rightarrow C^{\Leftarrow}_{\mathrm{MC}}$: Let w, w' be arbitrary worlds in a Chellas frame $\mathcal{F}_C = \langle W, R\rangle$ and let X be an arbitrary subset of W, such that $w' = w$ and $w \in X$. Then, C_{MP} implies that wR_Xw. Since $w = w'$ we obtain wR_Xw'.

(2) $C_{\mathrm{MC}} \Rightarrow C_{\mathrm{MP}}$: Let w be an arbitrary world in a Chellas frame $\mathcal{F}_C = \langle W, R\rangle$ and let X be an arbitrary subset of W, such that $w \in X$. Then, by C_{MC} we get wR_Xw.

$C_{\mathrm{MC}} \Rightarrow C_{\mathrm{VEQ}}$: Let w and w' be arbitrary worlds in a Chellas frame $\mathcal{F}_C = \langle W, R\rangle$ and let X be an arbitrary subset of W, such that wR_Xw'. Then, due to C_{MC} this implies $w' = w$.

$C_{\mathrm{MC}} \Rightarrow C_{\mathrm{EFQ}}$: Let w, w' be arbitrary worlds in a Chellas frame $\mathcal{F}_C = \langle W, R\rangle$ and let X be an arbitrary subset of W, such wR_Xw'. By C_{MC} we obtain $w \in X$ and, hence, $w \notin -X$ is the case. □

Proof Theory

My name for System **MC** ("(full) material collapse") presupposes that both the material implication and the material conditional are logically equivalent in that system. The following theorem shows that this assumption is justified:

Theorem 7.78. $(\alpha \rightarrow \beta) \leftrightarrow (\alpha \mathrel{\Box\!\!\rightarrow} \beta)$ *is a theorem of System* **MC**.

Proof. By Definition 7.76, logic **MC** is **CK** plus Principles MP, VEQ, and EFQ. MP gives us $(\alpha \mathrel{\Box\!\!\rightarrow} \beta) \rightarrow (\alpha \rightarrow \beta)$. Furthermore, VEQ and EFQ imply, on the basis of Lemma 7.79, Triv′, namely $(\alpha \rightarrow \beta) \rightarrow (\alpha \mathrel{\Box\!\!\rightarrow} \beta)$. □

In the proof for Theorem 7.78 we used the Principle Triv′:

Triv′ $(\alpha \rightarrow \beta) \rightarrow (\alpha \mathrel{\Box\!\!\rightarrow} \beta)$

Principle Triv′ is p.c. equivalent to the following principle:

Or-to-If: $\neg\alpha \vee \beta \rightarrow (\alpha \mathrel{\Box\!\!\rightarrow} \beta)$

The Principle Or-to-If is, for example, discussed in Bennett (2003, p. 139f, p. 142) and in Adams (1975, pp. 19–21). Principle Or-to-If is interesting insofar as it has some intuitive appeal at first glance, but is p.c.-equivalent to VEQ+EFQ, as Lemma 7.79 shows:

Lemma 7.79. *VEQ+EFQ* $\Leftrightarrow$ *Triv′*

Proof. Lemmata 7.80–7.82 yield 7.79. □

Lemma 7.80. *VEQ+EFQ* $\Rightarrow$ *Triv′*

Proof.

1.	$\alpha \rightarrow \beta$	given
2.	$\quad\neg\alpha$	assumption 1, proof by cases
3.	$\quad\alpha \mathrel{\Box\!\!\rightarrow} \beta$	2, EFQ

4.	$\neg\neg\alpha$	assumption 2, proof by cases
5.	β	1, 4, p.c.
6.	$\alpha \mathrel{\Box\!\!\rightarrow} \beta$	5, VEQ
7.	$\alpha \mathrel{\Box\!\!\rightarrow} \beta$	2–3, 4–6, proof by cases

□

Lemma 7.81. *Triv′ ⇒ VEQ*

Proof.

1.	β	given
2.	$\alpha \rightarrow \beta$	1, p.c.
3.	$\alpha \mathrel{\Box\!\!\rightarrow} \beta$	2, Triv′

□

Lemma 7.82. *Triv′ ⇒ EFQ*

Proof.

1.	$\neg\alpha$	given
2.	$\alpha \rightarrow \beta$	1, p.c.
3.	$\alpha \mathrel{\Box\!\!\rightarrow} \beta$	2, Triv′

□

Both Principles EFQ and VEQ are hardly plausible and correspond to S1′ and S2′ from Chapter 1, respectively. We saw in Chapter 1 that both principles are counter-intuitive for both indicative and counterfactual conditionals. Furthermore, as any conditional $\alpha \mathrel{\Box\!\!\rightarrow} \beta$ in System **MC** is equivalent to the corresponding material conditional $\alpha \rightarrow \beta$, all properties of the material conditional also hold for conditional formulas. As a result Mon, Trans, and CP are valid for all conditionals in System **MC**, which makes **MC** still a less plausible candidate for a conditional logic system.

Observe that CEM is derivable in **MC**, as Lemma 7.83 implies:

Lemma 7.83. *VEQ+EFQ ⇒ CEM*

Proof.

1.	α	assumption 1, proof by cases
2.	β	assumption 1.1, proof by cases

3.	$\alpha \boxright \beta$	2, VEQ
4.	$(\alpha \boxright \beta) \vee (\alpha \boxright \neg\beta)$	3, p.c.
5.	$\neg\beta$	assumption 1.2, proof by cases
6	$\alpha \boxright \neg\beta$	5, VEQ
7.	$(\alpha \boxright \beta) \vee (\alpha \boxright \neg\beta)$	6, p.c.
8.	$(\alpha \boxright \beta) \vee (\alpha \boxright \neg\beta)$	2–4, 5–7, proof by cases
9.	$\neg\alpha$	assumption 2, proof by cases
10.	$\alpha \boxright \beta$	9, EFQ
11.	$(\alpha \boxright \beta) \vee (\alpha \boxright \neg\beta)$	10, p.c.
12.	$(\alpha \boxright \beta) \vee (\alpha \boxright \neg\beta)$	1–8, 9–11, proof by cases

□

We finally show that System **MC** can be alternatively axiomatized by **CK** +MP+CS+EFQ, on the basis of the following theorem:

Theorem 7.84. *EFQ ⇒ (CS ⇔ VEQ)*

Proof. Lemmata 7.85 and 7.86 give us Theorem 7.84. □

Lemma 7.85. *EFQ+CS ⇒ VEQ*

Proof.

1.	β	given
2.	$\neg\alpha$	assumption 1, proof by cases
3.	$\alpha \boxright \beta$	2, EFQ
4.	$\neg\neg\alpha$	assumption 2, proof by cases
5.	$\alpha \wedge \beta$	1, 4, p.c.
6.	$\alpha \boxright \beta$	5, CS
7.	$\alpha \boxright \beta$	2–3,4–6, proof by cases

□

Lemma 7.86. *VEQ ⇒ CS*

Proof.

1.	$\alpha \wedge \beta$	given
2.	β	1, p.c.
3.	$\alpha \,\Box\!\!\rightarrow \beta$	2, VEQ

□

7.4 Summary

This concludes our investigation of conditional logic systems based on CS semantics. Let me summarize: In this chapter (i) I gave an objective and a subjective interpretation of basic CS semantics (Section 7.1), which should allow for a plausible interpretation of CS semantics for indicative and counterfactual conditionals, respectively. (ii) I then discussed a range of plausible principles from Table 5.1, such as Cautious Monotonicity (CM), Cautious Cut (CC), Or, and Rational Monotonicity (RM), which can be rendered valid if one strengthens the basic CS semantics by adding the respective frame conditions from Tables 5.3 to the basic CS semantics.

(iii) I then discussed how a range of indicative and counterfactual conditional logics can be modeled by CS semantics on the basis of these further frame conditions. Among these are D. Lewis' System **VC** (see Section 7.3.2) and Systems **C**, **CL**, and **P** of Kraus et al. (1990; see Sections 7.2.1–7.2.3). I also gave a direct characterization of the full material collapse systems with (System **MC**; see Section 7.3.4) and without bridge principles (System **M**; see Section 7.2.6).

I fully concede that the present book is only a first step in terms of the conditional logic project described in Chapter 1 and that a host of philosophical issues still have to addressed. The most pressing among these is the following question: While I argued that the basic CS semantics is plausible both as a basis for indicative and counterfactual conditionals, I did not explicitly address which system is "the" best system for indicative and counterfactual conditionals, respectively. I strongly believe that one must – in order to be able to fully answer this question – take not only the discussion in the area of philosophy of conditionals and conditional logics into account but also current results from linguistics and psychology of reasoning, both in much greater detail than it is possible here (cf. Section 1.1). Based on the limited discussion of these issues here, I would argue that the basic CS semantics, when augmented by frame conditions for CM, CC, and RM, might serve as a plausible "rough" candidate for such "best" indicative and counterfactual conditional systems.

Chapter 8

Concluding Remarks

In this final chapter I will first (i) give a short overview of the problems addressed in this book and (ii) provide a list of further problems that may be worth being investigated. I shall furthermore (iii) summarize the advantages of Chellas-Segerberg (CS) semantics compared to other types of conditional logic semantics.

Let us now start with point (i). I investigated a range of topics in this book related to conditional logic: I gave an argument for the conditional logic project (Chapter 1), investigated the interdisciplinary ramifications of a conditional logic approach and contrasted it with default logic approaches from the non-monotonic literature (Chapter 2). Then, I described important probabilistic and possible worlds semantics and discussed the difference between indicative and counterfactual conditionals. On that basis I defended possible worlds approaches, such as CS semantics (Chellas, 1975; Segerberg, 1989), against the criticism by Bennett (2003; see my Chapter 3). In Chapters 4–6 I provided a formal account of CS semantics – a type of possible worlds semantics for conditionals – in terms of soundness, completeness, and frame correspondence. Finally, in Chapter 7 I gave an objective and a subjective interpretation of CS semantics and described a range of indicative and counterfactual conditionals on the basis of CS semantics by the formal framework established in Chapters 4–6.

Let us now focus on point (ii). In my view this book marks the beginning of a project rather than its completion. In particular, the following topics might be worth further attention:

- To provide a more detailed argument for a conditional logic project,
- investigate further arguments against truth value based semantics for conditionals as, for example, described by Bennett (2003),
- prove logical dependence and independence of the axiom schemata from Tables 5.1 and 5.2,
- account for Adams' (1975) conditional default logic and AGM belief revision (Alchourrón et al., 1985) by a modification of CS semantics,
- investigate weaker possible worlds semantics for conditional logics, such as neighborhood set selection functions (e.g., Chellas, 1975, pp. 144–147; Arló-Costa, 2007, Section 3.1.2),
- describe the notion of conditional obligation by means of a conditional operator (as described herein) plus an obligation operator,
- provide correspondence and canonicity results for further conditional logic principles, such as Negation Rationality and Disjunctive Rationality (Lehmann & Magidor, 1992, p. 17),
- investigate for which parts of the lattice of conditional logic systems one can give completeness in terms of Chellas frames,
- describe predicate logic versions of CS semantics, and
- systematically investigate the relation of CS semantics with other conditional logic semantics, such as Stalnaker (1968), D. Lewis (1973b), Kraus et al. (1990), and Lehmann and Magidor (1992).

I shall now discuss point (iii), namely the advantages of CS semantics over alternative semantics. First, CS semantics allows to account for a broad class of existing conditional logic systems in a uniform framework. For example, Systems **S** (Stalnaker, 1968; Stalnaker & Thomason, 1970), **VC** (D. Lewis, 1973b; see Sections 7.3), **R** (Lehmann & Magidor, 1992; see Section 7.2),

P, **CL**, **C** (Kraus et al., 1990; see Section 7.2), and **CK** (Chellas, 1975; Segerberg, 1989; see Section 7.1) can be described by CS semantics. The formal results imply that the proof-theory of these systems is sound and complete w.r.t. the class of standard Segerberg models for which the respective conditions from Tables 5.3 and 5.4 hold. Furthermore, the proof-theory of the probabilistic Conditional Logic Systems **P** (Schurz, 2005b), $\mathbf{P}^*$ (Adams, 1965, 1966, 1977), and $\mathbf{P}^+$ (Adams, 1986; Schurz, 1998) can be accounted for by CS semantics (see Section 3.6).

Note that some conditional logics, such as the probabilistic Threshold System **O**, cannot be described in terms of CS semantics, since CS semantics validates inferences of type AND those which are not valid in System **O** (cf. Section 3.6.2). In constrast to the aforementioned probabilistic conditional logic systems, CS semantics can be described in the full language $\mathcal{L}_{\mathrm{CL}}$, which allows for both arbitrary Boolean combinations and nestings of conditionals and is thereby not prone to D. Lewis' (1976) triviality result (see Section 3.7).

Second, we saw that CS semantics can be interpreted both in terms of (A) objective alethic modality and (B) a modified Ramsey test (see Section 7.1). Since the difference between indicative and counterfactual conditionals seems to lie in the fact that indicative conditionals – in contrast to counterfactual conditionals – are interpreted relative to our subjective world knowledge (see Section 3.4), CS semantics can plausibly serve as basis for both indicative and counterfactual conditional logics. This conclusion is further supported by the fact that a range of indicative conditional logics (e.g., Systems **P** and **R**; Kraus et al., 1990; Lehmann & Magidor, 1992) and counterfactual conditional logics (e.g., System **VC**; D. Lewis, 1973b) can be formally described by CS semantics (see above).

Furthermore, I do not regard as a strong disadvantage of CS semantics that the semantics cannot account for all conditional logics (e.g., the probabilistic Threshold System **O**). CS semantics rather represents a plausible compromise in terms of strength and generality: CS semantics is on the one hand general enough to account for a wide range of conditional logic systems (see previous paragraph), while it is on the other hand strong enough to lend itself into objective and subjective interpretation in terms of (A) and (B), respectively.

Third, CS semantics relativizes the accessibility relation R_X in both Chellas models and Segerberg models to propositions X (sets of possible worlds) rather than formulas. This allows a characterization of CS semantics – in contrast to set-selection models and Stalnaker models (see Section 3.3.2) – in terms of frames and, thus, gives us a more natural and more flexible semantic characterization in terms of purely structural conditions (cf. Sections 3.3.2, 4.1, and 4.3).

I conclude that CS semantics has several pros which other conditional logic semantics do not have. Given the advantages of CS semantics, it should hence be regarded as a live option as a semantics for both indicative and counterfactual conditionals.

References

Adams, E. (1965). The Logic of Conditionals. *Inquiry*, *8*, 166–197.

Adams, E. (1966). Probability and the Logic of Conditionals. In J. Hintikka & P. Suppes (Eds.), *Aspects of Inductive Logic* (pp. 265–316). Amsterdam: North-Holland Publishing Company.

Adams, E. (1970). Subjunctive and Indicative Conditionals. *Foundations of Language*, *6*, 89–94.

Adams, E. (1975). *The Logic of Conditionals*. Dordrecht, Netherlands: D. Reidel Publishing Company.

Adams, E. (1977). A Note on Comparing Probabilistic and Modal Logics of Conditionals. *Theoria*, *18*, 186–194.

Adams, E. (1986). On the Logic of High Probability. *Journal of Philosophical Logic*, *15*, 255–279.

Adams, E. (1998). *A Primer of Probability Logic*. Stanford: Center for the Study of Language and Information.

Alchourrón, C. E., Gärdenfors, P., & Makinson, D. (1985). On the Logic of Theory Change. Partial Meet Contraction and Revision Functions. *The Journal of Symbolic Logic*, *50*, 510–530.

Anderson, A. R., Belnap, N., & Dunn, J. M. (1975). *Entailment. The Logic of Relevance and Necessity* (Vol. 1). Princeton, NJ: Princeton University Press.

Anderson, A. R., Belnap, N., & Dunn, J. M. (1992). *Entailment. The Logic of Relevance and Necessity* (Vol. 2). Princeton, NJ: Princeton University Press.

Arló-Costa, H. (2001). Bayesian Epistemology and Epistemic Conditionals.

On the Status of the Export–Import Laws. *The Journal of Philosophy*, *98*, 555–593.

Arló-Costa, H. (2007). The Logic of Conditionals. In E. N. Zalta (Ed.), *The Stanford Encyclopedia of Philosophy*. Retrieved from http://plato.stanford.edu/entries/logic-conditionals/.

Bacchus, F. (1990). *Representing and Reasoning with Probabilistic Knowledge*. Cambridge, Mass.: The MIT Press.

Baltag, A., & Smets, S. (2008). Probabilistic Dynamic Belief Revision. *Synthese*, *165*, 179–202.

Bamber, D. (1994). Probabilistic Entailment of Conditionals by Conditionals. *IEEE Transactions on Systems, Man, and Cybernetics*, *24*, 1714–1723.

Barwise, J., & Cooper, R. (1991). Generalized Quantifiers. *Linguistics and Philosophy*, *4*, 159–219.

Beck, S. R., & Guthrie, C. (2011). Almost Thinking Counterfactually. Children's Understanding of Close Counterfactuals. *Child Development*, *82*, 1189–1198.

Beck, S. R., Robinson, E. J., Carroll, D. J., & Apperly, I. A. (2006). Children's Thinking about Counterfactuals and Future Hypotheticals as Possibilities. *Child Development*, *77*, 413–426.

Bennett, J. (2003). *A Philosophical Guide to Conditionals*. Oxford: Oxford University Press.

Biazzo, V., Gilio, A., Lukasiewicz, T., & Sanfilippo, G. (2002). Probabilistic Logic under Coherence, Model-Theoretic Probabilistic Logic, and Default Reasoning in System P. *Journal of Applied Non-Classical Logic*, *12*, 189–213.

Blackburn, P., de Rijke, M., & Venema, Y. (2001). *Modal Logic*. Cambridge: Cambridge University Press.

Blau, U. (1978). *Die Dreiwertig Logik der Sprache* [The Three-Valued Logic of Language]. Berlin: Walter de Gruyter.

Bonati, R. (2005). *Nichtmonotone Systeme der deontischen Logik und ihre Bedeutung für die Behandlung bedingter Normen* [Non-Monotonic Systems of Deontic Logic and their Meaning for the Treatment of Conditional Norms] (Doctoral dissertation, University of Salzburg, Salzburg, Austria).

Bovens, L., & Hartmann, S. (2003). *Bayesian Epistemology*. Oxford: Oxford University Press.

Bradley, R. (2000). A Preservaton Condition for Conditionals. *Analysis*, *60*, 219–222.

Bremer, M. (2005a). *An Introduction to Paraconsistent Logics*. Frankfurt am Main: Peter Lang.

Bremer, M. (2005b). *Philosophische Semantik* [Philosophical Semantics]. Frankfurt am Main: Ontos-Verlag.

Brun, G. (2003). *Die Richtige Formel. Philosophische Probleme der logischen Formalisierung* [The Correct Formula. Philosophical Issues in Logical Formalization]. Frankfurt am Main: Dr. Hänsel-Hohenhausen.

Burgess, J. P. (1981). Quick Completeness Proofs for Some Logics of Conditionals. *Notre Dame Journal of Formal Logic*, *22*, 76–84.

Carnap, R. (1962). *Logical Foundations of Probability* (2nd ed.). Chicago: University of Chicago Press.

Cartwright, N. (2002). In Favor of Laws that are Not Ceteris Paribus after All. *Erkenntnis*, *57*, 425–439.

Chellas, B. F. (1975). Basic Conditional Logic. *Journal of Philosophical Logic*, *4*, 133–153.

Clocksin, W. F., & Mellish, C. S. (2003). *Programming in Prolog. Using the ISO Standard* (5th ed.). Berlin: Springer.

Connolly, A. C., Fodor, J. A., Gleitman, L. R., & Gleitman, H. (2007). Why Stereotypes Don't even Make Good Defaults. *Cognition*, *103*, 1–22.

Davis, W. A. (1979). Indicative and Subjunctive Conditionals. *The Philosophical Review*, *88*, 544–564.

Declerck, R., & Reed, S. (2001). *Conditionals. A Comprehensive Empirical Analysis*. Berlin: Mouton de Gruyter.

Delgrande, J. P. (1987). A First-Order Conditional Logic for Prototypical Properties. *Artificial Intelligence*, *33*, 105–130.

Delgrande, J. P. (1988). An Approach to Default Reasoning Based on a First-Order Conditional Logic. Revised Report. *Artificial Intelligence*, *36*, 63–90.

Delgrande, J. P. (1998). On First-Order Conditional Logics. *Artificial Intelligence*, *105*, 105–137.

Douven, I., & Romeijn, J.-W. (2011). A New Resolution of the Judy Benjamin Problem. *Mind*, *120*, 637–670.

Douven, I., & Verbrugge, S. (2010). The Adams Family. *Cognition*, *117*, 302–318.

Dunn, J. M., & Restall, G. (2002). Relevance Logic. In D. M. Gabbay & F. Guenthner (Eds.), *Handbook of Philosophical Logic* (2nd ed., Vol. 6, pp. 1–128). Dordrecht, Netherlands: Kluwer Academic Publishers.

Earman, J., Roberts, J., & Smith, S. (2002). Ceteris Paribus Lost. *Erkenntnis*, *57*, 281–301.

Edgington, D. (1995). On Conditionals. *Mind*, *104*, 235–329.

Edgington, D. (2007). On Conditionals. In D. M. Gabbay & F. Guenthner (Eds.), *Handbook of Philosophical Logic* (2nd ed., Vol. 14, pp. 127–221). Dordrecht: Springer.

Enderton, H. B. (2001). *A Mathematical Introduction to Logic* (2nd ed.). San Diego, CA: Harcourt/Academic Press.

Ereshefsky, M. (2007). Species. In E. N. Zalta (Ed.), *Stanford Encyclopedia of Philosophy.* Retrieved from http://plato.stanford.edu/entries /species/#SpePlu.

Evans, J. S., Handley, S. J., & Over, D. E. (2003). Conditionals and Conditional Probability. *Journal of Experimental Psychology. Learning, Memory, and Cognition*, *29*, 321–335.

Evans, J. S., & Over, D. E. (2004). *If.* Oxford: Oxford University Press.

Fagin, R., Halpern, J. Y., Moses, Y., & Vardi, M. Y. (2003). *Reasoning about Knowledge*. Cambridge, Mass.: MIT Press.

Feller, W. (1968). *An Introduction to Probability Theory and its Applications* (3rd ed., Vol. 1). New York: Wiley.

Fuhrmann, A. (1998). Nonmonotonic logic. In E. Craig (Ed.), *Routledge*

Encyclopedia of Philosophy (Vol. 7, pp. 30–35). London: Routledge.

Gabbay, D. M., Kurucz, A., Wolter, F., & Zakharyaschev, M. (2003). *Many-Dimensional Modal Logics. Theory and Applications*. Amsterdam: Elsevier.

Gärdenfors, P. (1979). Conditionals and Changes of Belief. *Acta Philosophica Fennica*, *30*, 381–404.

Gärdenfors, P. (1986). Belief Revision and the Ramsey Test for Conditionals. *The Philosophical Review*, *95*, 81–93.

Gärdenfors, P. (1988). *Knowledge in Flux. Modeling the Dynamics of Epistemic States*. Cambridge, Mass.: MIT Press.

Geffner, H., & Pearl, J. (1994). A Framework for Reasoning with Defaults. In H. E. Kyburg Jr., R. P. Loui, & G. N. Carlson (Eds.), *Knowledge Representation and Defeasible Reasoning* (pp. 69–87). Dordrecht, Netherlands: Kluwer Academic Publishers.

Gibbard, A. (1980). Two Recent Theories of Conditionals. In W. L. Harper, R. Stalnaker, & G. Pearce (Eds.), *Ifs. Conditionals, Belief, Decision, Chance, and Time* (pp. 211–247). Dordrecht: D. Reidel Publishing Company.

Gilio, A. (2002). Probabilistic Reasoning under Coherence in System P. *Annals of Mathematics and Artificial Intelligence*, *34*, 5–34.

Ginsberg, M. L. (1986). Counterfactuals. *Artificial Intelligence*, *30*, 35-79.

Giordano, L., Gliozzi, V., & Olivetti, N. (2005). Weak AGM Postulates and Strong Ramsey Test. A Logical Formalization. *Artificial Intelligence*, *168*, 1–37.

Goldszmidt, M., & Pearl, J. (1996). Qualitative Probabilities for Default Reasoning, Belief Revision, and Causal Modeling. *Artificial Intelligence*, *84*, 57–112.

Goodman, N. (1983). *Fact, Fiction, and Forecast* (4th ed.). Cambridge, Mass.: Harvard University Press.

Grice, H. P. (1975). Logic and Conversation. In P. Cole & J. L. Morgan (Eds.), *Syntax and Semantics* (Vol. 3, pp. 41–58). New York: Academic Press.

Grice, H. P. (1989). *Studies in the Way of Words*. Cambridge, Mass.: Havard University Press.

Grove, A. (1988). Two Modellings for Theory Change. *Journal of Philosophical Logic*, *17*, 157–170.

Haegeman, L. (2003). Conditionals Clauses. External and Internal Syntax. *Mind & Language*, *18*, 317–339.

Hawthorne, J. (1996). On the Logic of Nonmonotonic Conditionals and Conditional Probabilities. *Journal of Philosophical Logic*, *25*, 185–218.

Hawthorne, J., & Makinson, D. (2007). The Qantitative/Qualitative Watershed for Rules of Uncertain Inference. *Studia Logica*, *86*, 247–297.

Helfman, G. S., Collette, B. B., Facey, D. E., & Bowen, B. W. (2009). *The Diversity of Fishes. Biology, Evolution, and Ecology* (2nd ed.). Chichester, Sussex, U.K.: Wiley-Blackwell.

Hendry, H. E., & Pokriefka, M. L. (1985). Carnapian Extensions of S5. *Journal of Philosophical Logic*, *14*, 111–128.

Horty, J. F. (2001). Nonmonotonic Logic. In L. Goble (Ed.), *The Blackwell Guide to Philosophical Logic* (pp. 336–361). Malden, MA: Blackwell.

Howson, C., & Urbach, P. (1993). *Scientific Reasoning: the Bayesian Approach* (2nd ed.). Chicago: Open Court.

Hughes, G. E., & Cresswell, M. J. (1984). *A Companion to Modal Logic*. London: Methuen.

Hughes, G. E., & Cresswell, M. J. (2003). *A New Introduction to Modal Logic*. London: Routledge. (Original work published 1996)

Hüttemann, A. (1998). Laws and Dispositions. *Philosophy of Science*, *65*, 121–135.

Johnson-Laird, P. N., & Byrne, R. M. J. (2002). Conditionals. A Theory of Meaning, Pragmatics, and Inference. *Psychological Review*, *109*, 646–678.

Kleene, S. C. (1952). *Introduction to Meta-Mathematics*. Amsterdam: North-Holland Publishing Company.

Kolmogorov, A. N. (1950). *Foundations of the Theory of Probability* (N. Morrison, Trans.). New York: Chelsea.

Kratzer, A. (1977). What 'Must' and 'Can' Must and Can Mean. *Linguistics and Philosophy*, *1*, 337–355.

Kratzer, A. (1981). The Notional Category of Modality. In H.-J. Eikmeyer & H. Rieser (Eds.), *Words, Worlds, and Contexts* (pp. 38–74). Berlin: Walter de Gruyter.

Kraus, S., Lehmann, D., & Magidor, M. (1990). Nonmonotonic Reasoning, Preferential Models and Cumulative Logics. *Artificial Intelligence*, *44*, 167–207.

Lange, M. (1993). Natural Laws and the Problem of Provisos. *Erkenntnis*, *38*, 233–248.

Lehmann, D., & Magidor, M. (1992). What Does a Conditional Knowledge Base Entail? *Artificial Intelligence*, *55*, 1–60.

Leitgeb, H. (2004). *Inference on the Low Level*. Dordrecht, Netherlands: Kluwer Academic Publishers.

Leitgeb, H. (2010). On the Ramsey Test without Triviality. *Notre Dame Journal of Formal Logic*, *51*, 21–54.

Leitgeb, H., & Segerberg, K. (2007). Dynamic Doxastic Logic. Why, How, and Where to? *Synthese*, *155*, 167–190.

Leslie, S.-J. (2007). Generics and the Structure of the Mind. *Philosophical Perspectives*, *21*, 375–403.

Levesque, H. J. (1990). All I Know. A Study in Autoepistemic Logic. *Artificial Intelligence*, *42*, 263–309.

Levinson, S. C. (2000). *Presumptive Meanings. The Theory of Generalized Conversational Implicature*. Cambridge, Mass.: MIT Press.

Lewis, C. I. (1912). Implication and the Algebra of Logic. *Mind*, *21*, 522–531.

Lewis, C. I., & Langford, C. H. (1932). *Symbolic Logic*. New York: Dover Publications.

Lewis, D. (1970). General Semantics. *Synthese*, *22*, 18–67.

Lewis, D. (1971). Completeness and Decidability of Three Logics of Counterfactual Conditionals. *Theoria*, *37*, 74–85.

Lewis, D. (1973a). Causation. *The Journal of Philosophy*, *70*, 556–567.

Lewis, D. (1973b). *Counterfactuals*. Malden, MA: Blackwell.

Lewis, D. (1976). Probabilities of Conditionals and Conditional Probabilities. *The Philosophical Review*, *85*, 297–315.

Lewis, D. (1979). Counterfactual Dependence and Time's Arrow. *Nous*, *13*, 455–476.

Lewis, D. (1980). A Subjectivist's Guide to Objective Chance. In R. C. Jeffrey (Ed.), *Studies in Inductive Logic and Probability* (Vol. 2, pp. 263–293). Berkeley: University of California Press.

Lewis, D. (1997). Finkish Dispositions. *The Philosophical Quarterly*, *47*, 143–158.

Lewis, D. (2000). Causation as Influence. *The Journal of Philosophy*, *97*, 182–197.

Lindström, S., & Rabinowicz, W. (1995). The Ramsey Test Revisited. In G. Crocco, L. Fariñas del Cerro, & A. Herzig (Eds.), *Conditionals. From Philosophy to Computer Science* (pp. 147–191). Oxford: Oxford University Press.

Macdonald, D. W., Seifert, E. R., Balcome, K., Dunbar, R., Bininda-Emonds, O., Best, R., et al. (2007). What is a Mammal? In D. W. Macdonald (Ed.), *The Encyclopedia of Mammals*. Oxford University Press. Oxford Reference Online. Retrieved from http://www.oxfordreference.com/views/ENTRY.html?subview=Main&entry=t227.e156 via Univ- und Landesbibliothek Düsseldorf.

Makinson, D. (1993). Five Faces of Minimality. *Studia Logica*, *52*, 339–379.

Makinson, D. (1994). General Patterns in Nonmonotonic Reasoning. In D. M. Gabbay, C. J. Hogger, & J. A. Robinson (Eds.), *Handbook of Logic in Artificial Intelligence and Logic Programming* (Vol. 3, pp. 35–110). Oxford: Clarendon Press.

McCall, S. (1966). Connexive Implication. *The Journal of Symbolic Logic*, *31*, 415–433.

McCarthy, J. (1980). Circumscription – A Form of Non-Monotonic Reasoning. *Artificial Intelligence*, *13*, 27–39.

McCarthy, J. (1989). Artificial Intelligence, Logic and Formalizing Common Sense. In R. H. Thomason (Ed.), *Philosophical Logic and Artificial Intelligence* (pp. 161–190). Dordrecht, Netherlands: Kluwer Academic Publishers.

McCarthy, J., & Hayes, P. J. (1969). Some Philosophical Problems from the Standpoint of Artificial Intelligence. In B. Meltzer & D. Michie (Eds.), *Machine Intelligence 4* (pp. 463–502). Edinburgh: Edinburgh University Press.

McGee, V. (1985). A Counterexample to Modus Ponens. *The Journal of Philosophy*, *82*, 462–471.

McGee, V. (1989). Conditional Probabilities and Compounds of Conditionals. *The Philosophical Review*, *98*, 485–541.

McKenna, M. C., & Bell, S. K. (1997). *Classification of Mammals above the Species Level*. New York: Columbia University Press.

Meyer, J.-J. C., & van der Hoek, W. (1995). *Epistemic Logic for AI and Computer Science*. Cambridge: Cambridge University Press.

Nejdl, W. (1992). *The P-Systems. A Systematic Classification of Logics of Nonmonotonicity. Extended Report* (Tech. Rep.). Vienna, Austria: Technical University of Vienna.

Nickel, B. (2009). Generics and the Ways of Normality. *Linguistics and Philosophy*, *31*, 629–648.

Nickel, B. (2010). Ceteris Paribus Laws. Generics & Natural Kinds. *Philosphers' Imprint*, *10*, 1–25.

Nute, D. (1980). *Topics in Conditional Logic*. Dordrecht, Netherlands: D. Reidel Publishing Company.

Nute, D. (1992). Basic Defeasible Logic. In L. Fariñas del Cerro & M. Penttonnen (Eds.), *Intentional Logics for Programming* (pp. 125–154). Oxford: Oxford University Press.

Nute, D., & Cross, C. B. (2001). Conditional Logic. In D. M. Gabbay & F. Guenthner (Eds.), *Handbook of Philosophical Logic* (2nd ed., Vol. 4, pp. 1–98). Dordrecht, Netherlands: Kluwer Academic Publishers.

Oberauer, K., & Wilhelm, O. (2003). The Meaning(s) of Conditionals. Conditional Probabilities, Mental Models, and Personal Utilities. *Journal of Experimental Psychology. Learning, Memory, and Cognition*, *29*, 680–693.

Pearl, J. (1988). *Probabilistic Reasoning in Intelligent Systems*. San Francisco: Morgan Kaufmann. (Revised Second Printing)

Pearl, J. (1990). System Z. A Natural Ordering of Defaults with Tractable Applications to Nonmonotonic Reasoning. In R. Parikh (Ed.), *Proceedings of the Third Conference on Theoretical Aspects of Reasoning about Knowledge* (pp. 121–135). San Francisco: Morgan Kaufmann.

Pelletier, F. J. (1997). Generics and Defaults. In J. van Benthem & A. ter Meulen (Eds.), *Handbook of Logic and Language* (pp. 1125–1177). Amsterdam: Elsevier.

Pelletier, F. J., Elio, R., & Hanson, P. (2008). Is Logic All in Our Heads? From Naturalism to Psychologism. *Studia Logica*, *88*, 3–66.

Peters, S., & Westerståhl, D. (2008). *Quantifiers in Language and Logic*. Oxford: Oxford University Press.

Pfeifer, N., & Kleiter, G. D. (2010). The Conditional in Mental Probability Logic. In M. Oaksford & N. Chater (Eds.), *Cognition and Conditionals. Probability and Logic in Human Thinking* (pp. 153–173). Oxford: Oxford University Press.

Poole, D. (1988). A Logical Framework for Default Reasoning. *Artificial Intelligence*, *36*, 27–47.

Portner, P. (2009). *Modality*. Oxford: Oxford University Press.

Priest, G. (2008). *An Introduction to Non-Classical Logics* (2nd ed.). Cambridge: Cambridge University Press.

Quine, W. v. O. (1960). *Word & Oject*. Cambridge, Mass.: The MIT Press.

Rafetseder, E., Cristi-Vargas, R., & Perner, J. (2010). Counterfactual Reasoning. Developing a Sense of "Nearest Possible World". *Child Development*, *81*, 376–389.

Rafetseder, E., & Perner, J. (2010). Is Reasoning from Counterfactual Antecedents Evidence for Counterfactual Reasoning? *Thinking & Reasoning*, *16*, 131–155.

Ramsey, F. P. (1965). *The Foundations of Mathematics and Other Logical Essays* (R. B. Braithwaite, Ed.). London: Routledge & Kegan Paul LTD.

Åqvist, L. (2002). Deontic Logic. In D. Gabbay & F. Guenther (Eds.), *Handbook of Philosophical Logic* (2nd ed., Vol. 8, pp. 147–264). Dordrecht, Netherlands: Kluwer.

Reichenbach, H. (1949). *The Theory of Probability*. Berkeley: University of California Press.

Reiter, R. (1980). A Logic for Default Reasoning. *Artificial Intelligence*, *13*, 81–132.

Reutlinger, A., Schurz, G., & Hüttemann, A. (2011). Ceteris Paribus Laws. In E. N. Zalta (Ed.), *Stanford Encyclopedia of Philosophy*. http://plato.stanford.edu/entries/ceteris-paribus/.

Rott, H. (2001). *Chance, Choice and Inference. A Study of Belief Revision and Nonmonotonic Reasoning*. Oxford: Oxford University Press.

Routley, R., & Montgomery, H. (1968). On Systems Containing Aristotle's Thesis. *The Journal of Symbolic Logic*, *33*, 82–96.

Schrenk, M. (2008). A Lewisian Theory for Special Science Laws. In D. Bohse, K. Dreimann, & S. Walter (Eds.), *Selected Papers Contributed to the Sections of GAP.6, 6th International Congress of the Society for Analytic Philosophy*. Paderborn: Mentis.

Schurz, G. (1991). Relevant Deduction. From Solving Paradoxes towards a General Theory. *Erkenntnis*, *35*, 391–437.

Schurz, G. (1994). Probabilistic Justification of Default Reasoning. In B. Nebel & L. Dreschler-Fischer (Eds.), *KI-94. Advances in Artificial Intelligence* (pp. 248–259). Berlin: Springer.

Schurz, G. (1996). The Role of Negation in Nonmonotonic Logic and Defeasible Reasoning. In H. Wansing (Ed.), *Negation. A Notion in Focus* (pp. 197–231). Berlin: Walter de Gruyter.

Schurz, G. (1997a). *The Is-Ought Problem*. Dordrecht, Netherlands: Kluwer Academic Publishers.

Schurz, G. (1997b). Probabilistic Default Logic Based on Irrelevance and Relevance Assumptions. In D. M. Gabbay, R. Kruse, A. Nonnengart, &

H. J. Ohlbach (Eds.), *Qualitative and Quantitative Practical Reasoning* (pp. 536–553). Berlin: Springer.

Schurz, G. (1998). Probabilistic Semantics for Delgrande's Conditional Logic and a Counterexample to His Default Logic. *Artificial Intelligence*, *102*, 81–95.

Schurz, G. (2001a). Rudolf Carnap's Modal Logic. In W. Stelzner & M. Stöckler (Eds.), *Zwischen traditioneller und moderner Logik. Nichtklassische Ansätze* (pp. 365–380). Paderborn: Mentis.

Schurz, G. (2001b). What is 'Normal'? An Evolution-Theoretic Foundation for Normic Laws and Their Relation to Statistical Normality. *Philosophy of Science*, *68*, 476–497.

Schurz, G. (2002). Ceteris Paribus Laws. Classification and Deconstruction. *Erkenntnis*, *57*, 351–372.

Schurz, G. (2004). Logic, Matter of Form, and Closure under Substitution. In L. Běhounek & M. Bílková (Eds.), *The Logica Yearbook 2004* (pp. 33–46). Prague: Filosofia.

Schurz, G. (2005a). Alethic Modal Logics and Semantics. In D. Jacquette (Ed.), *A Companion to Philosophical Logic* (pp. 442–477). Malden, MA: Blackwell. (Original work published 2002)

Schurz, G. (2005b). Non-Monotonic Reasoning from an Evolution-Theoretic Perspective. Ontic, Logical and Cognitive Foundations. *Synthese*, *146*, 37–51.

Schurz, G. (2007). Human Conditional Reasoning Explained by Non-Monotonicity and Probability. An Evolutionary Account. In S. Vosniadou, D. Kayser, & A. Protopapas (Eds.), *Procceedings of the EuroCogSci 2007* (pp. 628–633). Hove, East Sussex, U.K.: Lawrence Erlbaum Associates.

Schurz, G. (2008). *Einführung in die Wissenschaftstheorie* [Introduction to Philosophy of Science] (2nd ed.). Darmstadt, Germany: Wissenschaftliche Buchgesellschaft.

Schurz, G. (2011). Tweety, or Why Probabilism and even Bayesianism Need Objective and Evidential Probabilities. In D. Dieks, W. J. Gonzales,

S. Hartmann, M. Stöltzner, & M. Weber (Eds.), *Probabilities, Laws and Structures* (pp. 57–74). New York: Springer.

Schurz, G. (2012). Prototypes and their Composition from an Evolutionary Point of View. In M. Werning, W. Hinzen, & E. Machery (Eds.), *The Oxford Handbook of Compositionality* (pp. 530–553). Oxford: Oxford University Press.

Schurz, G., & Thorn, P. D. (2012). Reward versus Risk in Uncertain Inference. Theorems and Simulations. *Review of Symbolic Logic*, *5*, 574–612.

Segerberg, K. (1989). Notes on Conditional Logic. *Studia Logica*, *48*, 157–168.

Sperber, D., & Wilson, D. (1995). *Relevance. Communication and Cognition* (2nd ed.). Malden, MA: Blackwell.

Spohn, W. (1975). An Analysis of Hansson's Dyadic Deontic Logic. *Journal of Philosophical Logic*, *4*, 237–252.

Spohn, W. (2012). *The Laws of Belief. Ranking Theory and Its Philosophical Applications*. Oxford: Oxford University Press.

Stalnaker, R. C. (1968). A Theory of Conditionals. In N. Rescher (Ed.), *Studies in Logical Theory* (pp. 98–112). Oxford: Basil Blackwell.

Stalnaker, R. C. (1970). Probability and Conditionals. *Philosophy of Science*, *37*, 64–80.

Stalnaker, R. C. (1975). Indicative Conditionals. *Philosophia*, *5*, 269–286.

Stalnaker, R. C., & Thomason, R. H. (1970). A Semantical Analysis of Conditional Logic. In S. Körner (Ed.), *Practical Reason* (pp. 23–42). Oxford: Basil Blackwell.

Stein, E. (1996). *Without Good Reason. The Rationality Debate in Philosophy and Cognitive Science*. Oxford: Oxford University Press.

Strevens, M. (in press). Ceteris Paribus Hedges. Causal Voodoo that Works. *Journal of Philosophy*.

Talbott, W. (2008). Bayesian Epistemology. In E. N. Zalta (Ed.), *The Stanford Encyclopedia of Philosophy*. Retrieved from http://plato.stanford.edu/entries/epistemology-bayesian/.

Thorn, P. D. (in press). Defeasible Conditionalization. *Journal of Philosophical Logic*.

Unterhuber, M., & Schurz, G. (in press). The New Tweety Puzzle. Arguments against Monistic Bayesian Approaches in Epistemology and Cognitive Science. *Synthese*.

van Benthem, J. (2001). Correspondence Theory. In D. M. Gabbay & F. Guenthner (Eds.), *Handbook of Philosophical Logic* (2nd ed., Vol. 3, pp. 325–408). Dordrecht, Netherlands: Kluwer Academic Publishers.

Wansing, H. (2010). Connexive Logic. In E. N. Zalta (Ed.), *Stanford Encyclopedia of Philosophy*. Retrieved from http://plato.stanford.edu/entries/logic-connexive/.

Weingartner, P., & Schurz, G. (1986). Paradoxes Solved by Simple Relevance Criteria. *Logique et Analyse*, *113*, 3–40.

Werning, M. (2012). Non-Symbolic Compositional Representation and Its Neuronal Foundation. Towards an Emulative Semantics. In M. Werning, W. Hinzen, & E. Machery (Eds.), *The Oxford Handbook of Compositionality* (pp. 633–654). Oxford: Oxford University Press.

Bei Fragen zur Produktsicherheit wenden Sie sich bitte an:
If you have any questions regarding product safety,
please contact:

Walter de Gruyter GmbH
Genthiner Straße 13
10785 Berlin
productsafety@degruyterbrill.com